OBSERVING MARINE INVERTEBRATES

DONALD P. ABBOTT

OBSERVING MARINE INVERTEBRATES

Drawings from the Laboratory

Edited by Galen Howard Hilgard

With 26 drawings by Professor Abbott's students

STANFORD UNIVERSITY PRESS
Stanford, California 1987

The photographs on pp. xvi and 376 are by Galen Howard Hilgard. The photograph on p. 374 is by the News and Publications Office, Stanford University; used with permission.

Stanford University Press
Stanford, California

© 1987 by the Board of Trustees of the
Leland Stanford Junior University
Printed in the United States of America

CIP data appear at the end of the book

CONTENTS

Foreword by Lawrence Blinks   xv
Author's Preface   xvii
Editor's Introduction   xix
Abbreviations Used in the Drawings   xxiii

THE DRAWINGS

Phylum Cnidaria (=Coelenterata)   1–49
   Class Hydrozoa   1–21
      Order Anthomedusae   1–11
         Syncoryne eximia   1, 2
         Tubularia marina   3, 4
         Eudendrium californicum   5, 6, 7
         Polyorchis montereyensis   8
         Velella sp.   9
         Rathkea sp.   10
         Cladonema sp.   11
      Order Leptomedusae   12–20
         Aglaophenia sp.   12, 13
         Plumularia sp.   14
         Eucopella sp.   15, 16, 17
         Aequorea sp.   18
         Eutonina sp.   19, 20
      Order Limnomedusae   21
         Vallentinia adhaerens   21
   Class Scyphozoa   22–36
      Order Stauromedusae   22
         Manania sp. (?)   22
      Order Semaeostomeae   23–31
         Chrysaora melanaster   23, 24
         Pelagia sp.   25, 26, 27
         Aurelia aurita   28
         Phacellophora camtshatica   29, 30, 31
      Order Coronatae   32–36
         Atolla sp.   32, 33
         Periphylla sp.   34, 35, 36
   Class Anthozoa   37–49
     Subclass Alcyonaria   37–42

Order Gorgonacea 37
    Psammogorgia arbuscula 37
Order Alcyonacea 38–39
    Anthomastus ritteri 38, 39
Order Pennatulacea 40–42
    Stylatula elongata 40
    Stylatula gracilis 41
    Renilla sp. 42
Subclass Zoantharia 43–49
  Order Actiniaria 43–45
    Metridium sp. 43
    Epiactis prolifera 44, 45
  Order Corallimorpharia 46–47
    Corynactis sp. 46, 47
  Order Madreporaria (=Scleractinia) 48–49
    Balanophyllia elegans 48, 49
Phylum Ctenophora 50–54
  Class Nuda 50–51
    Beroë cucumis 50, 51
  Class Tentaculata 52–54
    Pleurobrachia sp. 52, 53, 54
Phylum Platyhelminthes 55–57, (214), (351)
  Class Turbellaria 55–57, (214)
    Order Acoela 55
      Polychoerus carmelensis 55
    Order Polycladida 56–57
      Notoplana sp. 56, 57
    Order Tricladida (unidentified triclad, with *Ischnochiton*) (214)
  Class Cestoda (unidentified cestode, with *Pagurus*) (351)
Phylum Ectoprocta (=Bryozoa) 58–78
  Order Cheilostomata 58–73
    Suborder Anasca 58–67
      Membranipora sp. 58, 59, 60
      Membranipora tuberculata 61
      Thalamoporella californica 62, 63
      Bugula californica 64, 65
      Tricellaria occidentalis 66
      Scrupocellaria sp. 67
    Suborder Ascophora 68–73
      Phidolopora sp. 68, 69

          Hippodiplosia insculpta   70
          Celleporaria (=*Holoporella*) *brunnea*   71, 72, 73
      Order Cyclostomata   74
          Crisia sp.   74
      Order Ctenostomata   75–78
          Flustrella corniculata   75
          Bowerbankia sp.   76, 77, 78

Phylum Entoprocta   79–80
          Pedicellina sp.   79
          Barentsia sp.   80

Phylum Brachiopoda   81–84
          Terebratalia or *Terebratulina* sp.   81, 82, 83
          Laqueus californicus   84

Phylum Phoronida   85–86
          Phoronis sp.   85, 86

Phylum Echinodermata   87–117
  Class Asteroidea   87–98
      Order Paxillosida   87–89
          Astropecten sp.   87, 88, 89
      Order Platyasterida   90
          Luidia sp.   90
      Order Spinulosida   91–93
          Henricia sp.   91
          Patiria miniata   92, 93
      Order Forcipulatida   94–98
          Leptasterias sp.   94, 95, 96
          Pisaster giganteus   97
          Pycnopodia helianthoides   98
  Class Holothuroidea   99–107
          Cucumaria curata   99
          Eupentacta quinquesemita   100, 101, 102, 103
          Leptosynapta sp.   104, 105, 106, 107
  Class Ophiuroidea   108–115
          Amphipholis squamata with *Rhopalura,* an orthonectid mesozoan   108
          Ophioplocus esmarki   109
          Amphiodia occidentalis   110, 111, 112, 113, 114, 115
  Class Echinoidea   116–117
          Dendraster excentricus   116, 117

Phylum Chordata   118–126
  Subphylum Tunicata   118–126

Class Ascidiacea   118–125
    Polyclinum planum   118, 119
    Synoicum parfustis   120
    Perophora annectens   121, 122
    Botryllus sp.   123
    Ascidia ceratodes   124, 125
Class Thaliacea   126
    unidentified sp. (Family Doliolidae)   126

Phylum Mollusca   127–219
  Class Gastropoda   127–175
    Subclass Prosobranchia   127–155
      Order Archaeogastropoda   127–136
        Haliotis sp. (Superfamily Zygobranchia) and *Tegula funebralis* (Superfamily Trochacea)   127
        Fissurella volcano   128, 129
        Megatebennus bimaculatus   130, 131
        Notoacmea persona   132, 133
        Tegula sp. (Superfamily Trochacea)   134
        Acmaea or *Lottia* sp. (Superfamily Patellacea)   135
        Acmaea or *Lottia* sp. (with *Littorina*, Order Mesogastropoda; see below)   136
      Order Mesogastropoda   136–143
        Littorina sp. (with *Acmaea* or *Lottia*, Order Archaeogastropoda; see above)   136
        Littorina planaxis   137, 138, 139
        Polinices sp.   140
        Lamellaria sp.   141, 142, 143
      Order Neogastropoda   144–155
        Acanthina spirata   144, 145
        Nucella emarginata   146, 147, 148
        Nassarius sp.   149
        Olivella biplicata   150, 151, 152, 153
        Granulina sp.   154
        Conus californicus   155
    Subclass Opisthobranchia   156–175
      Order Cephalaspidea   156–163
        Rictaxis (=*Actaeon*) *punctocaelatus*   156, 157, 158, 159
        Aglaja (=*Navanax*) sp.   160, 161, 162
        Haminoea sp.   163
      Order Notaspidea   164–165
        Pleurobranchaea sp.   164, 165

    Order Nudibranchia 166–174
     Suborder Doridacea 166–170
      Anisodoris nobilis 166
      Triopha catalinae (= *T. carpenteri*) 167
      Triopha maculata 168, 169
      Archidoris montereyensis 170
     Suborder Dendronotacea 171
      Melibe leonina 171
     Suborder Aeolidacea 172–174
      Hermissenda crassicornis 172, 173
      Flabellinopsis iodinea 174
    Order Gymnosomata 175
      Clione sp. (?) (a pteropod) 175
Class Bivalvia 176–207
  Subclass Palaeotaxodonta 176
    Order Nuculoida 176
      Nuculana (= *Nucula*) sp. 176
  Subclass Pteriomorpha 177–185
    Order Mytiloida 177–182
      Adula (= *Botula*) *californiensis* 177, 178
      Lithophaga sp. 179, 180
      Mytilus californianus 181
      Mytilus edulis 182
    Order Pterioida 183–185
      Hinnites giganteus 183
      Pododesmus cepio 184, 185
  Subclass Heterodonta 186–205
    Order Veneroida 186–202
      Clinocardium (= *Cardium*) *nuttallii* 186, 187, 188, 189
      Irus lamellifer 190, 191
      Kellia laperousii 192
      Protothaca sp. 193
      Petricola (= *Rupellaria*) sp. 194
      Macoma secta 195, 196, 197
      Tresus nuttallii 198, 199
      Siliqua lucida 200, 201
      Cryptomya californica 202
    Order Myoida 203–205
      Penitella penita 203, 204
      Hiatella arctica 205
  Subclass Anomalodesmata 206–207

       Order Pholadamyoida   206–207
           Lyonsia californica   206
           Pandora sp.   207
  Class Polyplacophora   208–214
       Nuttallina californica   208, 209
       Cyanoplax hartwegi   210
       Mopalia muscosa   211
       Katharina tunicata   212, 213
       Ischnochiton (= *Lepidozona*) *mertensii* (with triclad flatworm)   214
  Class Aplacophora   215
       Chaetoderma erudita   215
  Class Scaphopoda   216–219
       Cadulus fusiformis   216, 217, 218
       Dentalium sp. (*D. neohexagonum*?)   219
Phylum Annelida   220–248
  Class Polychaeta   220–248
      Family Polynoidae   220–223
        Halosydna brevisetosa   220, 221, 222
        Hesperonoë sp.   223
      Family Hesionidae   224–225
        unidentified hesionid   224, 225
      Family Dorvilleidae   226–227
        Dorvillea moniloceras   226, 227
      Family Lumbrinereidae   228–229
        Lumbrinereis sp.   228, 229
      Family Syllidae   230
        unidentified syllid   230
      Family Cirratulidae   231–233
        Cirriformia sp.   231
        Dodecaceria fewkesi   232, 233
      Family Opheliidae   234–235
        Ophelia sp. (?)   234, 235
      Family Sternaspidae   236–238
        Sternaspis sp.   236, 237, 238
      Family Terebellidae   239–241
        unidentified terebellid   239, 240, 241
      Family Sabellariidae   242–243
        Phragmatopoma californica   242, 243
      Family Pectinariidae   244–245
        Pectinaria   244, 245

Contents

        Family Sabellidae   246–248
           unidentified sabellid   246, 247
           Myxicola sp.   248
Phylum Sipuncula   249–258
        Dendrostomum (= *Themiste*) *zostericola*   249, 250, 251
        Phascolosoma agassizii   252, 253, 254, 255, 256, 257, 258
Phylum Priapulida   259
        Priapulus sp.   259
Phylum Arthropoda   260–373
  Class Crustacea   260–362
    Subclass Branchiopoda   260–268
      Order Anostraca   260–264
        Artemia sp.   260, 261, 262, 263, 264
      Order Conchostraca   265–266
        unidentified conchostrachan   265, 266
      Order Cladocera   267–268
        Ceriodaphnia sp. (?)   267, 268
    Subclass Ostracoda   269–270
      unidentified ostracod (Podocopa)   269
      unidentified ostracod   270
    Subclass Copepoda   271–273
      Tigriopus californicus   271, 272, 273
    Subclass Branchiura   274
      unidentified branchiuran on sand dab   274
    Subclass Cirripedia   275–283
      Suborder Lepadomorpha   275–278
        Pollicipes polymerus   275, 276, 277, 278
      Suborder Balanomorpha   279–283
        Catophragmus sp.   279
        Tetraclita squamosa   280
        Balanus glandula   281, 282, 283
    Subclass Malacostraca   284–362
      Superorder Phyllocarida   284–287
        Order Leptostraca   284–287
          Nebalia pugettensis   284, 285, 286
          Nebalia sp.   287
      Superorder Hoplocarida   288
        Order Stomatopoda   288
          unidentified stomatopod   288
      Superorder Peracarida   289–322

Order Mysidacea 289–292
    Acanthomysis (=*Neomysis*) sp. 289, 290, 291, 292
Order Isopoda 293–306
    Cirolana harfordi 293, 294, 295, 296, 297
    Idotea (=*Pentidotea*) *stenops* 298, 299, 300, 301
    Idotea (=*Pentidotea*) *resecata* 302, 303, 304
    unidentified isopod parasitic on lingcod 305
    Microcerberus abbotti 306
Order Tanaidacea 307–310
    Synapseudes intumescens (probably) 307
    Pagurapseudes sp. 308
    Leptochelia dubia (?) 309
    Tanais sp. (?) 310
Order Amphipoda 311–322
    Orchestoidea sp. 311, 312, 313, 314, 315, 316, 317
    unidentified caprellid 318, 319, 320, 321
    Corophium sp. 322
Superorder Eucarida 323–362
  Order Euphausiacea 323–327
    Euphausia pacifica 323, 324, 325, 326, 327
  Order Decapoda 328–362
    Section Caridea 328–335
      Hippolyte sp. 328, 329, 330
      Spirontocaris sp. 331, 332, 333, 334, 335
    Section Anomura 336–356
      Superfamily Hippidea 336–340
        Emerita analoga 336, 337, 338, 339, 340
      Superfamily Paguridea 341–351
        Cryptolithodes sitchensis 341, 342
        Pagurus samuelis 343, 344, 345, 346, 347, 348
        Pagurus granosimanus 349, 350, 351 (with cestode flatworm)
      Superfamily Galatheidea 352–356
        Petrolisthes sp. 352, 353, 354, 355
        Pleuroncodes planipes 356
    Section Brachyura 357–362
      Loxorhynchus crispatus 357
      Pachygrapsus crassipes 358, 359, 360, 361, 362
Class Pycnogonida 363–368
    Tanystylum duospinum 363, 364, 365, 366
    Halosoma viridintestinale 367
    Lecythorhynchus marginatus 368

Class Arachnida 369–372
  Order Pseudoscorpionida 369–370
    Garypus californicus 369, 370
  Order Acarina 371–372
    Neomolgus littoralis 371
    unidentified mesostigmatid mite 372
Class Tardigrada 373
    unidentified tardigrade 373
Phylum (?) Mesozoa (108)
    Rhopalura ophiocomae (an orthonectid, from *Amphipholis*) (108)

List of Drawings by Professor Abbott's Students 375
Index 377

FOREWORD

A Remembrance of Don Abbott

Biology 111–112H, "Marine Invertebrates," was one of the most successful and most popular courses ever offered at the Hopkins Marine Station. There must have been over 500 earnest students enrolled in it over the 31 years it was taught by Donald Abbott, and the course was undoubtedly responsible (along with the famous "Spring Class") for Don's receipt of Stanford University's Dinkelspiel Award for teaching in 1982. Of course, there was more to this course than the innards and behavior of invertebrates: there were 5:00 A.M. risings to meet the low tides of summer, and there were long, thorough considerations of the interphyletic relationships of animals, from protozoa to chordata, including tunicates (Don's beloved sea squirts).

But my own intense memories of Don, seen through the classroom door, are of his drawing the chalk versions of these pictures on the blackboard, clarifying the complex anatomy of these endlessly fascinating and amazingly diverse animals without backbones. Now these drawings are spread out here for all students and professionals (even amateurs!) to see and understand; they make available, for neurobiologists, embryologists, immunologists, and others, an unexplored wealth of material particularly pertinent to their own research. May these drawings help us avoid the overlooking of critical body parts, as had occurred, for example, when the squid axon was assumed to be a blood vessel prior to J. Z. Young's setting the record straight in 1936. Who knows what gonad, looking very much like a kidney, remains to be discovered out there?

A few attentive hours with these drawings will encourage the mindset necessary to the perceptive observation of these animals. To all of us who seek a richer understanding and enjoyment of marine invertebrates, then, this book is a great boon.

Lawrence Blinks
Hopkins Marine Station
Pacific Grove, California
August 1986

Get the experience of looking at fresh things. If you watch live animals, you gain clearer insights in shorter time than you would watching dead animals for much longer.

Best keep taking notes and drawing diagrams—you won't find these pictures in any books.

There's no substitute for fine forceps. None.

Classification is a very human thing to do. By grouping, we can generalize and predict. Classification helps us organize what we know, and it makes us better biologists.

You'll be tempted to grouse about the instability of taxonomy; but stability occurs only where people stop thinking and stop working.

Cultivate a suspicious attitude toward people who do phylogeny.

I like this photograph. So many pictures of professors make them look as if they've been stuffed by bad taxidermists.

AUTHOR'S PREFACE

Most of my professional life was spent as a faculty member at the Hopkins Marine Station of Stanford University at Pacific Grove, California, a laboratory devoted to the research and teaching of biology at the seashore. Located on a peninsula of several acres, projecting into Monterey Bay just west of the new Monterey Bay Aquarium, the Marine Station and the adjacent coastline offer habitats as different as wave-swept exposed rocky shores, protected sandy beaches, and tidepools. Harbor pilings and sandy mud are available nearby. Reflecting this diversity of habitat is an immense variety of invertebrates potentially available for study by students and researchers.

Classes in marine invertebrates were offered in the Spring (as Biology 176H, primarily research) and Summer (as Biology 111–112H, primarily academic and survey), and it was very soon apparent that for most of the abundant and convenient invertebrates immediately available, adequate accounts of structure, physiology, and behavior did not exist in English-language publications. The drawings reproduced in this volume, therefore, often represent fairly common but not well-known organisms.

In my very first year at Hopkins, my colleague Professor Rolf Bolin gave me an extraordinarily good piece of advice: "Always save yourself a seat and a microscope in the classroom; this way you can work right along with the students." Many of the drawings here represent hours of enjoyment as I carried out that advice.

The summer classes in invertebrates lasted all day, often starting at dawn with field trips at low tide, when the most beautiful and fragile forms were exposed to those equipped with hip boots and wool sweaters. The remainder of the day was devoted to lecture on the groups collected, and on lab study of living or freshly anaesthetized forms. Generally, the emphasis was on the gross anatomy and body plan of the major groups (class, order, and sometimes family), and on functional anatomy and behavior. I rarely called attention to species, since taxonomy was not one of the aspects of invertebrates that were important to the course.

As students became more competent at dissection and at conducting their own studies in the lab, there was more time for me to work on my own, sometimes on the creatures we had chosen for lab work that day, sometimes on animals collected previously that had looked promising for future study. None of the drawings I made were done with publication in mind.

My own sketches served two purposes for me. Many were diagrammed still further and used as blackboard drawings to introduce students to new animals, or for comparisons of "systems" in different animals. Others were photocopied and used to help introduce students to forms for which English accounts were not readily available. These drawings were never complete in either the form shown or the parts labeled. I always detested the idea of a formal "lab manual" where a student is confronted with complete artistic drawings of the animal to be studied (leaving little for the student but to identify the parts), and a detailed set of instructions on what to do. Detailed instructions are the death of originality and imagination. I always regarded the sketches of my own that I put out for students as incomplete and imperfect road maps, or incomplete "plumbing and wiring" diagrams; they were useful in introducing a form, but students would soon see that they were incomplete and improve upon them—as I had hoped they would.

I believe that students should be judged on their observation and understanding of the organisms they study. The lab notebooks that I required to be turned in at the end of the course were evaluated on those factors and not on the artistic or technical quality of the drawings. Most students showed improvement in their drawings as the course went along, a reflection that they were coming to understand what they drew.

Certain general rules are expressed in the drawings presented here, and it seems best to make them explicit:

1. Reality is too complex to express in mere drawings. One need not be an artist. What you need to do is to decide *what sorts of details* are to be stressed or omitted, and then in the sketch to try to simplify reality without distorting it too much in the process.

2. Clean, *simple* lines, all connected, are almost always better than a lot of short, sketchy lines that give everything a fuzzy appearance and conceal rather than reveal detail.

3. In anatomy, the *relationships* of parts to one another and to the whole are so important that nearly everything else should be sacrificed to show relationships. It may be necessary in some cases to make "exploded" drawings where parts are displayed as pulled apart so that all *connections* between parts are clear.

4. A certain degree of neatness in drawing, format, labeling, etc., is helpful both to yourself and to others using your drawings, but one should not make a fetish of it. "Freehand" neatness often serves the purpose and saves endless time in the rendering of picayune, unnecessary details. On the other hand, unlabeled drawings are not useful at all.

I thank Galen H. Hilgard, who received a Master's degree working with me on barnacles many years ago, and who twice came to Honolulu to help choose, discuss, and edit the drawings presented in this volume. I am pleased to have included in this volume drawings by some of the former Biology 111–112H students, and I thank them, also. Their illustrations are testimony to their interest in the animals, their attention to relationships, and their skill in translating their observations.

Donald P. Abbott
Honolulu, Hawaii
December 1985

EDITOR'S INTRODUCTION

It is a privilege to have worked with Donald P. Abbott (known affectionately to his students as DPA) in compiling and editing the drawings in this volume. As one of his students whose life was strongly influenced by his inspirational teaching, I have felt the impact of his vision, his great knowledge and drive, his scholarly habits and curiosity, and his kindness and humor. It was in his sharing of these things and his deep involvement with the animals themselves that he lit the way for his students.

Donald Abbott's 347 pages of drawings express his remarkable ability to combine intellectual, observational, and visual elements in his teaching; they exemplify his undaunted wet-hands approach to the study of marine zoology and reflect the profusion of original work that he carried out in the lab. Through his own example, he taught the rewards of reaching into the aquarium, tidepool, or jar and "asking the *animals*" for answers to questions. Emphasizing a comparative approach to invertebrates, he said "The key thing is, never lose sight of the interrelationships that exist between animals." He made these sketches with comparative purposes in mind.

Each of his drawings that Donald Abbott approved for this project, during the final weeks of his life, is included in its original form. None has been touched up or altered, though in some cases some simple additions (e.g., titles, classifications) have been made in the *annotations* of the drawings to make them more understandable. Most of these changes were executed by Dr. Abbott himself. However, six of the original sketches were not in condition for reproduction and with DPA's permission they were retraced and redrawn in more finished form: *Cucumaria curata* (p. 99); *Ascidia ceratodes* (p. 125); *Nuttallina californica* (p. 209, drawings in upper right and lower left); *Cirriformia* sp. (p. 231, lower left); *Idotea stenops* (p. 298, lower right); and *Pagurus granosimanus* (p. 350). Don's hand-written annotations are reproduced just as he set them down, for we realized that to have set them in type would have defeated the essence of the drawings—which is to inspire and encourage the taking of notes, not to intimidate.

The book includes 26 pages of drawings by Dr. Abbott's students, made—essentially—at the time they were enrolled in his course "Marine Invertebrates." He always encouraged students to make their own observations and often found great merit in their work; it is in this spirit that he suggested some students' drawings be included. Author and editor collaborated in their selection, and the final editorial decisions were made by the editor. These drawings show people's individual styles, and also clearly reflect the guidelines that Dr. Abbott established in his lectures. They were added where further coverage of particular animal groups would complement DPA's drawings. He particularly liked simplified sketches that catch movement and show behavior, and some of the contributions are of this sort.

With a few exceptions, the classification and sequence of animals reflected in this book's Contents pages and in the running heads on the pages of plates follow those in Morris, Abbott, and Haderlie, *Intertidal Invertebrates of California* (Stanford University Press, 1980). Other main sources referred to for classification include Ralph Smith and James Carlton, *Light's Manual* (University of California Press, 1975), Isabella Abbott and George Hollenberg, *Marine Algae of California* (Stanford University Press, 1976) and D. P. Abbott's "Outline of Classification" (1982; handed out and used in the "Marine Invertebrates" class of that year). Within the drawings themselves, classification and nomenclature sometimes reflect DPA's understanding of 15 or 20 years ago, but with the running heads as a guide the reader should find this nomenclature more interesting than troublesome. As is often the case with invertebrates, the animals sometimes were not identified to species or even to genus or family; as Dr. Abbott mentions in his Preface, he rarely called attention to species. We can be content to leave detailed identification to those who specialize in each taxonomic group.

These editorial notes would be incomplete without mention of Donald Abbott's lectures themselves. They have become one of the major tools for the study of marine invertebrates in the United States. In 1982, Stanford University's Dinkelspiel Award was won by Dr. Abbott "for the extraordinary scientific mastery that enabled him—through the creation of a justly famous summer program and through his writings—to provide national direction and shape to an entire discipline." From the present collection of his drawings, he developed his well-known lecture diagrams (seven of which are included and labeled as such in this book). All are based on his personal observations of the animals themselves, and on his own decipherings of their structures. Most of these drawings also illustrate some movement and activity, reflecting his functional approach to studies of morphology.

In addition to continuously updating his lectures, Don Abbott meticulously coached 115 student projects to publication and completed 50 published works drawn from his personal research, over and above his writing and editing of *Intertidal Invertebrates of California* (Stanford University Press, 1980; with Robert H. Morris and Eugene C. Haderlie) and his editing of *Intertidal Invertebrates of the Central California Coast* (University of California Press, 1954; with Ralph I. Smith, Frank A. Pitelka, and Frances M. Weesner). Two appreciations of Don's career and legacy have appeared: Lawrence Blinks, Arthur C. Giese, and Colin Pittendrigh, "Memorial Resolution: Donald Putnam Abbott, 1920–1986," *The Stanford University Campus Report*, vol. 18, no. 31, May 7, 1986, p. 18; and Todd Newberry and Michael G. Hadfield, "Donald Putnam Abbott (1920–1986)," *The Veliger*, vol. 29, no. 2, October 1, 1986, pp. 138–41.

The remarks quoted on pp. xvi, 374, and 376 are from notes taken during my working times with DPA from 1982 to 1985.

Many people have been involved in the planning and shaping of this book. Though I have served as coordinator and editor, others have given generously of their ideas, wisdom, and support. I would like especially to thank Dr. Ilze K. Berzins for initially sorting through many hundreds of drawings with me. My gratitude also goes to Dr. Vicki Buchsbaum Pearse, Dr. Andrew Todd Newberry, and Dr. John S. Pearse of the University of California at Santa Cruz, who have been sources of encouragement and have given their time and expertise in consultation. Charles H. Baxter and Dr. Lani West, both of whom taught and worked alongside Donald Abbott in recent years at Hopkins Marine Station, were involved in this project in its early stages and gave their valued opinions as the book approached completion; to them my sincere thanks. I am grateful to Dr. Lawrence Blinks of Hopkins Marine Station for his warm interest in the project and for his Foreword to this volume. Generous thanks go to Dr. Michael G. Hadfield of the University of Hawaii for his many helpful deeds and evaluations; to Donald Abbott's wife, Dr. Isabella A. Abbott of the University of Hawaii (and formerly of Hopkins Marine Station) for her backing, and for sharing her botanical expertise; and to Dr. Margaret G. Bradbury of San Francisco State University, who flew to Honolulu with me the second time, and with whom I have had valuable exchanges. I also thank Adrian Wilson, book designer, who gave me excellent advice about the preparation of the original drawings. I extend my gratitude to the following: Ann Abbott, Dr. William Austin, Alan Baldridge, Henrietta Bensussen, Dr. Anne Cohen, Susan Harris, Dotty Hollinger, Gretchen Lambert, Ron Marian, Margaret Sowers, and the staff at Hopkins Marine Station. My deep appreciation goes to members of my family—Henry R., Galen R., Addie J., and Julia H. Hilgard—for their sustaining strength and assistance, and to William W. Carver of Stanford University Press, who so heartily welcomed this project. Bill Carver has patiently and kindly contributed his fine art of editing to the book. I am much indebted to him and to Kathleen Szawiola, also of the Press, who designed the book, including its cover which, at DPA's request, incorporates a big jellyfish. In lieu of paying royalties, the Press has agreed to keep the list price as low as

Editor's Introduction

possible, and thus to make the book available to more of those who will make the most of it.

Thanks go to the former "Marine Invertebrates" students of DPA who allowed us to review their notebooks and then assisted in editing their drawings for this publication: Eva Aladjem, Ilze K. Berzins, Michael Imperato, Vicki Buchsbaum Pearse, James Watanabe, and Lani West. Their drawings are listed on p. 375.

Most of all, I gratefully acknowledge the wonderful cooperative spirit and tremendous effort Donald Abbott put to the task of preparing his own drawings for publication in the fall of 1985. On January 18th, 1986, Don died in Honolulu. It is unfortunate that he did not live long enough to see the book published, but we feel sure he would have liked it.

G.H.H.
Santa Cruz, California
August 1986

ABBREVIATIONS USED IN THE DRAWINGS

0600 6 A.M., etc. (ship's time)
1300 1 P.M., etc. (ship's time)

A auricle
A$_1$, A$_2$ first antenna, second antenna
A.A.M. or AAM anterior adductor muscle
Ab$_1$, Ab$_2$, etc. first abdominal, second abdominal, etc.
abdom. abdominal
adduc. adductor
amt. amount
Ant$_1$, Ant$_2$ first antenna, second antenna
ant. anterior
A.P. axial polyp
ap or app. appendage
approx. approximately, about
assoc. associated
asym. asymmetrical

bldg. building

C carina
ca. or Ca. approximately, about; or California
CaCO$_3$ calcium carbonate
Calif. California
carp. carpus
cav. cavity
circ. circular (as in muscle) or circulatory (as in system)
CL carinolateral
cm. centimeters
C.N.S. or CNS central nervous system
Co. County
coll. collector, or collected at
comp. or cpd. compound
c.s. cross section
c.t or C.T. connective tissue
ct. filament ctenidial filament

D or d. dorsal
det. determined
devel. developing or developed
diag. diagrammatic
diam. diameter
diff. different
dig. div. digestive diverticulum
digest. digestive
dig. gl. digestive gland

E east
ea. each
ed. editor
ejac. ejaculatory
end. or endop. endopodite
epip. epipodite
esoph. esophagus
esp. especially
est. estimated
ex. or exop. exopodite

Fert. fertilized
fig. or Fig. figure
fil. filament(s); or filamentary, in Cirripedia
fms. fathoms
ft. foot or feet

g. or gang. or gng. ganglion
gt. great
G.Z. gastrozooid

H head
H$_2$O water, usually sea water
Hbr. Harbor
Hd. head
H.M.S. or HMS Hopkins Marine Station
hr(s) hour(s)

I. Island
i.e. that is
I.L.V. internal longitudinal vessel
incr. increase
indiv. individual
irreg. irregular
isch ischium

juv. or Juv. juvenile

K kidney
KO "knock out" or anaesthetize

L left; or lateral, in Cirripedia
L.A. left auricle
lat. lateral
lg. large
L.K. left kidney
longitud. longitudinal
l-sec or l-section longitudinal section

m. muscle or muscles
mag. magnified
mand. mandible or mandibular
mat. maturity
m.c. mantle cavity
md. or Md. mandible or mandibular
mech. mechanism
med. medial or medium
mer. merus
merid. meridional
MgCl$_2$ magnesium chloride
mi mile or miles
min. minute or minutes
mm. millimeters
mt. mount
musc. muscle
Mx$_1$, Mx$_2$ first maxilla, second maxilla
Mxp maxilliped
Mxpd$_1$, Mxpd$_2$ first maxilliped, second maxilliped

N north
No. or no. number
n.s. nervous system

O order
obs. observer
occ. occasional
operc. operculum
ovig. ovigerous

P.A.M. or PAM or P.Ad.M. posterior adductor muscle
P.G. Pacific Grove, California
poss. possibly or possibility
post. posterior
pp'tate precipitate
pr(s) pair(s)
prep. preparation
pro protopodite
prob. probably
protop. protopodite
Pt. Point

R right; or rostrum, in Cirripedia
R.A. right auricle
rad. radial
r.c. radial canal
reg. regular
Riv. River

RL rostrolateral
R/V Research Vessel

S south; or scutum, in Cirripedia
sec(s). second(s); or section
seg(s). segment(s)
SO suborder
sp. or spp. species (single or plural)
spec. specimen
Sta. Station
stom. stomach
superfam. superfamily
SW or sw sea water
sys. system

T telson; tergum, in Cirripedia
temp. temperature
Th$_1$, Th$_2$, etc. first thoracic, second thoracic, etc.
th. or tho. thorax, or thoracic
T/S tergum/scutum

USNPG United States Naval Postgraduate School, Monterey, California

V ventral, or ventricle of heart
var. varying
VH visceral hump
V.N.C. ventral nerve cord

W west
w/ with
wks. weeks
w-v water vascular

x-sec cross section

♀ female
♂ male
⚥ hermaphrodite
± more or less
> more than
∴ therefore
∡ angle
A → P anterior to posterior
~ approximately, about
∝ varying

THE DRAWINGS

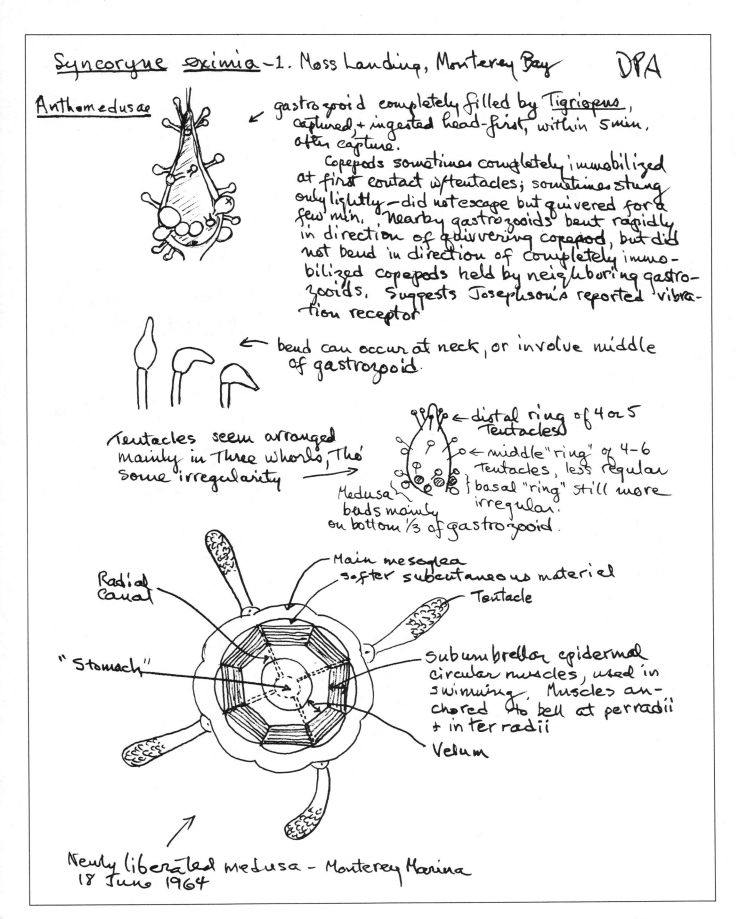

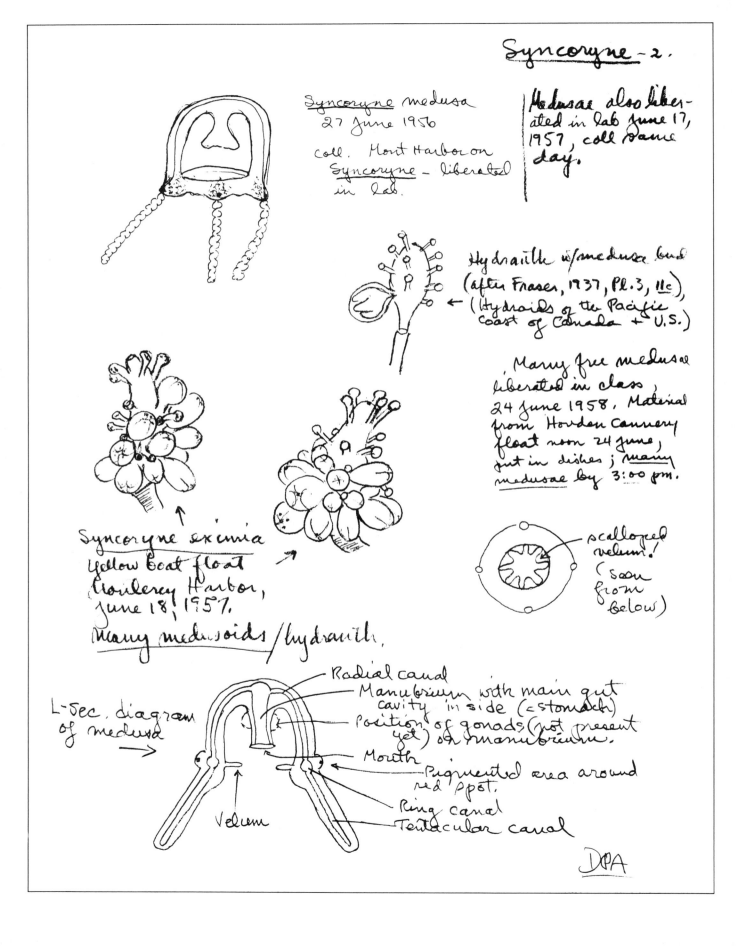

Cnidaria/Hydrozoa/Anthomedusae

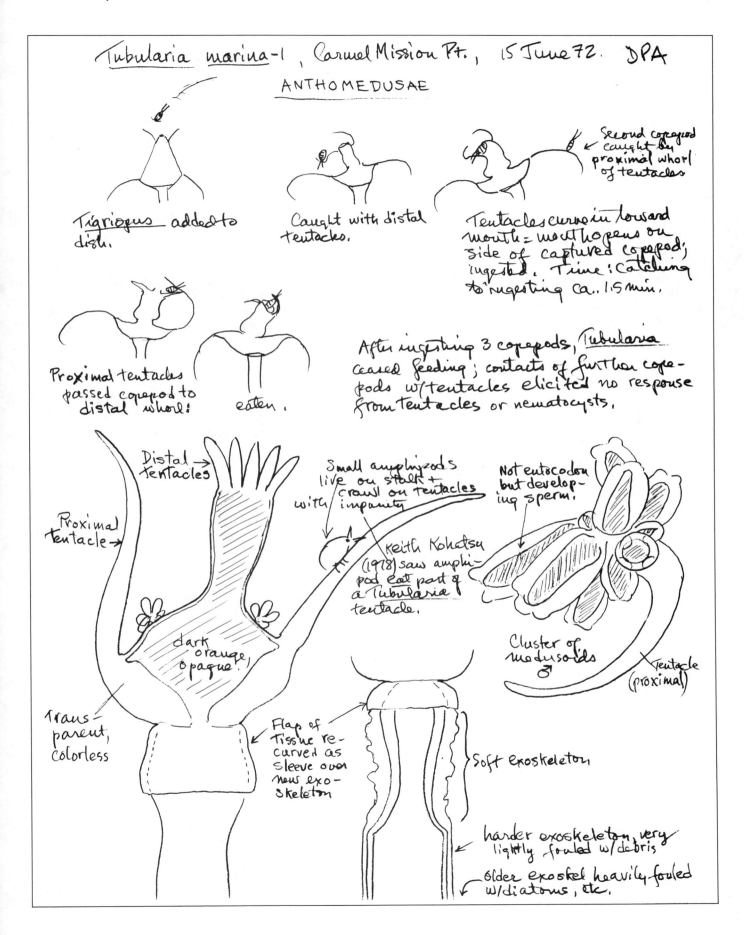

Cnidaria/Hydrozoa/Anthomedusae

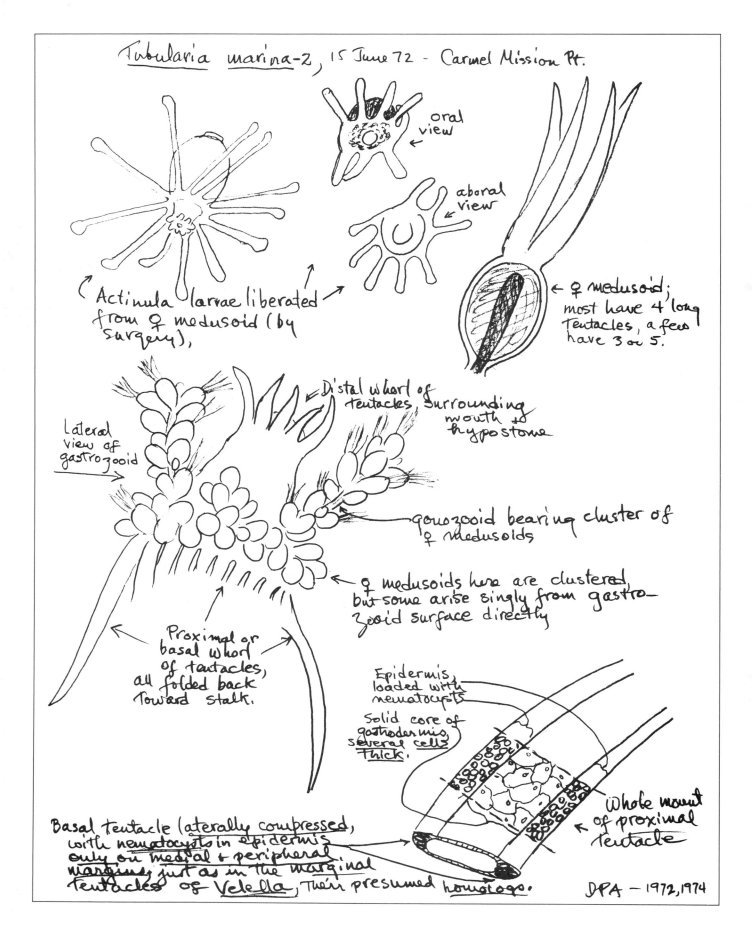

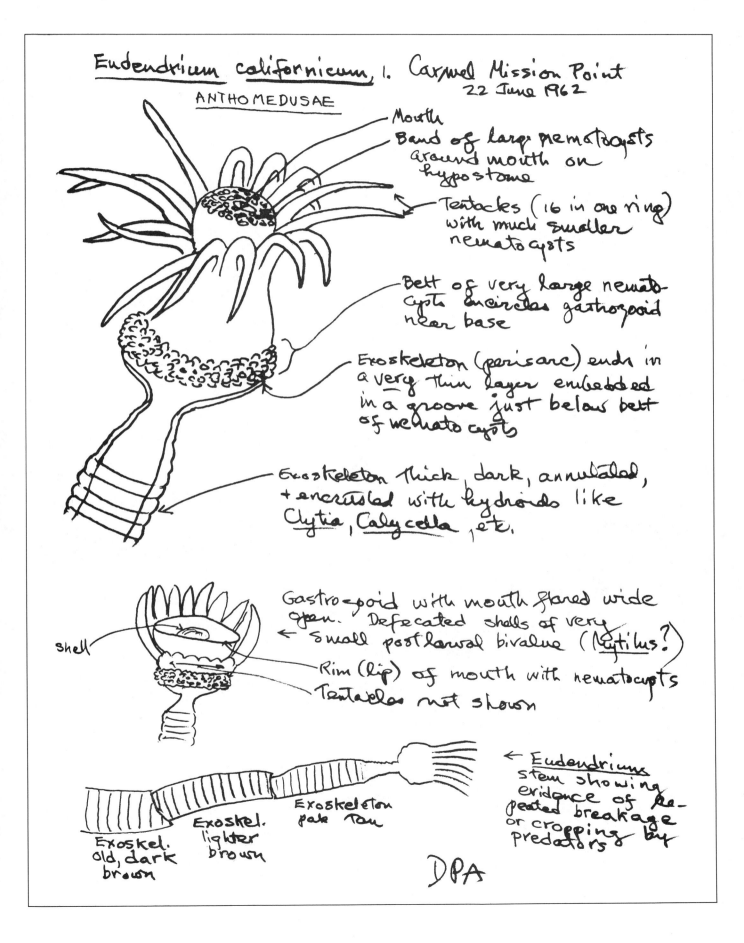

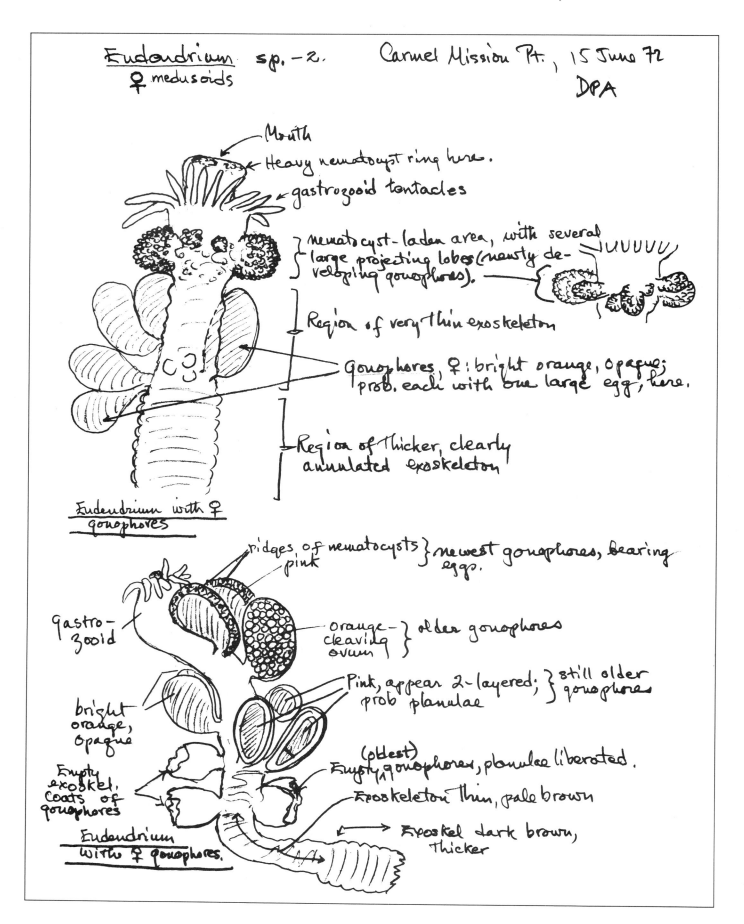

Cnidaria/Hydrozoa/Anthomedusae

Eudendrium (3), on rope, 10 ft depth, suspended under Wharf #2, Monterey, Calif.

DPA

gastrozooid with 18 tentacles

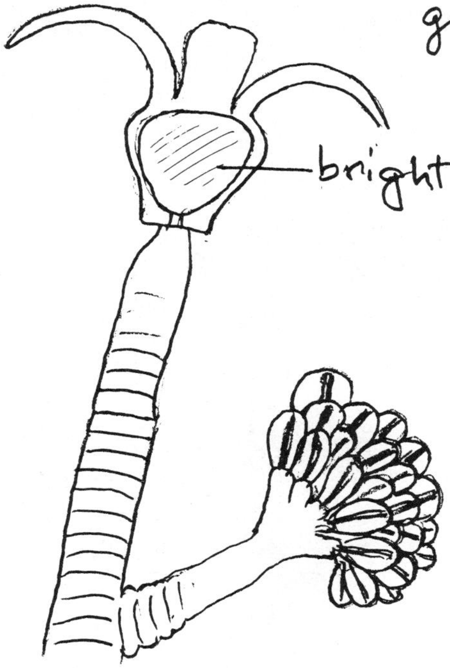

bright orange interior

Cluster of ♂ medusoids, tightly packed together. No gastrozooid seen in center.

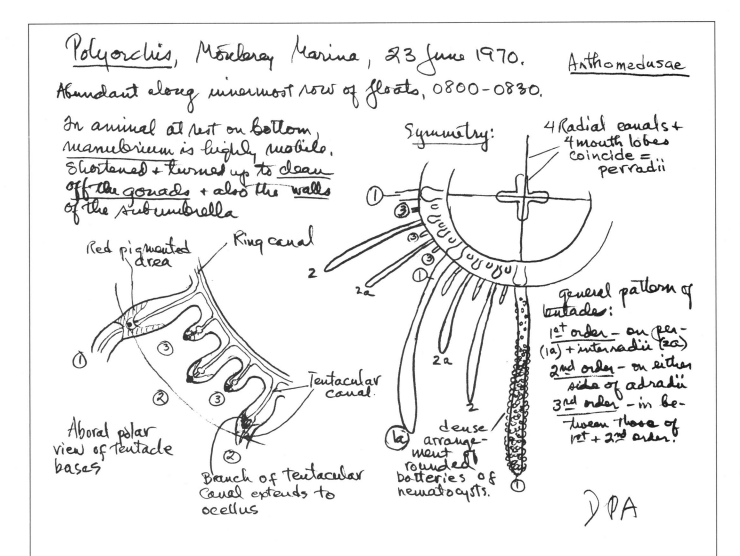

Cnidaria/Hydrozoa/Anthomedusae

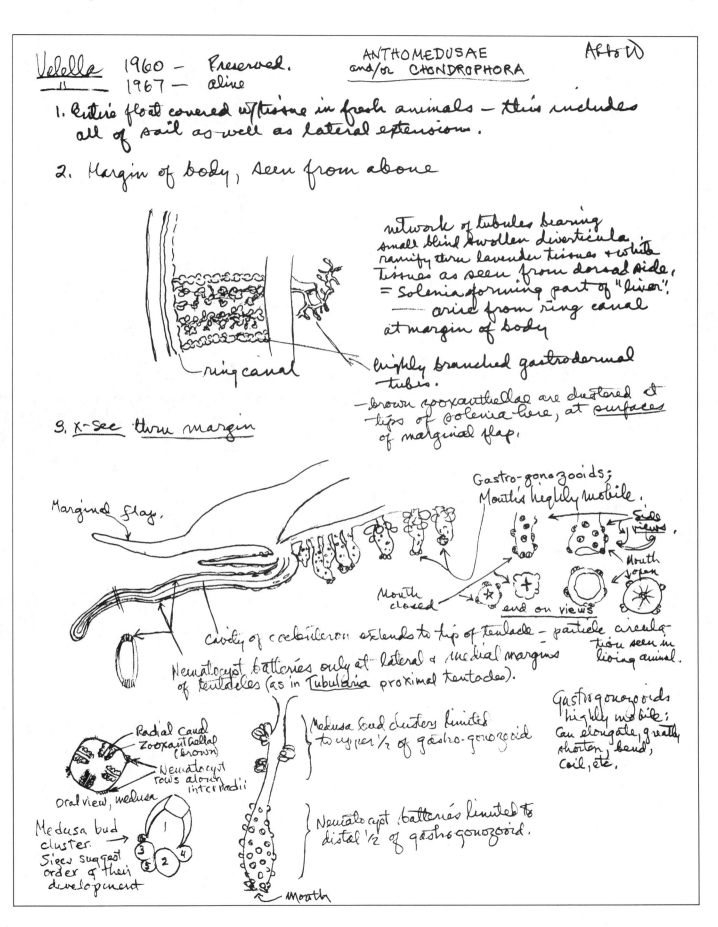

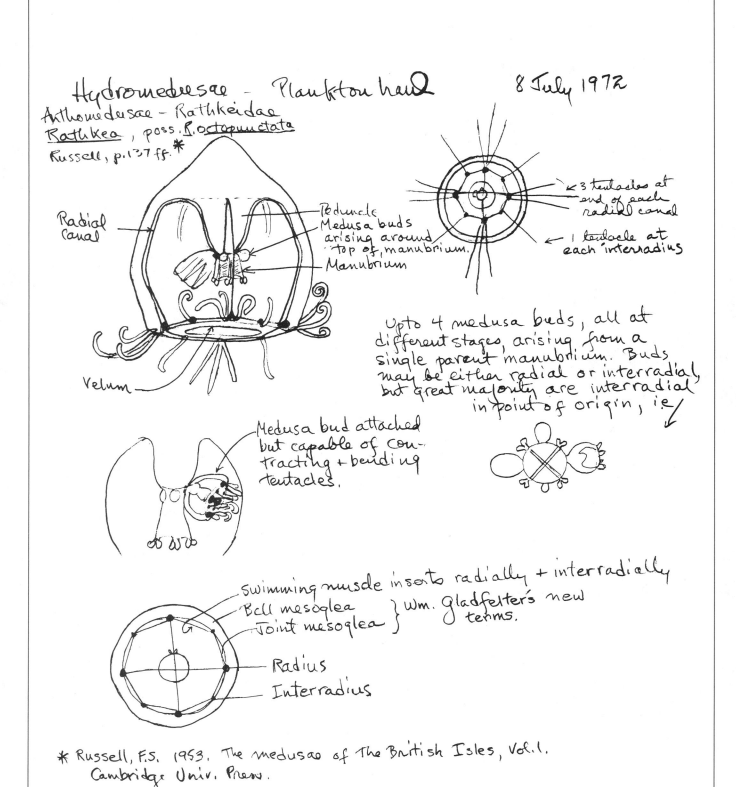

Cnidaria/Hydrozoa/Anthomedusae

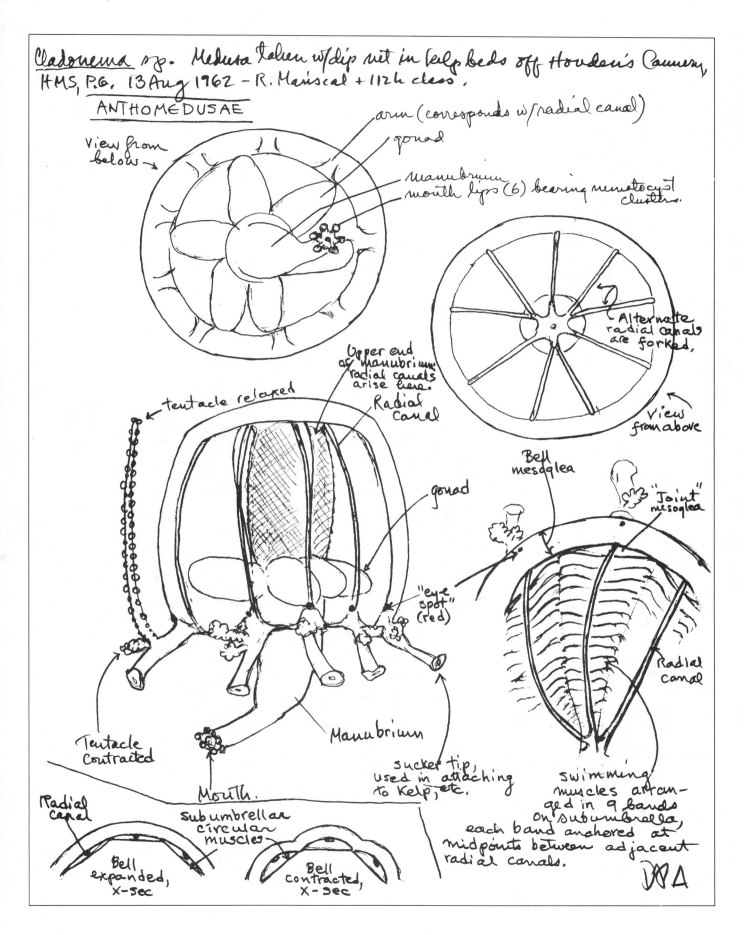

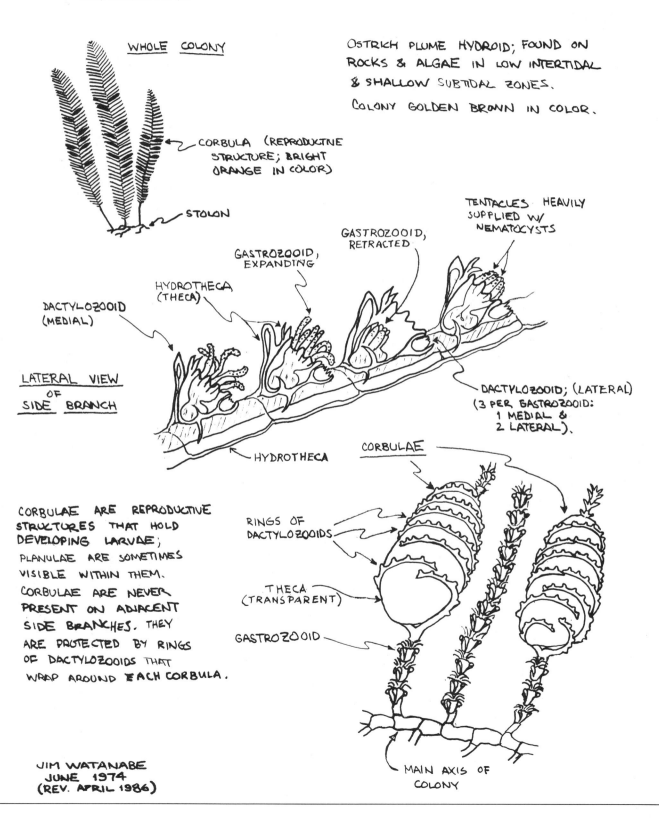

Cnidaria/Hydrozoa/Leptomedusae

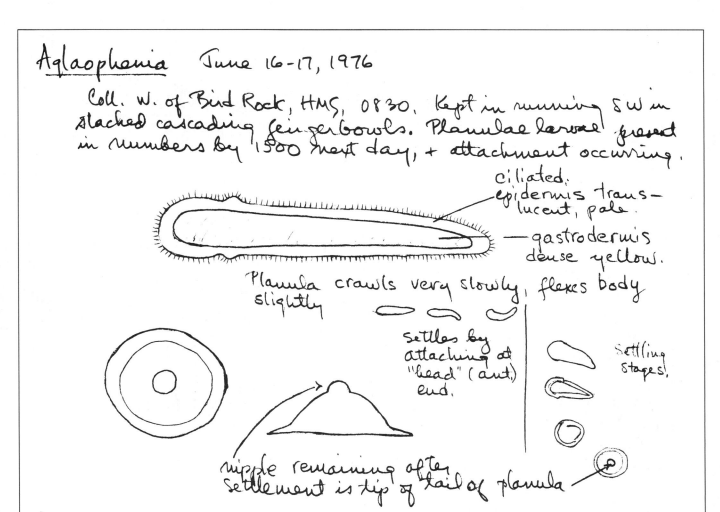

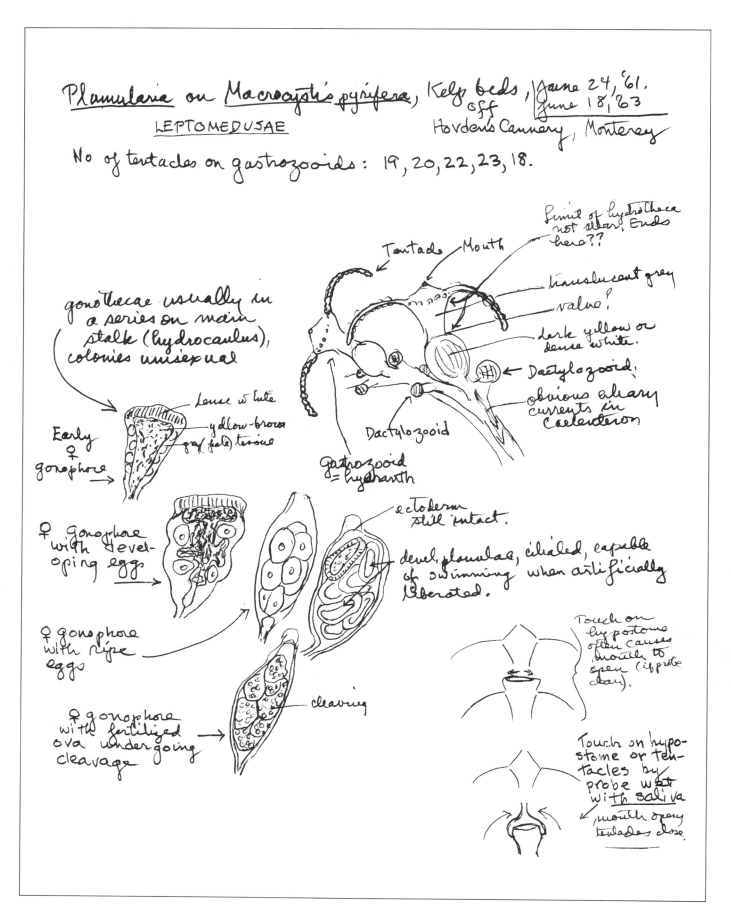

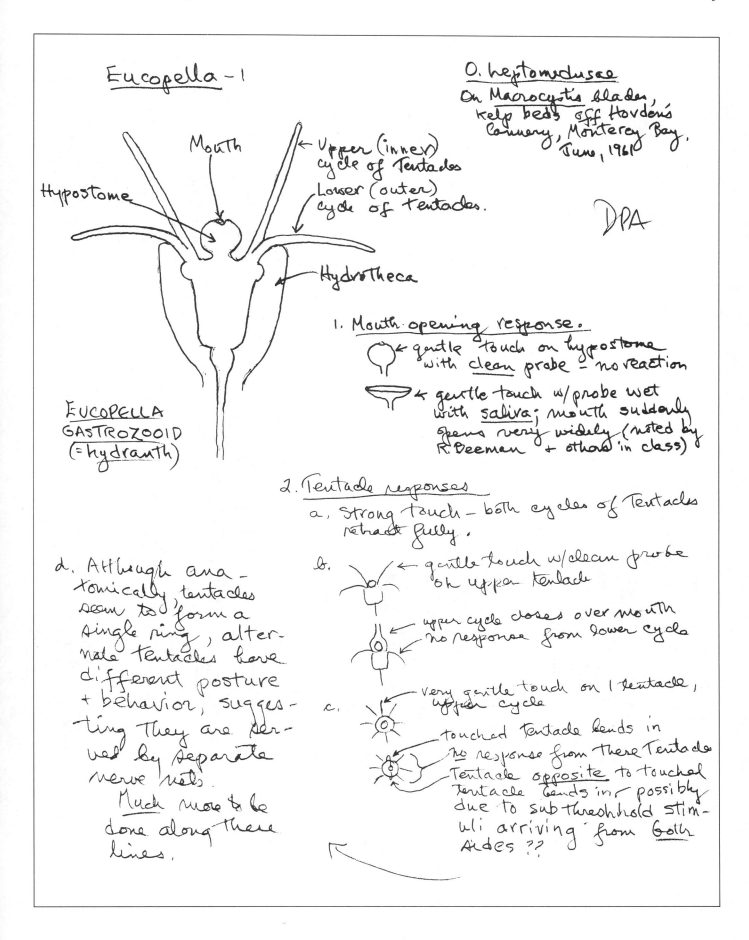

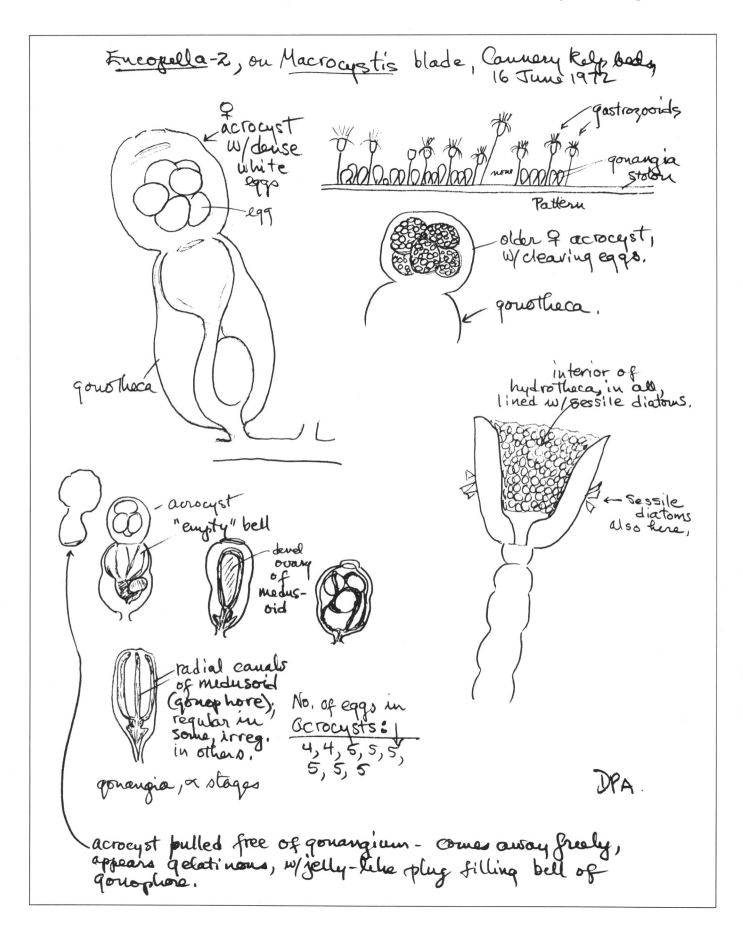

Cnidaria/Hydrozoa/Leptomedusae

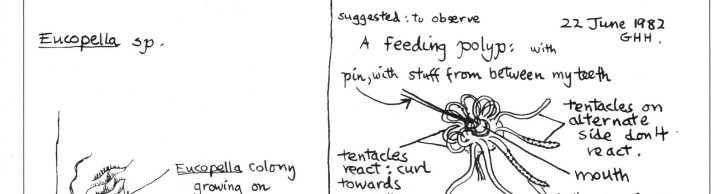

Eucopella sp.

Eucopella colony growing on Alga (red)
Life size

suggested: to observe   22 June 1982
                              GHH.
A feeding polyp: with pin, with stuff from between my teeth
- tentacles on alternate side don't react.
- mouth
- batteries of nematocysts (present on all tentacles)
- tentacles react: curl towards mouth, and right into mouth.
× 18

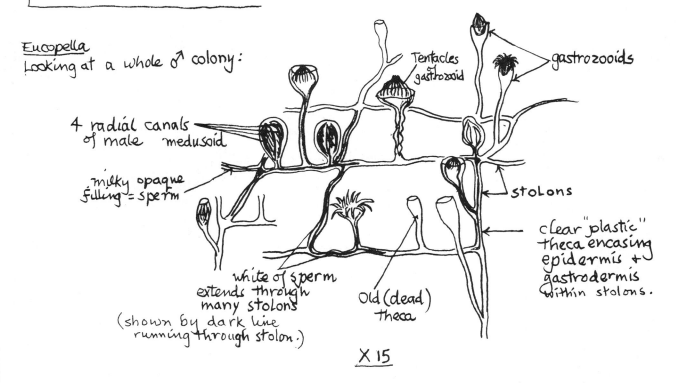

Eucopella
Looking at a whole ♂ colony:

- 4 radial canals of male medusoid
- milky opaque filling = sperm
- Tentacles gastrozooid
- gastrozooids
- stolons
- clear "plastic" theca encasing epidermis + gastrodermis within stolons.
- white of sperm extends through many stolons (shown by dark line running through stolon.)
- Old (dead) theca

× 15

This is a male colony growing on the blade of a red alga. Caprellids are clambering all over it. Several parasitic pycnogonids are present in or around thecae.

I find it interesting that the branches of stolons are often branching off at <u>close</u> to a right angle (L) from "original" stolon — it makes for good use of space. Is this related to the texture of the blade of alga?

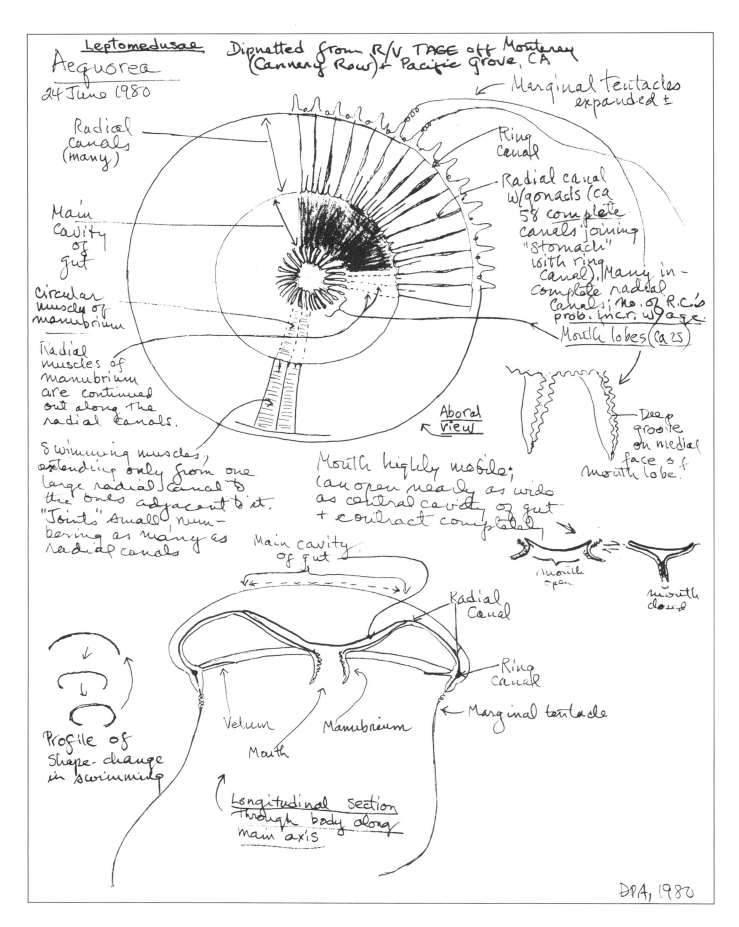

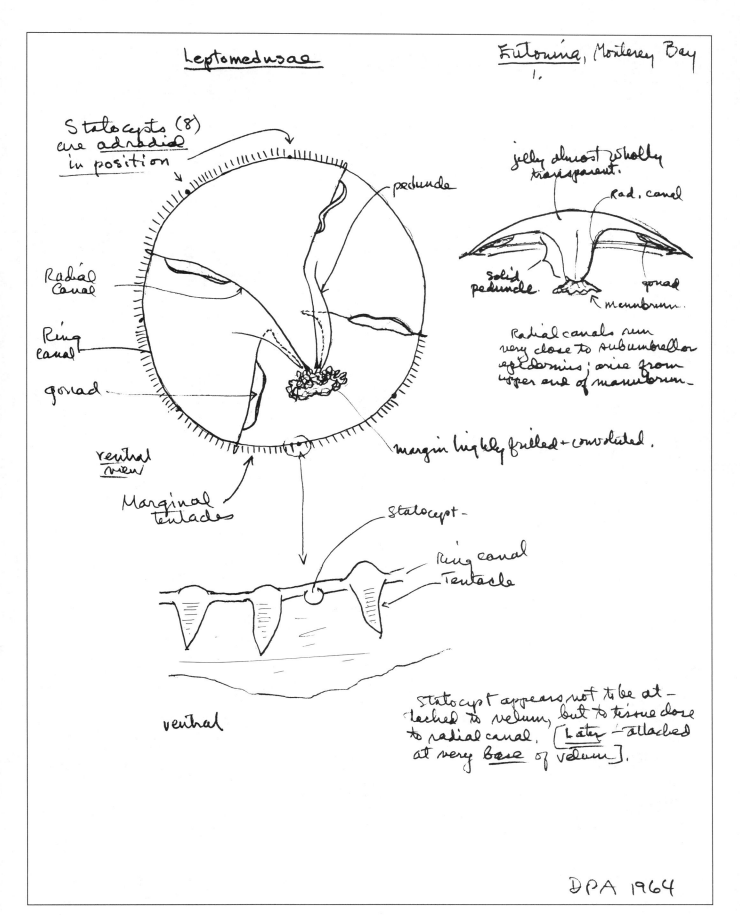

Cnidaria/Hydrozoa/Leptomedusae

Eutonina 2. — dipnetted outside *Macrocystis* beds at HMS, 19 July 1965.

Leptomedusae

— gametes liberated by a few specimens.
♂'s had either active sperm, or balls of spermatids w/ tails pointing outward
♀'s w/ eggs.

Peduncle mobile, apparently moved by radial muscles, running along radial canals.

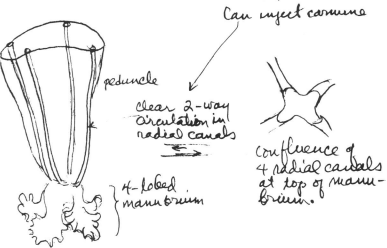

- peduncle
- Can inject carmine
- clear 2-way circulation in radial canals
- 4-lobed manubrium
- confluence of 4 radial canals at top of manubrium.

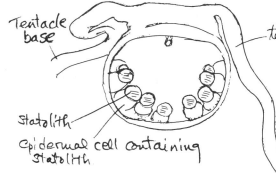

- Tentacle base
- tentacle base (ciliated)
- Statolith
- Epidermal cell containing statolith
- Statocyst.

Epidermal cells bearing statoliths remain broadly attached basally; Ends bearing statoliths are free in inner cavity of vesicle.
Specimens w/ 8, 9, + 10 statoliths seen in class.

DWA 1965

Cnidaria/Hydrozoa/Limnomedusae

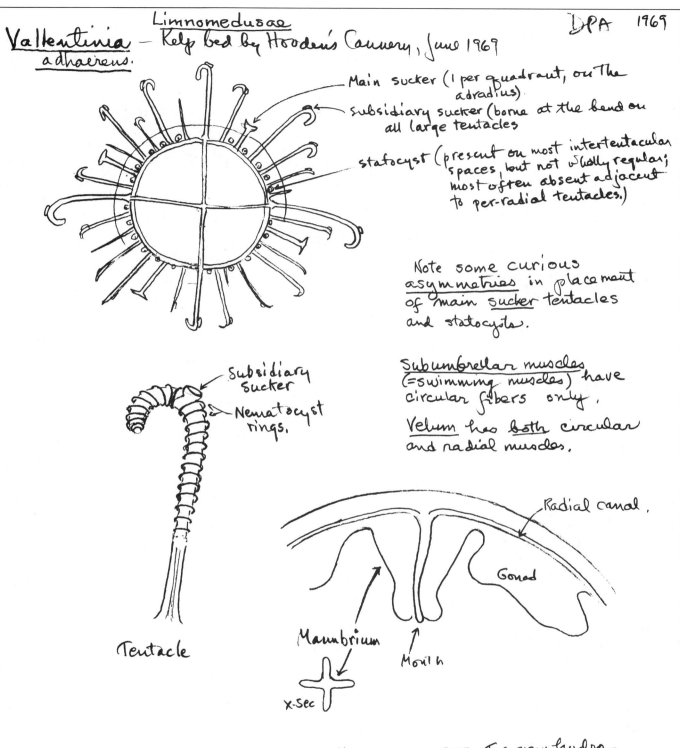

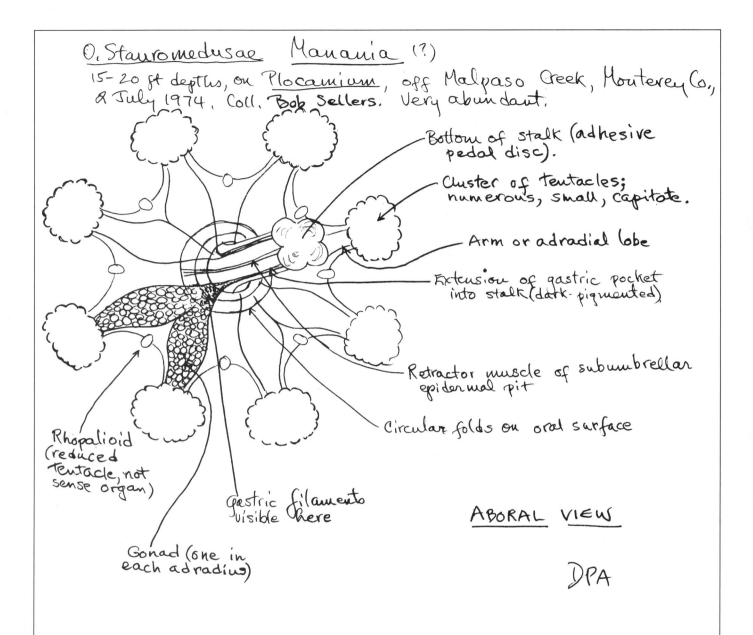

Cnidaria/Scyphozoa/Semaeostomeae

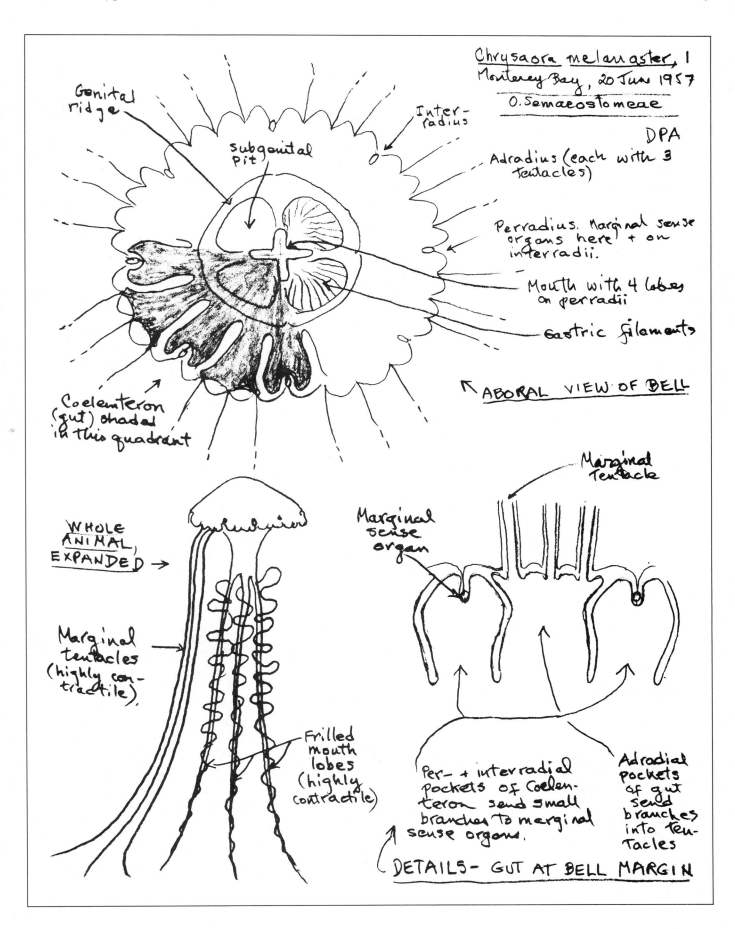

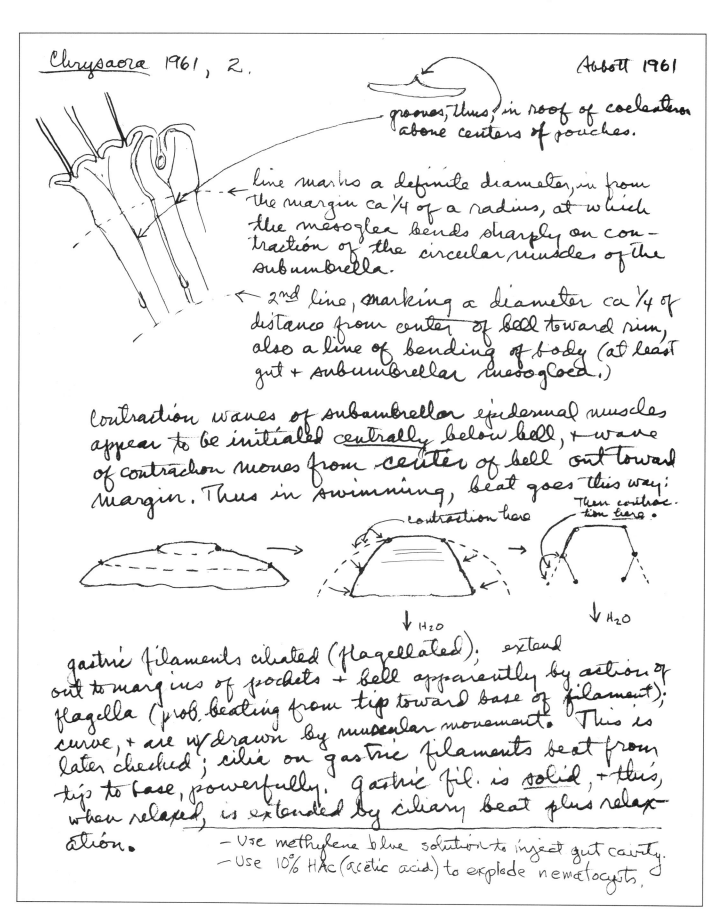

Chrysaora 1961, 2.   Abbott 1961

→ grooves, thus, in roof of coelenteron above centers of pouches.

← line marks a definite diameter, in from the margin ca ¼ of a radius, at which the mesoglea bends sharply on contraction of the circular muscles of the subumbrella.

← 2nd line, marking a diameter ca ¼ of distance from center of bell toward rim, also a line of bending of body (at least gut + subumbrellar mesoglea.)

Contraction waves of subumbrellar epidermal muscles appear to be initiated centrally below bell, + wave of contraction moves from center of bell out toward margin. Thus in swimming, beat goes this way:

contraction here → Then contraction here.

↓ H₂O   ↓ H₂O

Gastric filaments ciliated (flagellated); extend out to margins of pockets + bell apparently by action of flagella (prob. beating from tip toward base of filament); curve, + are w/drawn by muscular movement. This is later checked; cilia on gastric filaments beat from tip to base, powerfully. Gastric fil. is solid, + thus, when relaxed, is extended by ciliary beat plus relaxation.

— Use methylene blue solution to inject gut cavity.
— Use 10% HAc (acetic acid) to explode nematocysts.

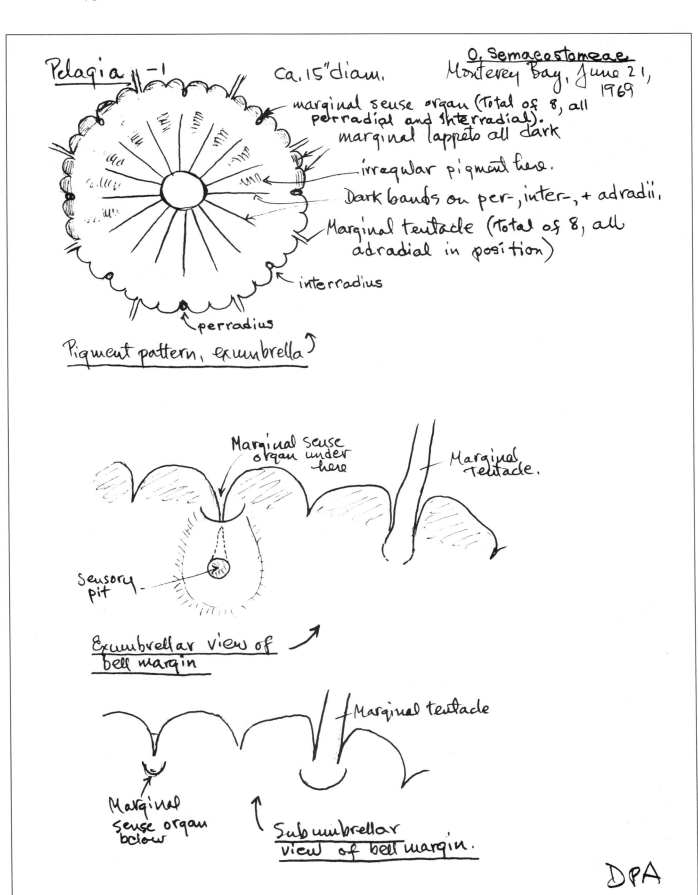

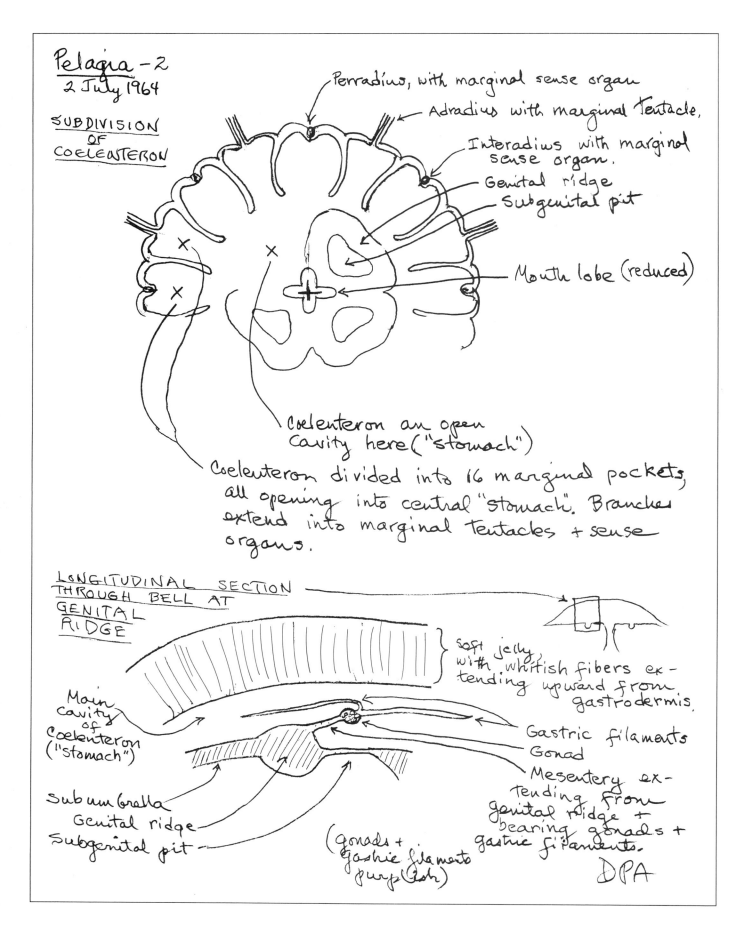

Cnidaria/Scyphozoa/Semaeostomeae

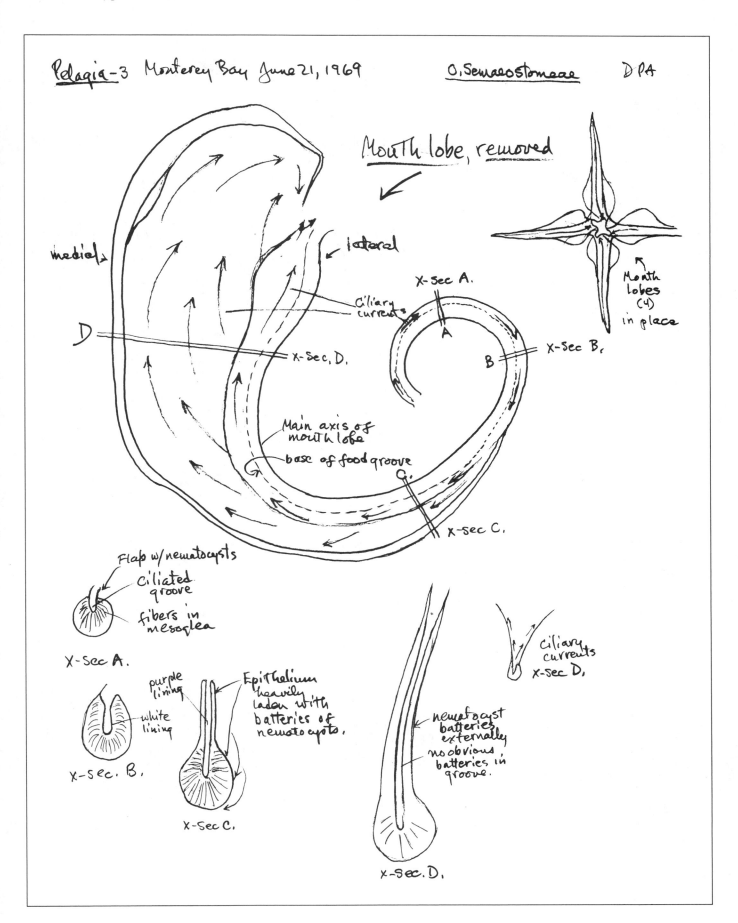

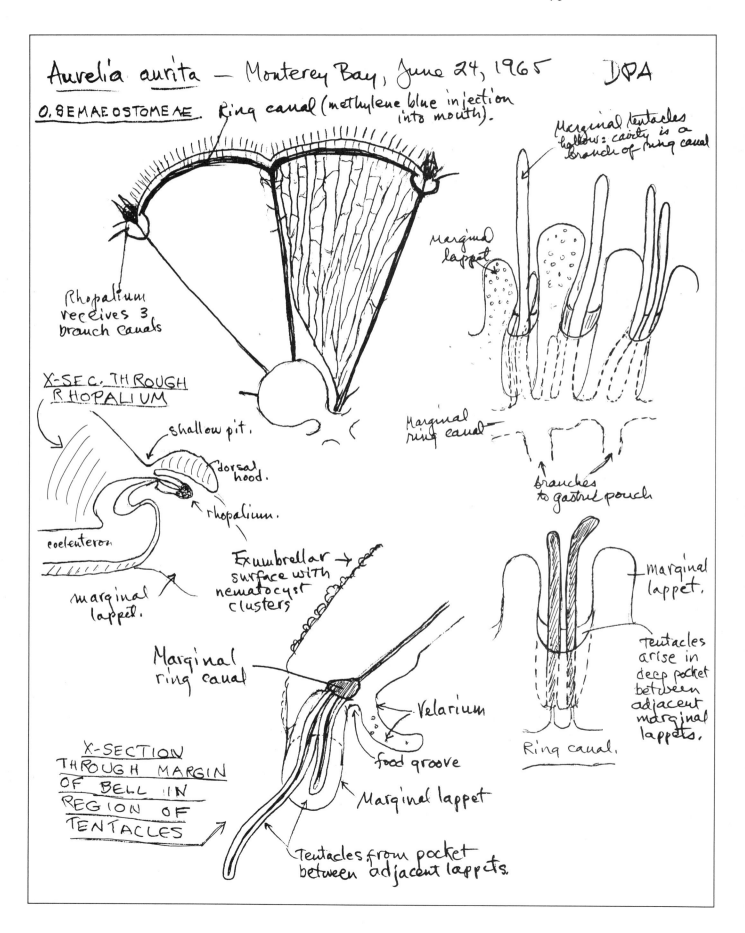

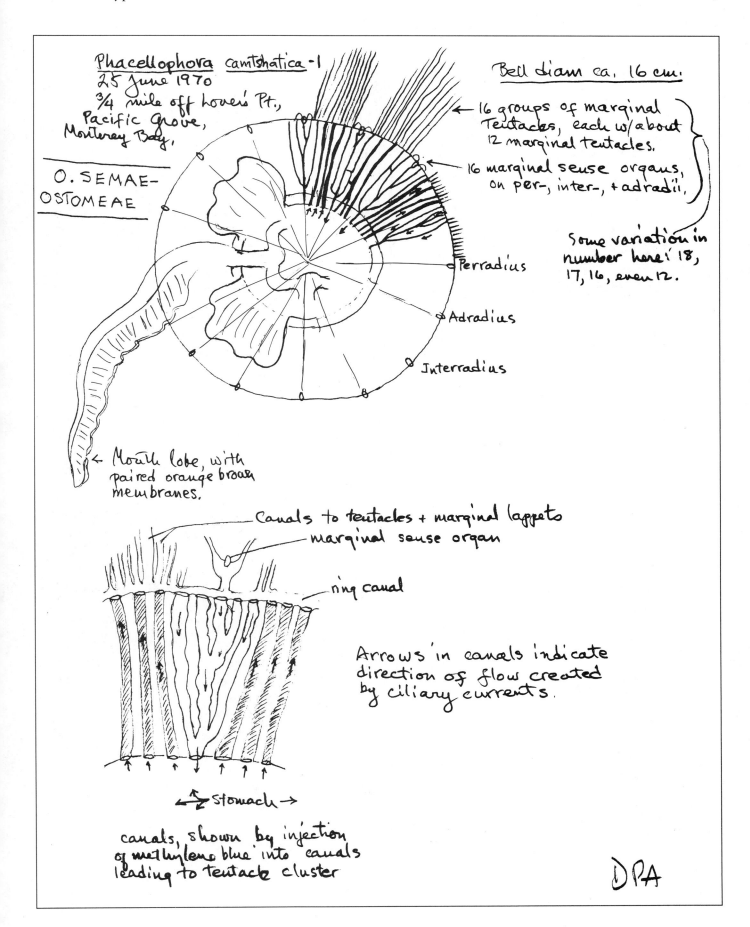

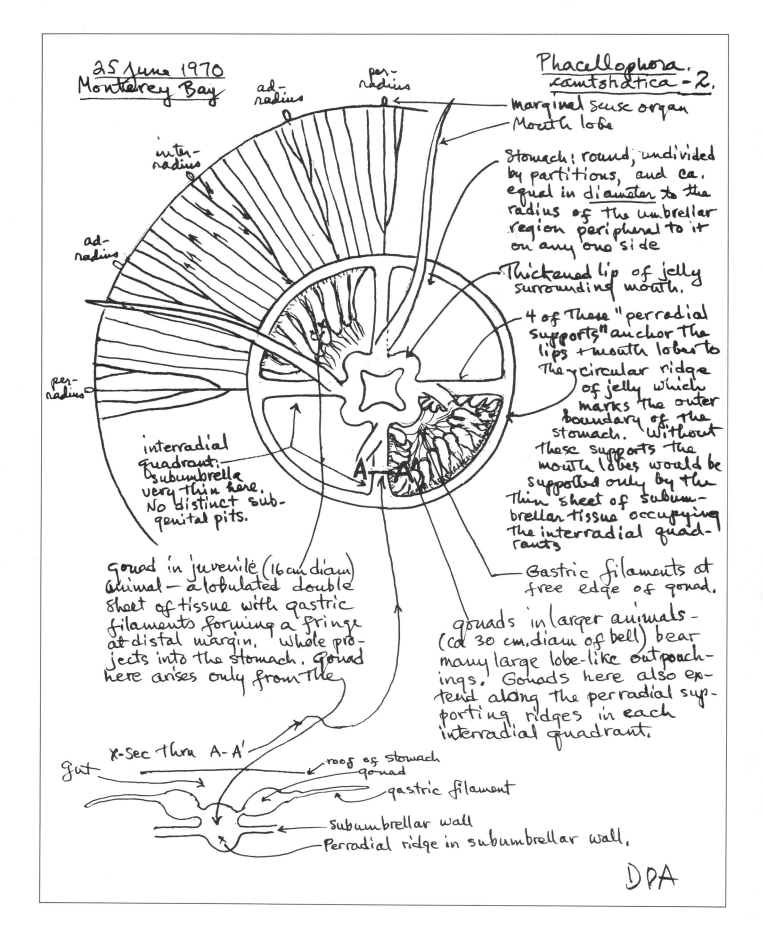

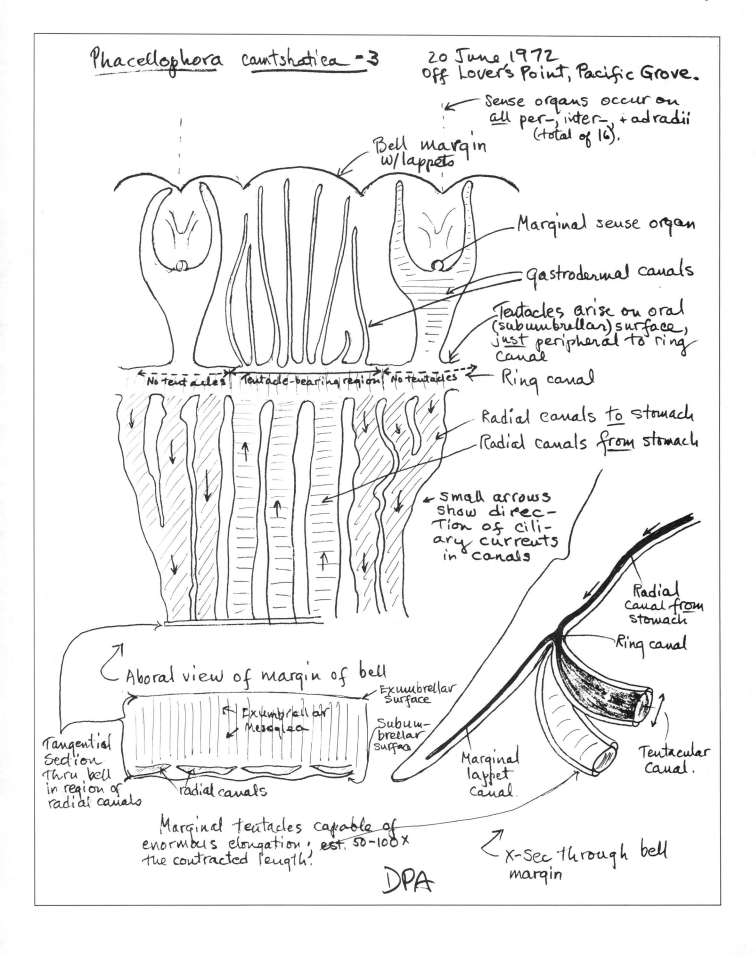

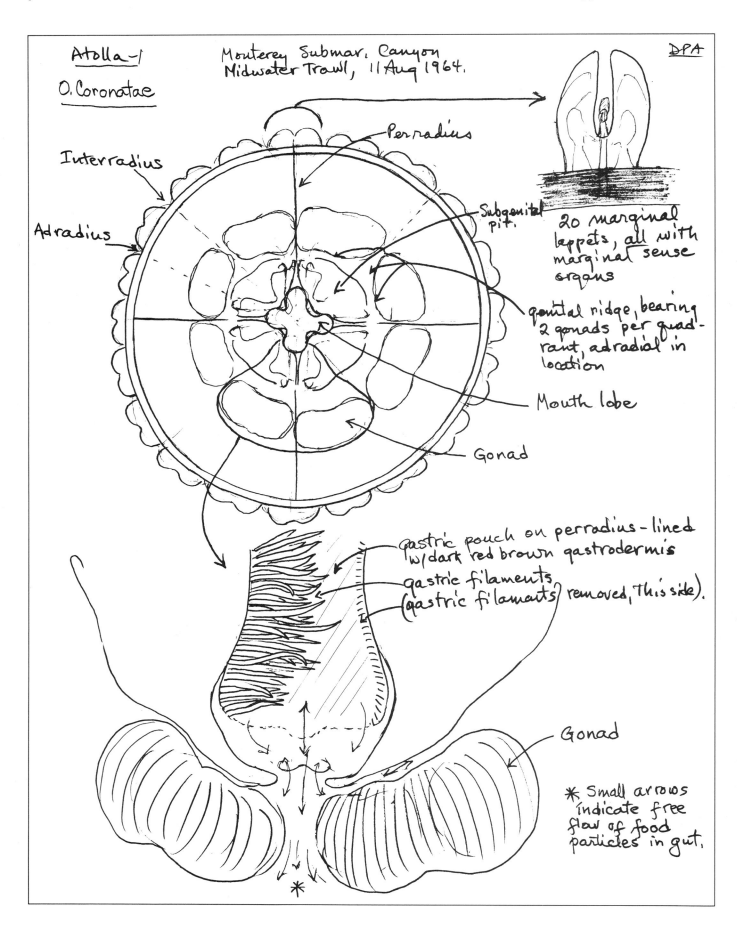

Cnidaria/Scyphozoa/Coronatae

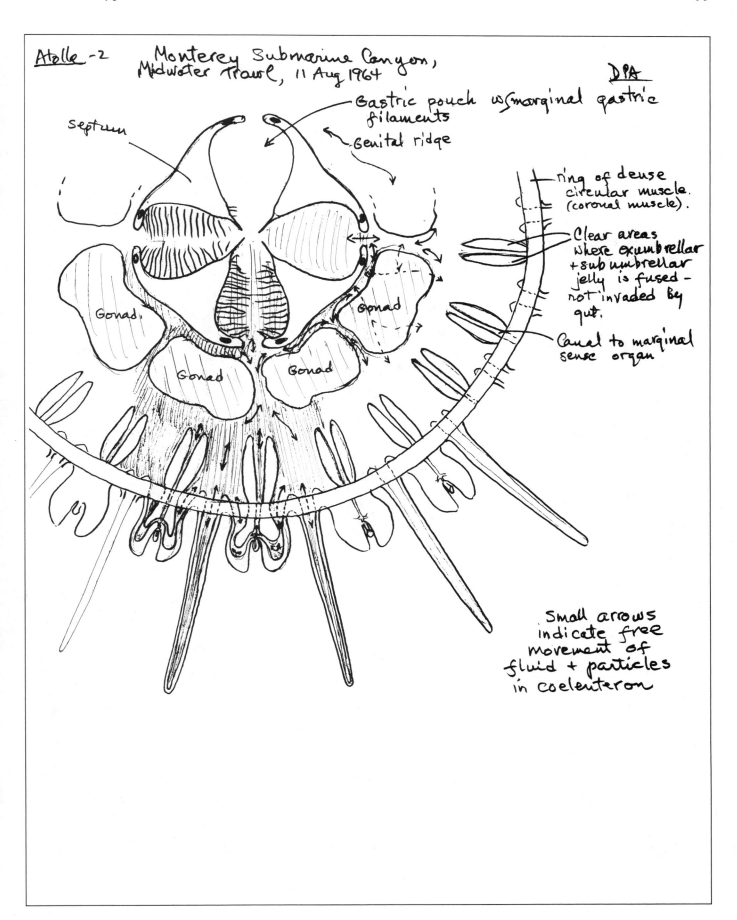

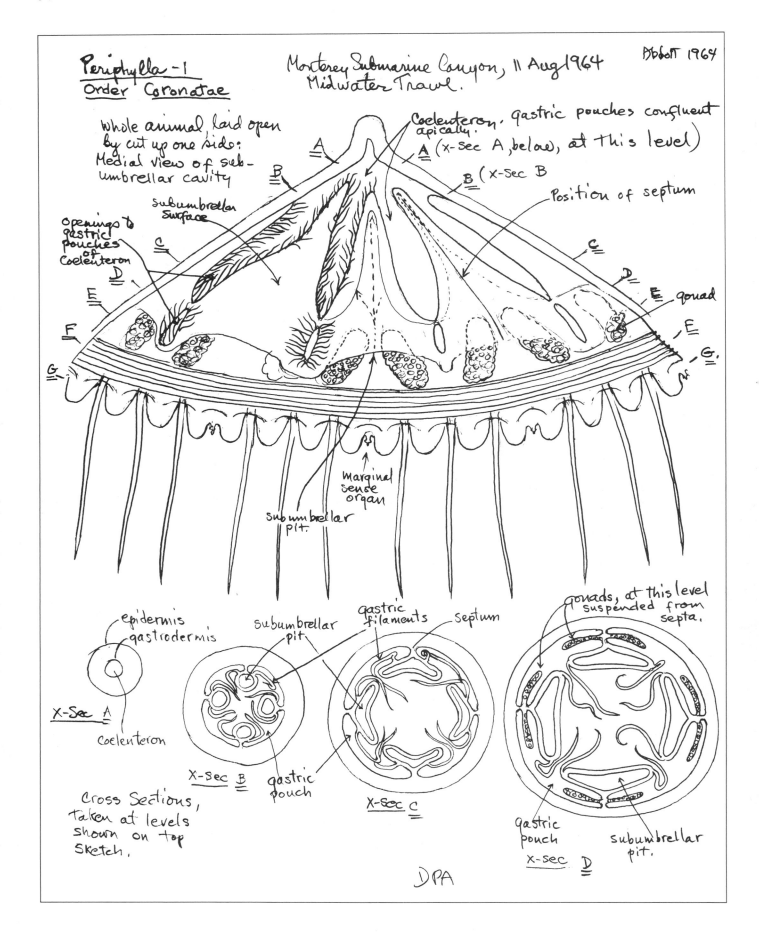

Cnidaria/Scyphozoa/Coronatae

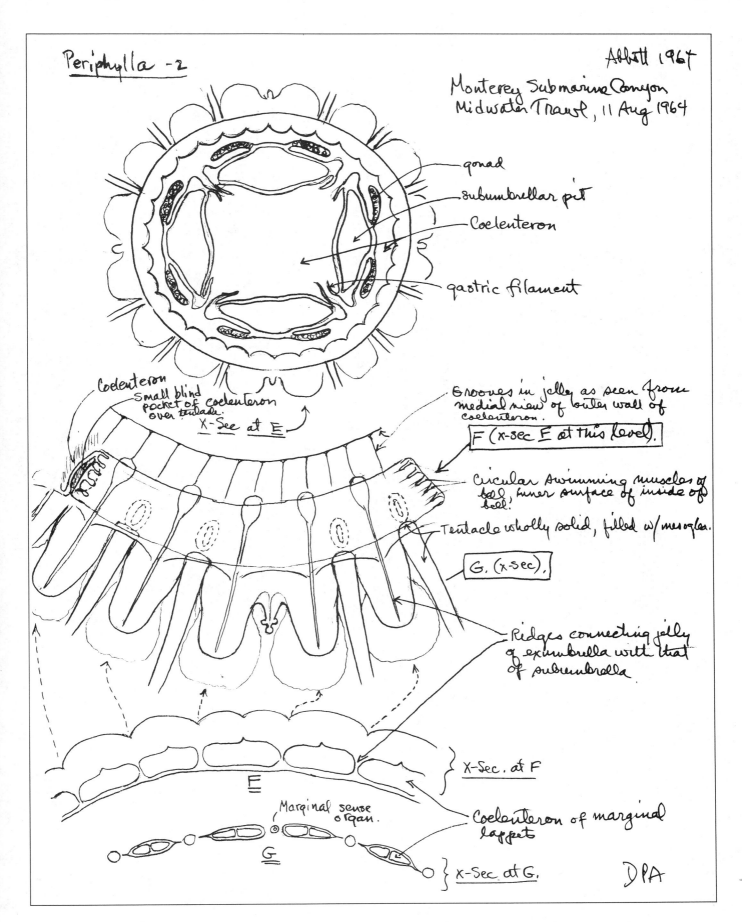

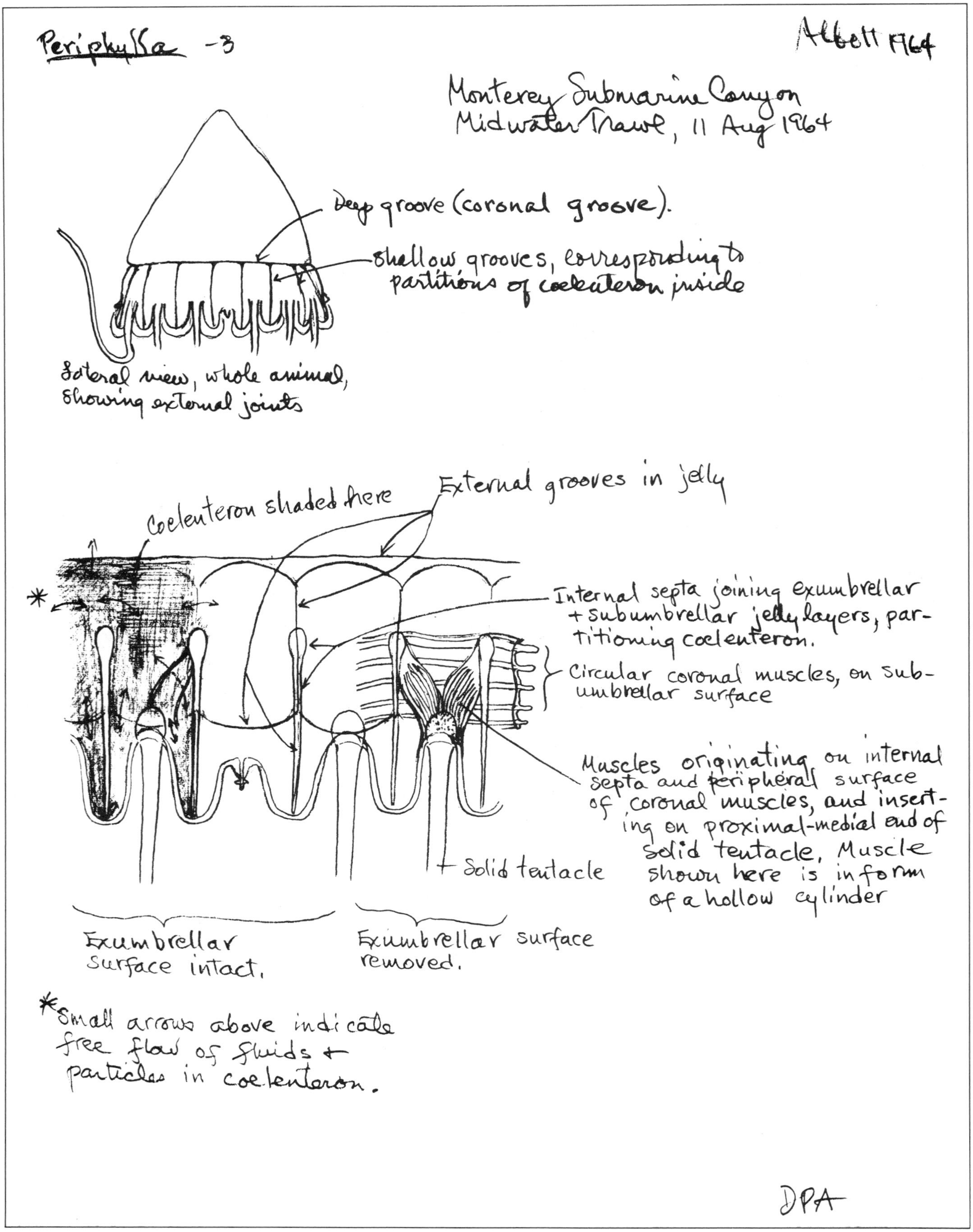

Periphylla – 3   Abbott 1964

Monterey Submarine Canyon
Midwater Trawl, 11 Aug 1964

Deep groove (coronal groove).

Shallow grooves, corresponding to partitions of coelenteron inside

Lateral view, whole animal, showing external joints

Coelenteron shaded here — External grooves in jelly

Internal septa joining exumbrellar + subumbrellar jelly layers, partitioning coelenteron.

Circular coronal muscles, on subumbrellar surface

Muscles originating on internal septa and peripheral surface of coronal muscles, and inserting on proximal-medial end of solid tentacle. Muscle shown here is in form of a hollow cylinder

Solid tentacle

Exumbrellar surface intact.   Exumbrellar surface removed.

*Small arrows above indicate free flow of fluids + particles in coelenteron.

DPA

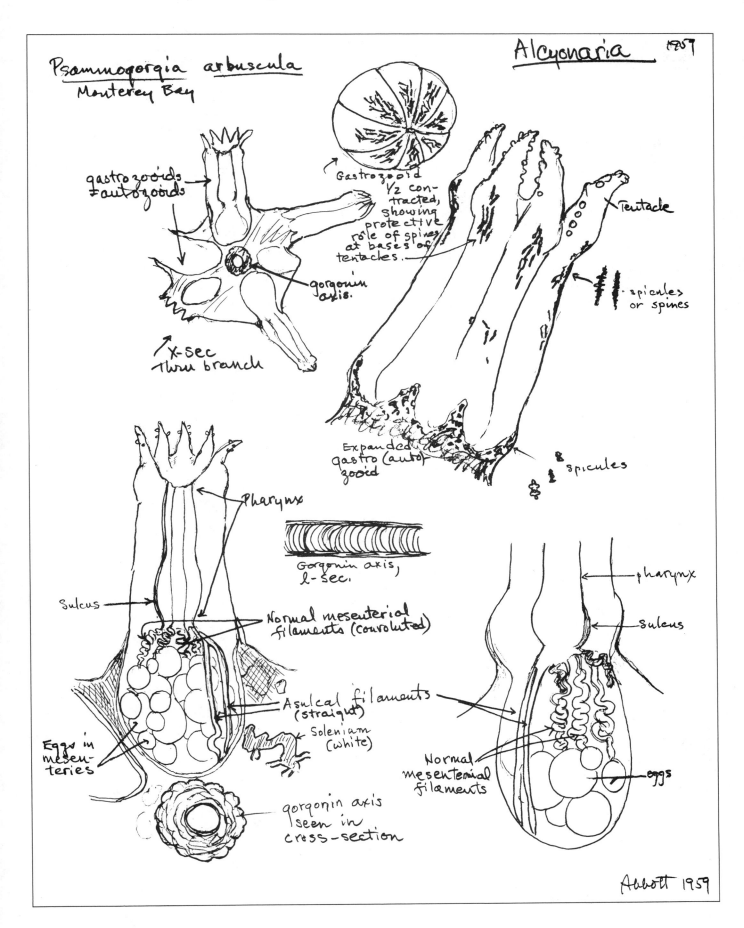

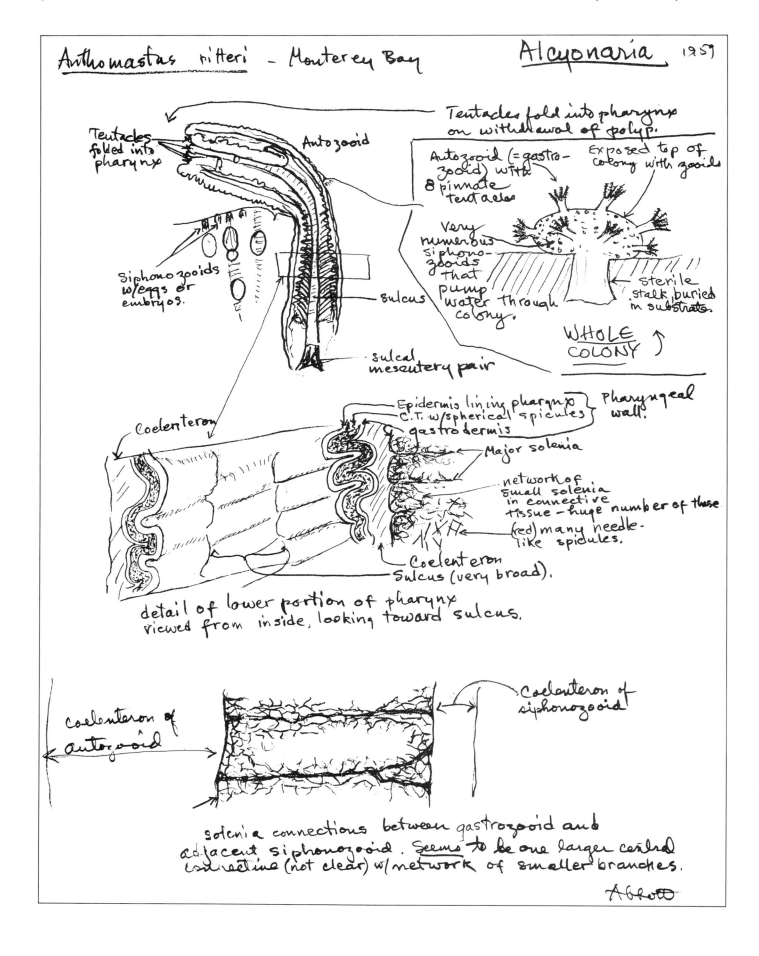

Cnidaria/Anthozoa/Alcyonaria/Alcyonacea

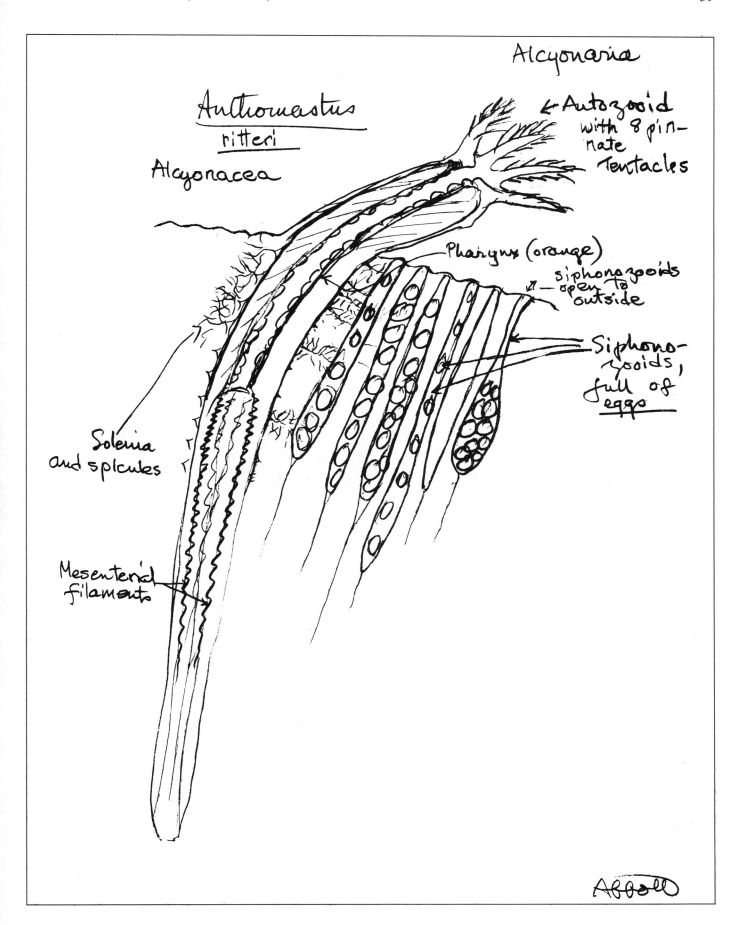

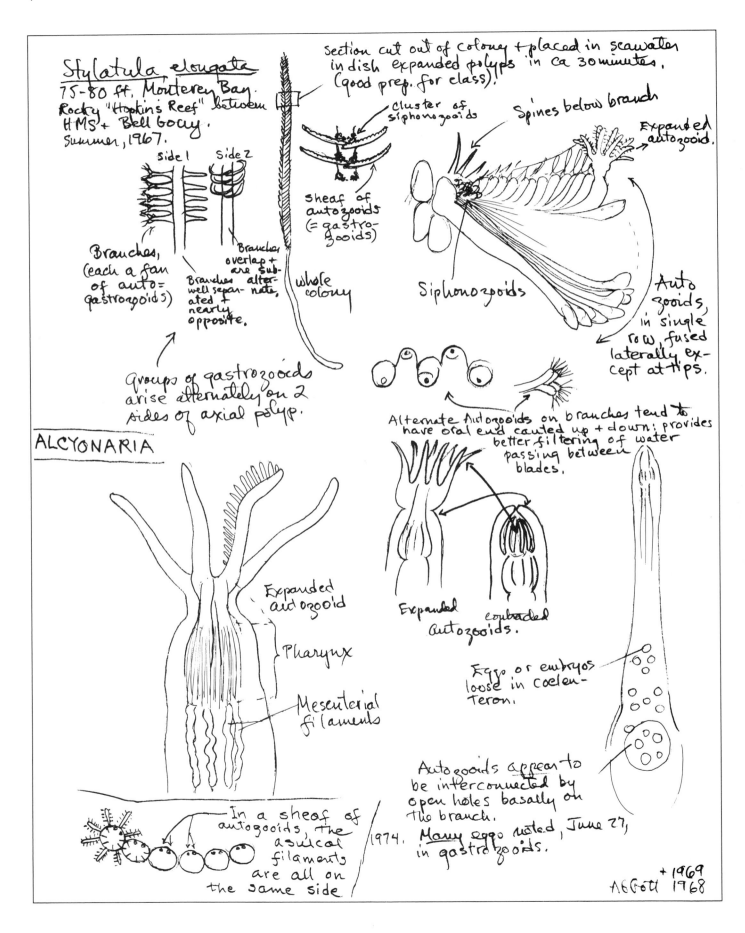

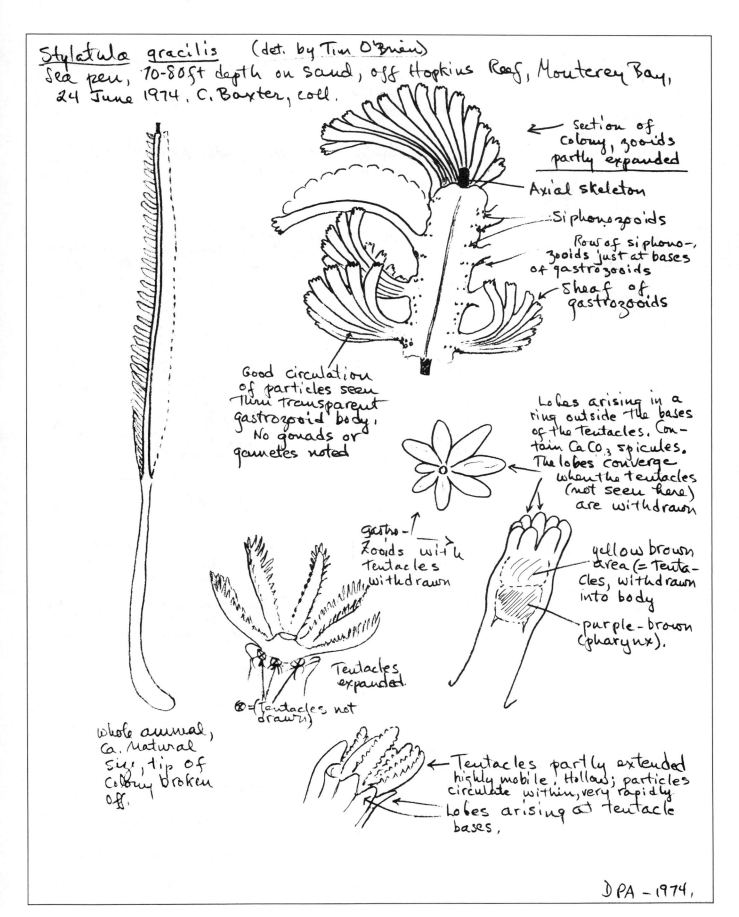

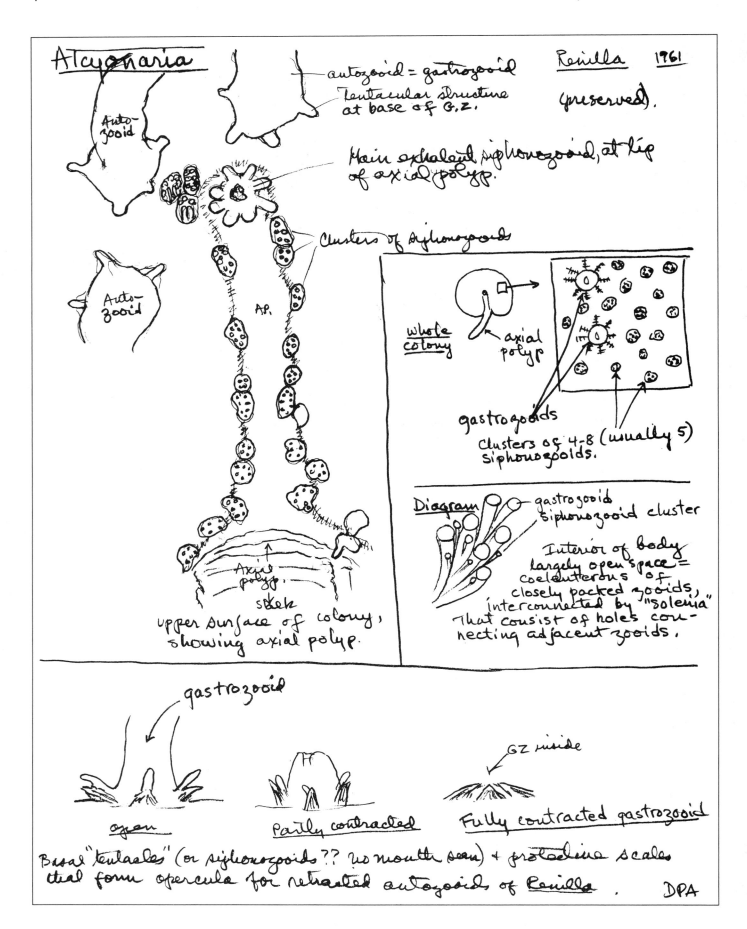

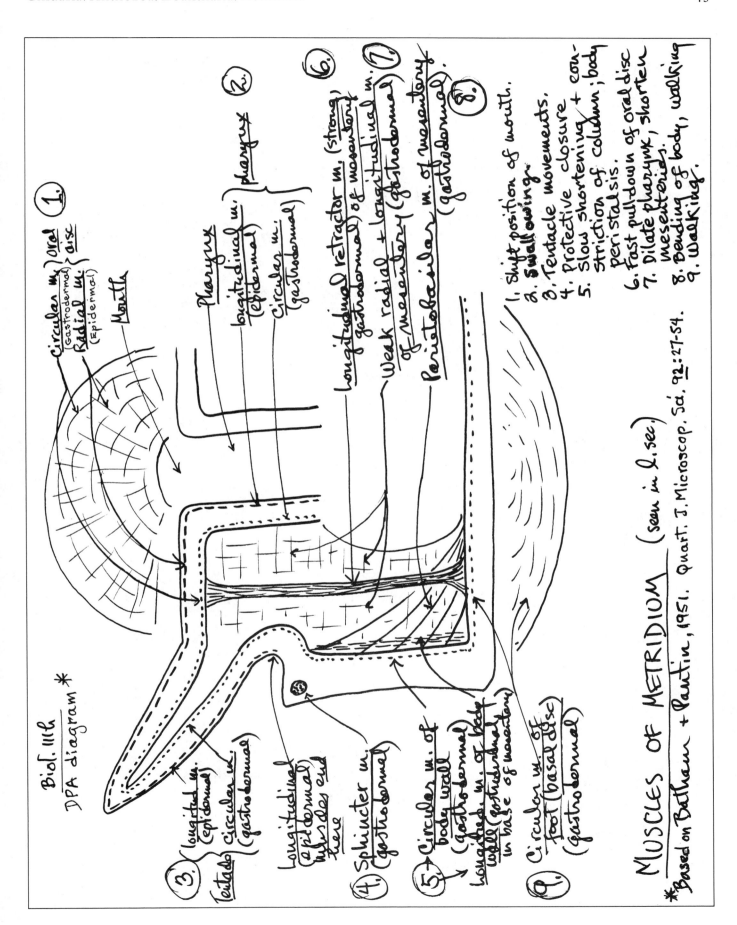

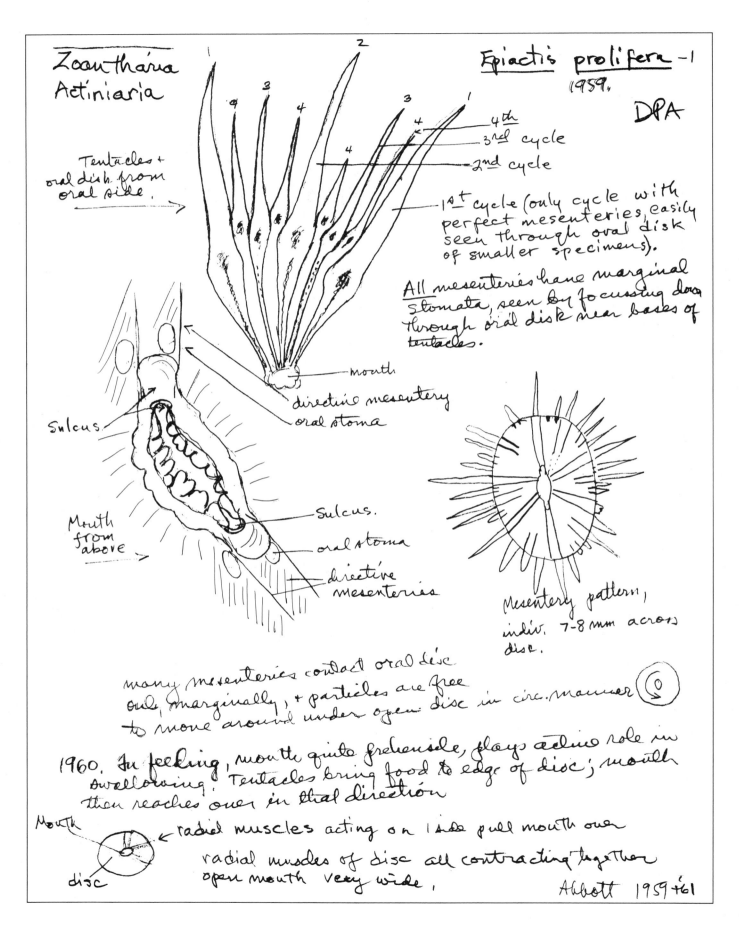

Epiactis prolifera - 2.
Zoantharia, Actiniaria

DPA 1961

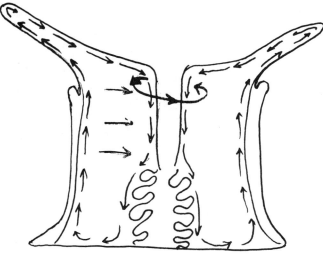

Circulation in Coelenteron of juv., disc diam. 2.5mm, removed from parent.

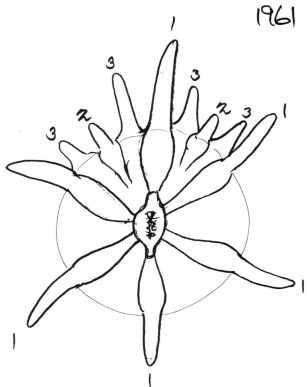

Oral view of juvenile *Epiactis*, showing different orders of tentacles.

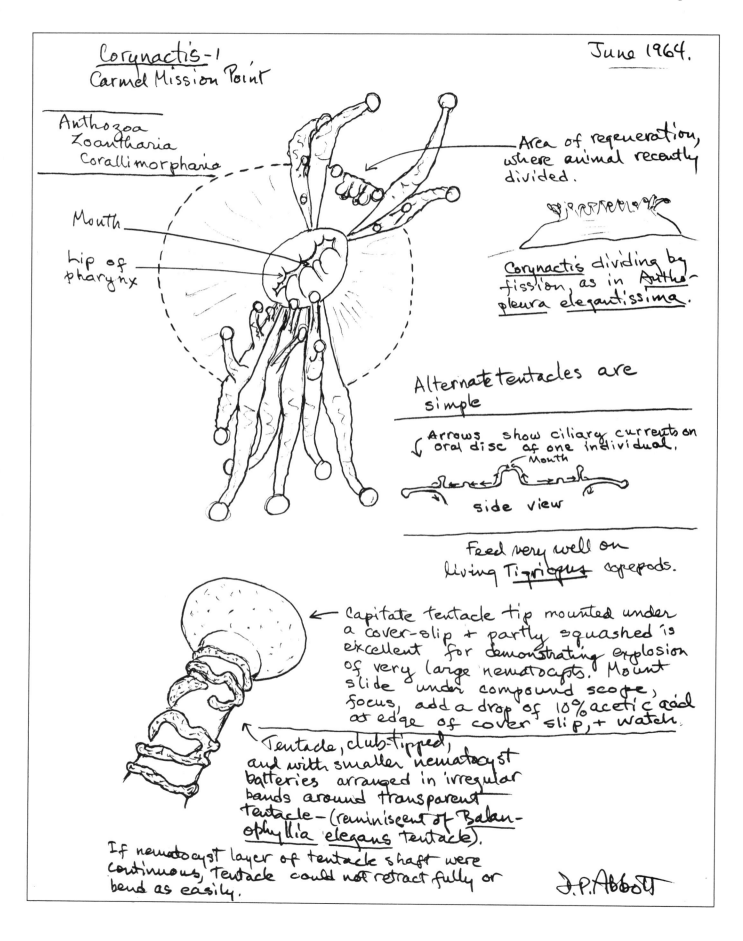

Corynactis-2
Carmel Mission Point
25 June 1964

Corallimorpharia   DPA

— 2 *Corynactis* in dish of seawater.
— Few *Tigriopus* copepods added to dish.

#1, fully expanded
#2, ca 1/3 expanded.

#1: instantly stung + immobilized copepods that touched tentacles.

#2: copepods touching + even clinging to tentacles elicited no response.

— 1 min. later, piece of egg yolk placed on tentacles on one side:

yolk added, stimulating nematocyst discharge.

Less than one second later, two copepods that were crawling on tentacles on opposite side were also stung + immobilized.

Only one observation but leaves impression of a feeding response in #2 "waked up" on contact with yolk; as though something happened to lower threshold of nematocysts to mechanical/chemical stimulation, + this spread rapidly to opposite side of body resulting in capture of copepods.

(Later, 1968, Fred McCollum (U.C. Berkeley) reports essentially same thing).

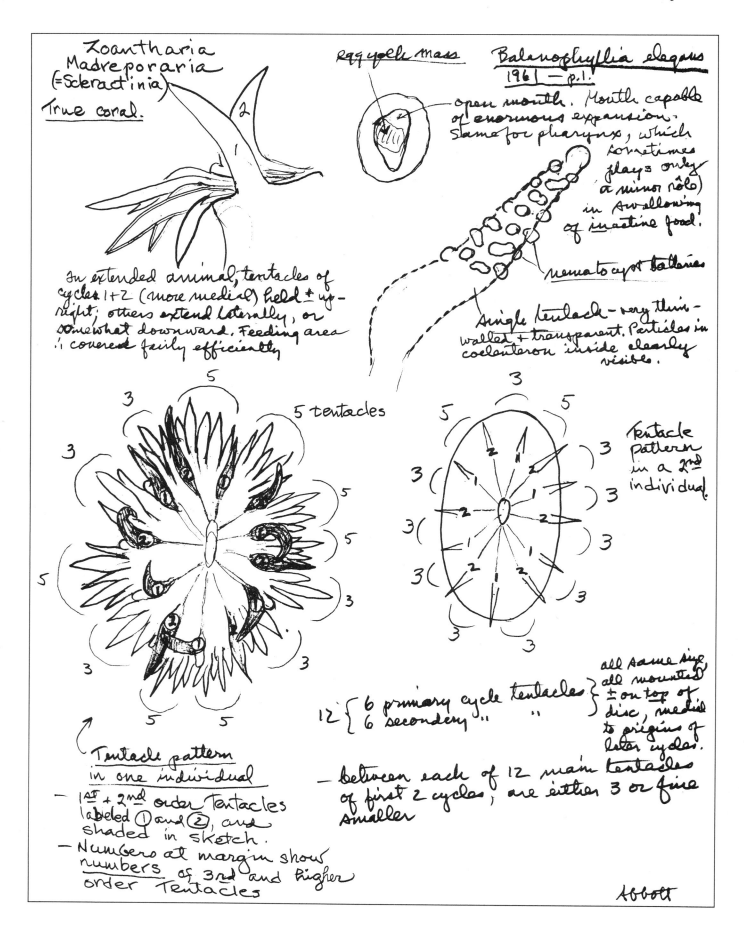

Balanophyllia elegans — 2, 1962
DPA

Feeding

B. elegans readily catches active *Tigriopus*. Reaction of copepod on touching tentacle is instantaneous paralysis + adhesion to tentacle. Few copepods immobilized later struggled a bit, but majority showed no further motion.

Tentacle(s) w/ copepod retract + curve toward mouth; other tentacles remain extended. Mouth, usually at this stage upraised → ⌒ above oral disc, bends toward tentacle till copepod lies within the lips. No muscular action seen in swallowing. Copepod moved into throat by flagellar action — acting apparently in mucus layer. Copepod is drawn smoothly inward, tentacle + copepod parting with a discernible jerk as nematocysts are pulled out of either tentacle or copepod.

Beautiful directional coordination along oral disc between mouth + tentacles.

1. Tentacle catches food
2. Within a few seconds (10± sec) mouth bends and/or elongates directly toward tentacle w/ food, + lips open as mouth nears food.
3. On one occasion this happened

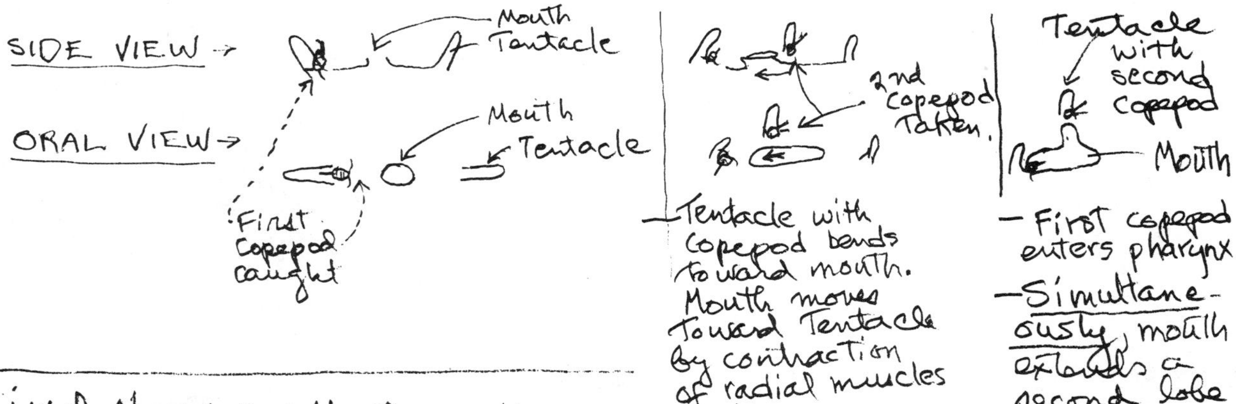

SIDE VIEW → [diagram: Mouth, Tentacle]
ORAL VIEW → [diagram: Mouth, Tentacle, First copepod caught]

[diagram: 2nd copepod taken]
— Tentacle with copepod bends toward mouth. Mouth moves toward tentacle by contraction of radial muscles of disc.
— Meanwhile a 2nd copepod is caught.

[diagram: Tentacle with second copepod, Mouth]
— First copepod enters pharynx
— Simultaneously, mouth extends a second lobe to pick up 2nd copepod

Animal shows excellent coordination along radii between mouth + tentacles, but perhaps little or no circular conduction.

(Later tried experiment with three copepods. Coral seemed "confused")

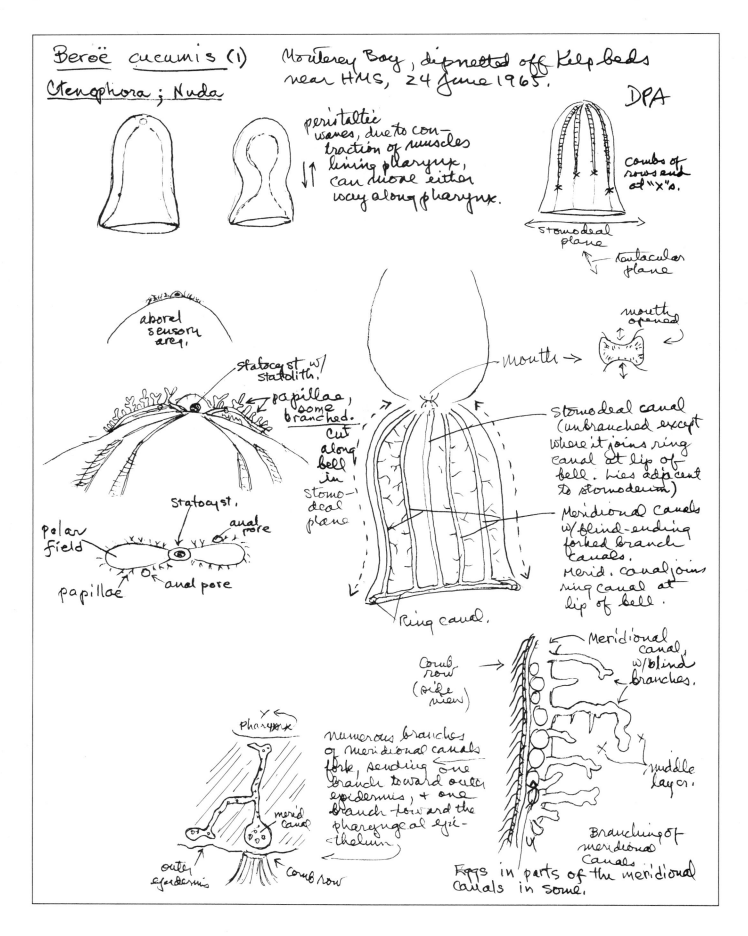

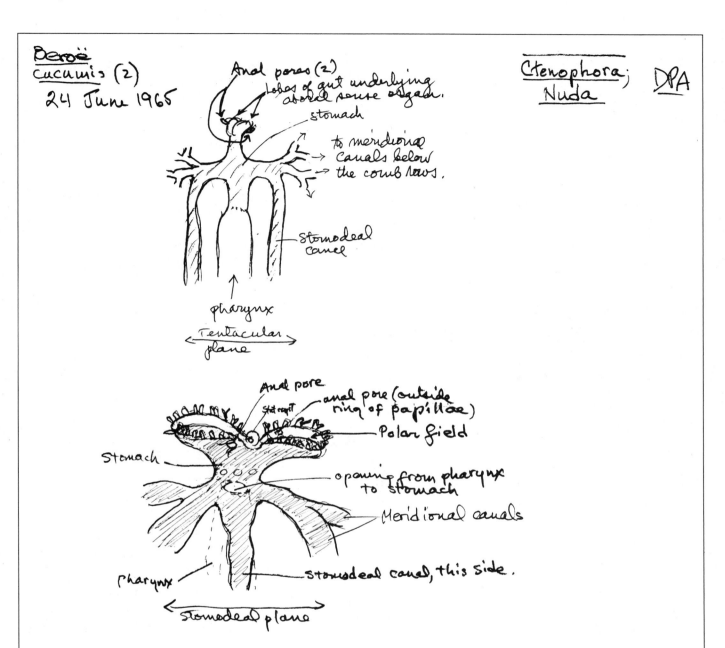

Pleurobrachia — 1
Ctenophora; Tentaculata.

Monterey Bay, June 21, 1969.
DPA

Feeding responses.

Placed with *Tigriopus* in bowl of seawater. Fed voraciously.

1. Presence of copepods in near vicinity appeared to stimulate extrusion of tentacles by contraction of tentacle sheath.

2. Copepods caught + held very effectively by colloblasts, against force of struggling animal.

3. Food brought to mouth thus (observed by Nat Howe, Bob Engel; + DPA confirmed).

① Copepod caught by colloblasts.

② Tentacle with copepod shortened by contraction till almost level with mouth.

③ Body *rotated* by opposite beat of comb rows on two sides of body, so mouth is brought up to copepod on tentacle.

④ Mouth opens + copepod is ingested mainly by strong inward ciliary action in pharynx. Part of tentacle may be drawn in with food, + must be pulled out again.

⑤ Copepod retained in pharynx for preliminary digestion.

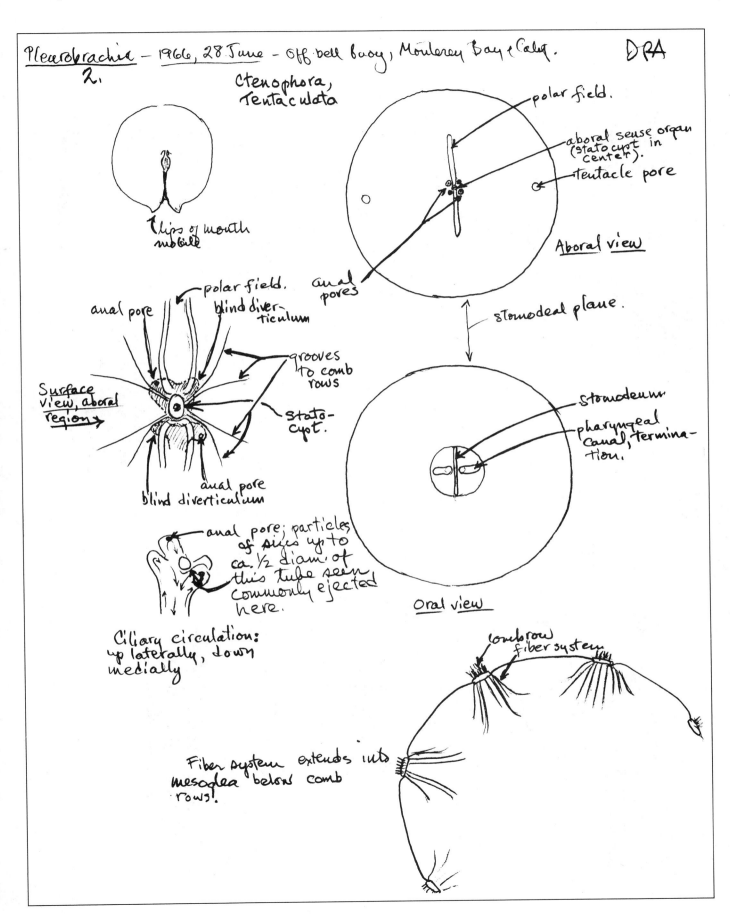

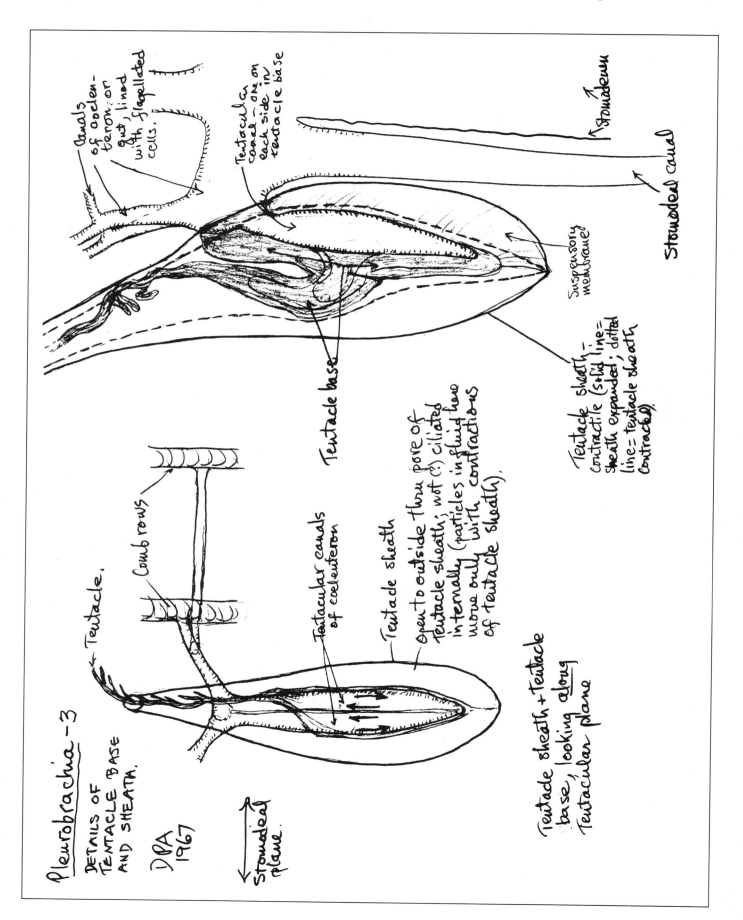

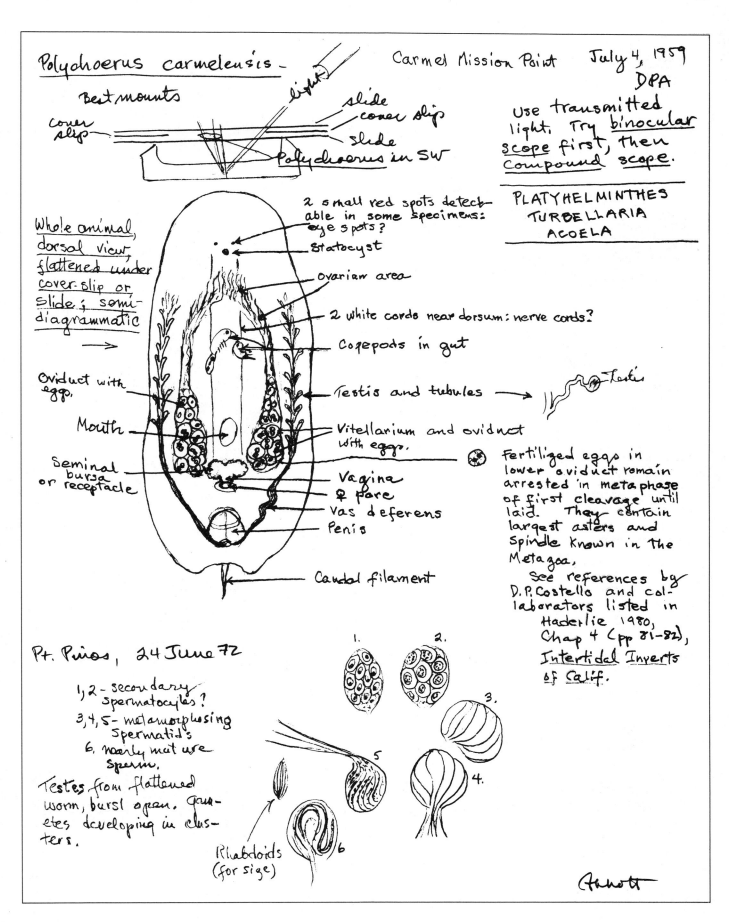

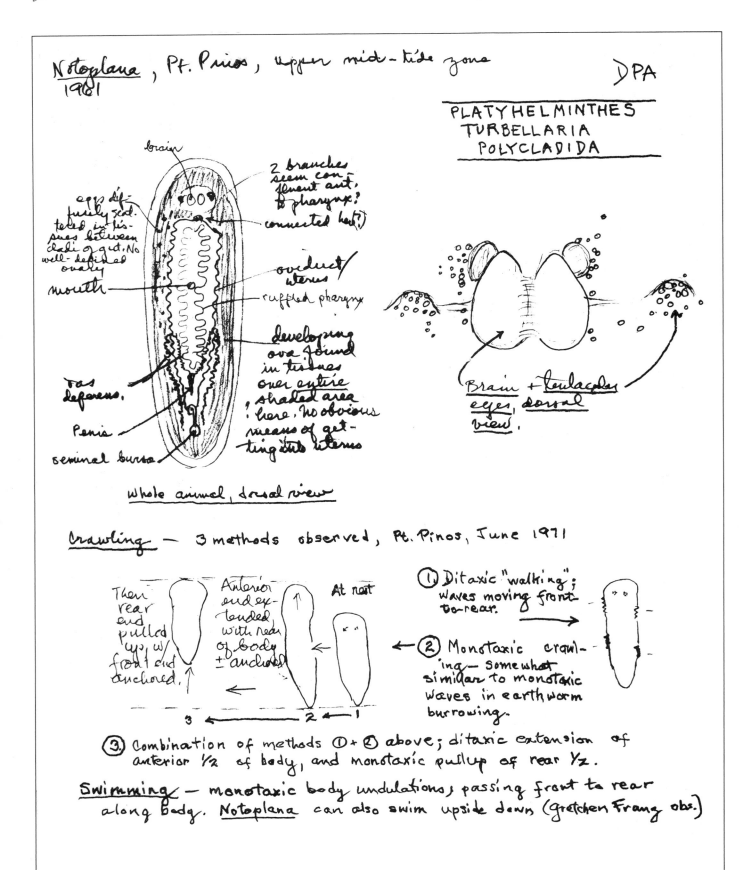

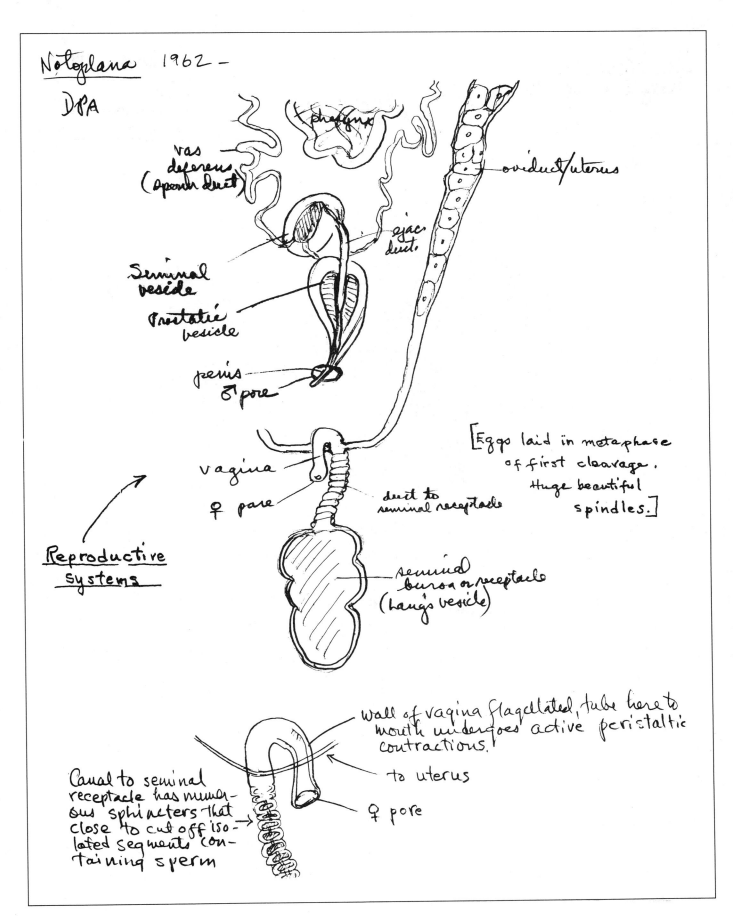

Membranipora — 1
Kelp beds off Cannery Row, Monterey

DPA 1967
ECTOPROCTA
CHEILOSTOMATA
ANASCA

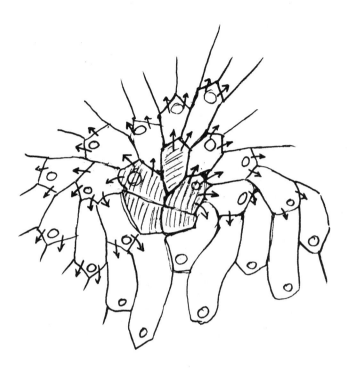

Budding pattern in vicinity
of ancestrula (later shaded)

On metamorphosis, larva generates 2 zooids
(a 3rd zooid?)
[see Hyman diagrams of colony]

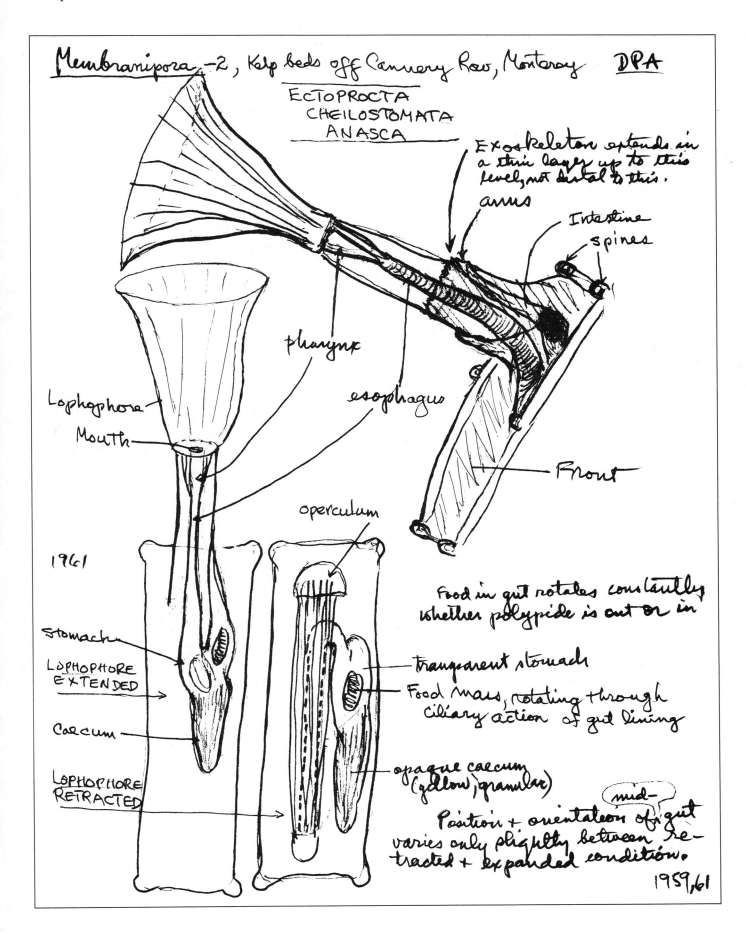

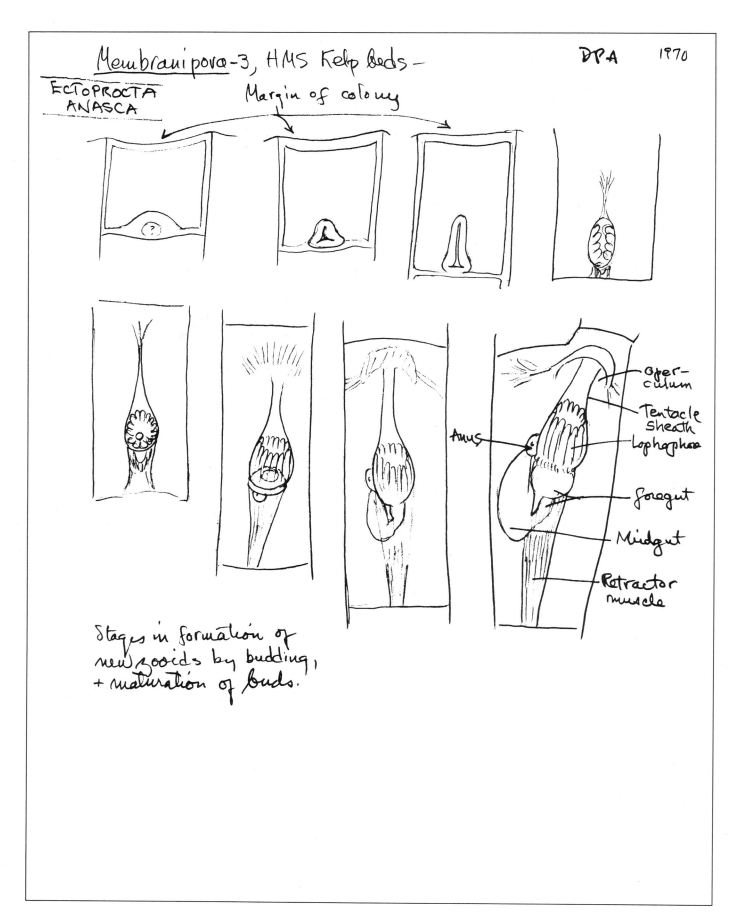

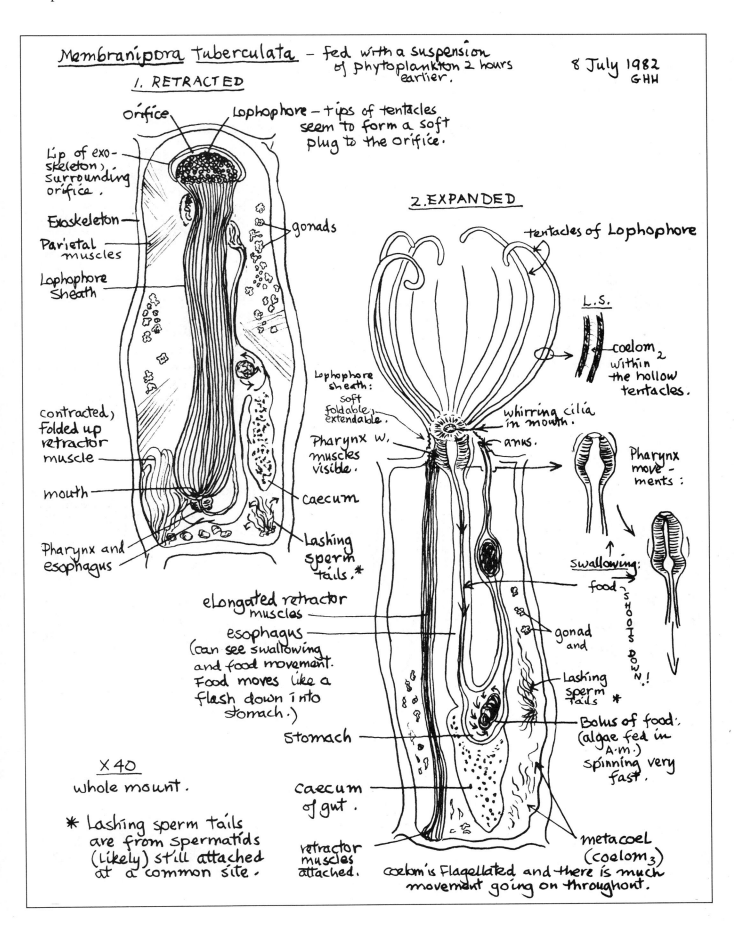

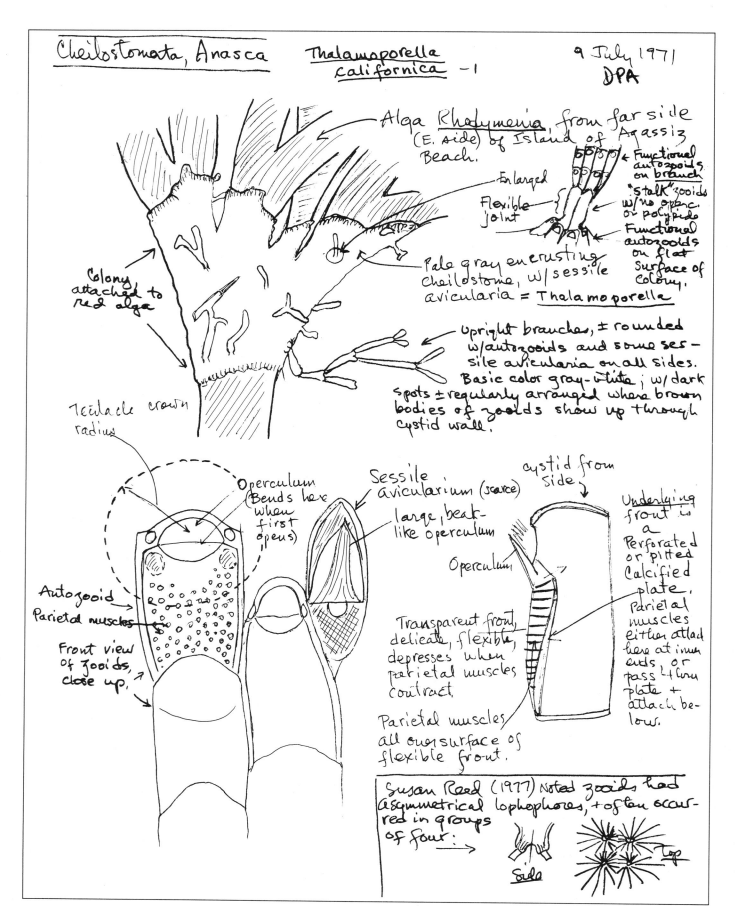

Ectoprocta/Cheilostomata/Anasca

Thalamoporella - 2
Seal Island area, HMS, among *Phyllospadix* blades.   DPA

30 June 77

Colony growing on eel grass effectively binds adjacent grass blades together. Clearly, growing tips of branches can attach to any substratum.

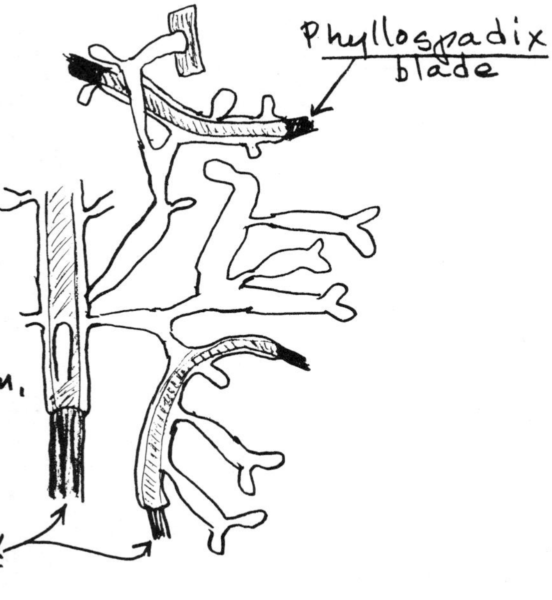

Phyllospadix blade

Phyllospadix blades

63

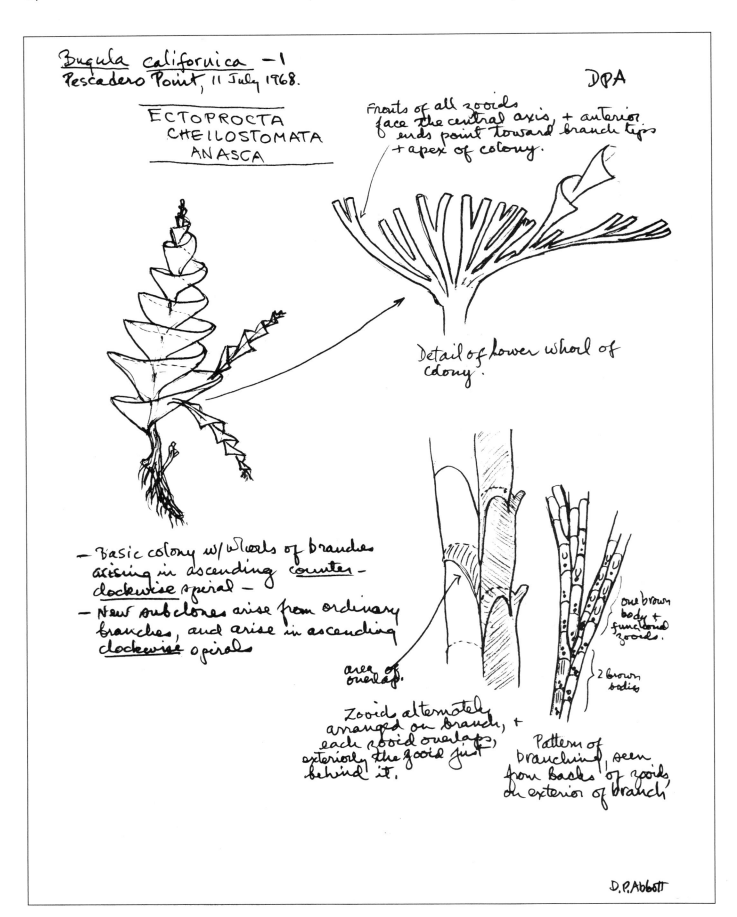

Ectoprocta/Cheilostomata/Anasca 65

Bugula californica - 2

Pescadero Pt, 27 June '72
DPA

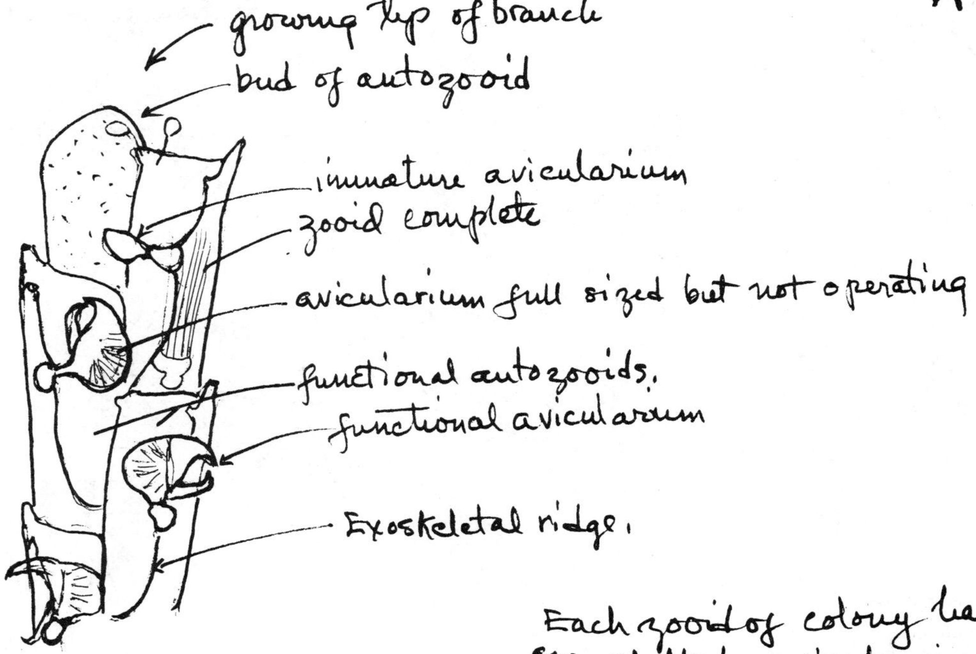

- growing tip of branch
- bud of autozooid
- immature avicularium
- zooid complete
- avicularium full sized but not operating
- functional autozooids.
- functional avicularium
- Exoskeletal ridge.

Tip of branch, 2nd whorl from colony apex, medial view.

Each zooid of colony has one stalked avicularium, about ½ way back on zooecium, borne along an exoskeletal ridge.

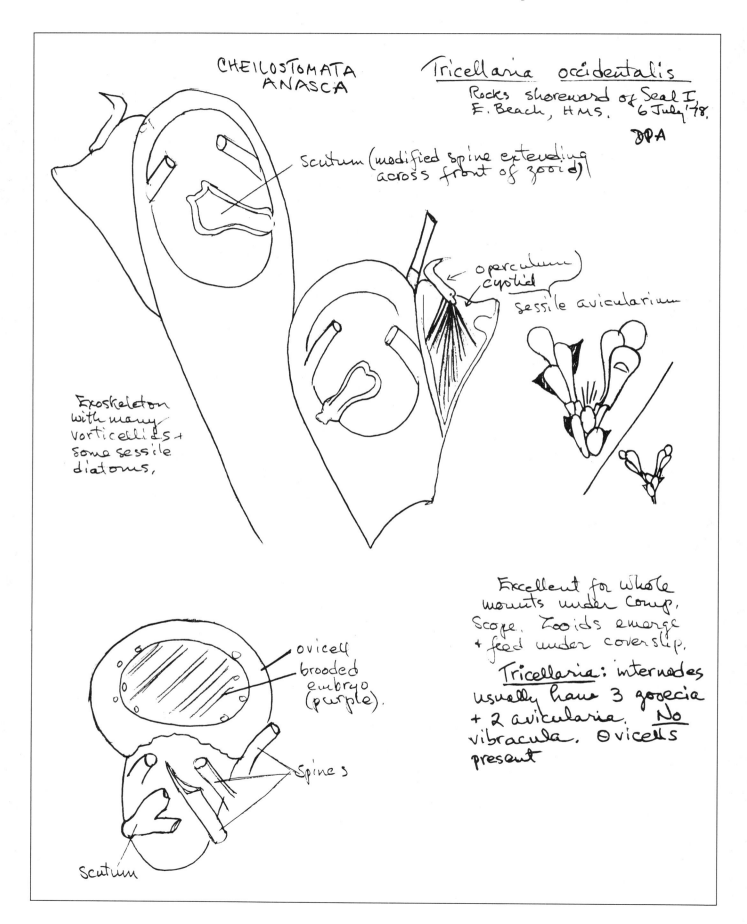

Scrupocellaria   CHEILOSTOMATA ANASCA   DPA

Pattern of branching — basically dichotomous in colony

Rhizoids arising from lower branches of colony — can apparently arise from any branch in contact with an available surface. They serve to anchor colony

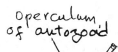

- operculum of autozooid
- Operculum or seta of vibraculum — sweeps back & forth across backs of zooecia, in a motion which should scrape off any organisms which might try to attach there.
- Cystid of vibraculum
- Frontal membrane of autozooid
- Spine on frontal membrane
- operculum / cystid } sessile avicularium

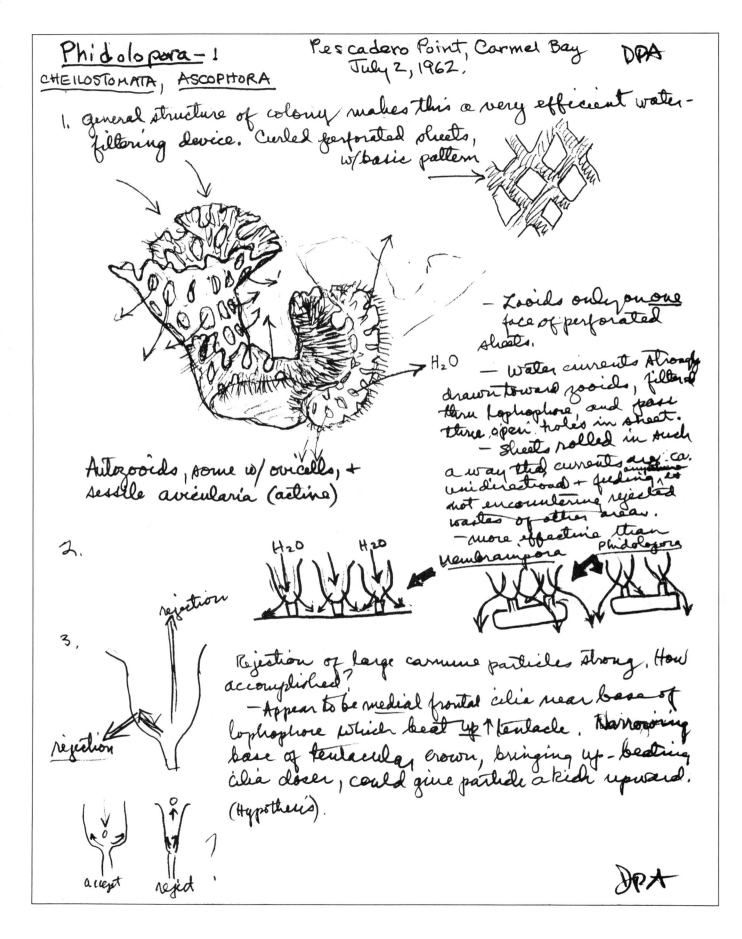

Phidolopora – 1
CHEILOSTOMATA, ASCOPHORA

Pescadero Point, Carmel Bay
July 2, 1962.   DPA

1. General structure of colony makes this a very efficient water-filtering device. Curled perforated sheets, w/ basic pattern →

   Autozooids, some w/ ovicells, + sessile avicularia (active)

   – Zooids only on one face of perforated sheets.
   – Water currents strongly drawn toward zooids, filtered thru lophophore, and pass thru open holes in sheet.
   – Sheets rolled in such a way that currents are ca. unidirectional + feeding animals do not encountering rejected wastes of other areas.
   – more effective than Membranipora   Phidolopora

2.

3. Rejection of large carmine particles strong. How accomplished?
   – Appear to be medial frontal cilia near base of lophophore which beat up ↑ tentacle. Narrowing base of tentacular crown, bringing up-beating cilia closer, could give particle a kick upward. (Hypothesis).

   accept   reject

DPA

Ectoprocta/Cheilostomata/Ascophora

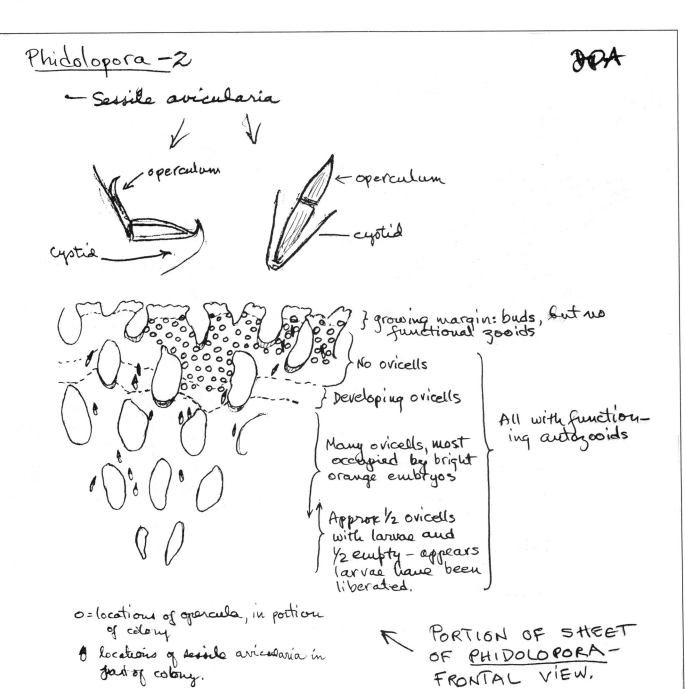

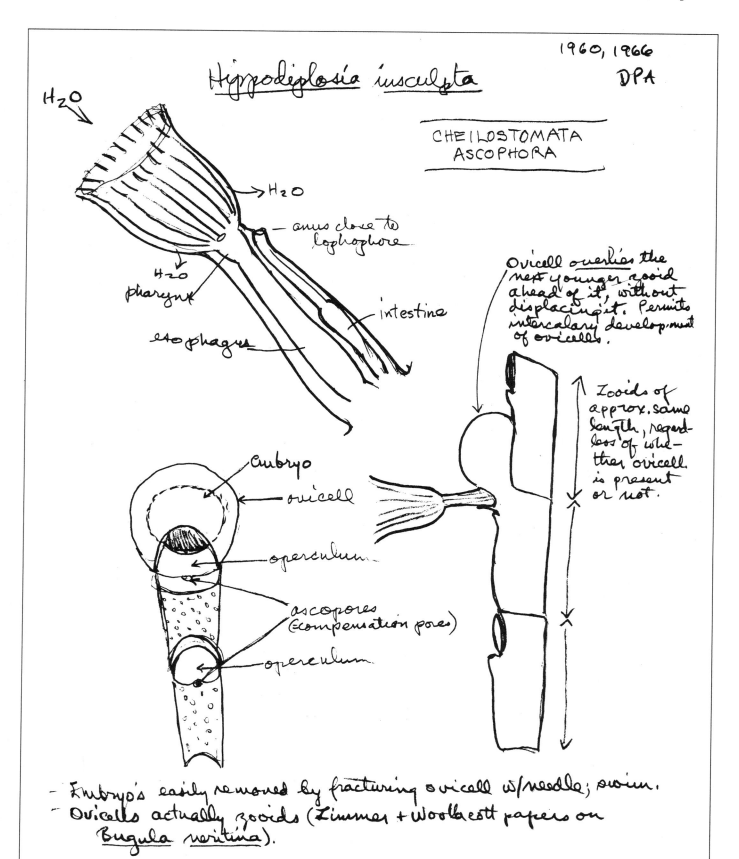

Ectoprocta/Cheilostomata/Ascophora

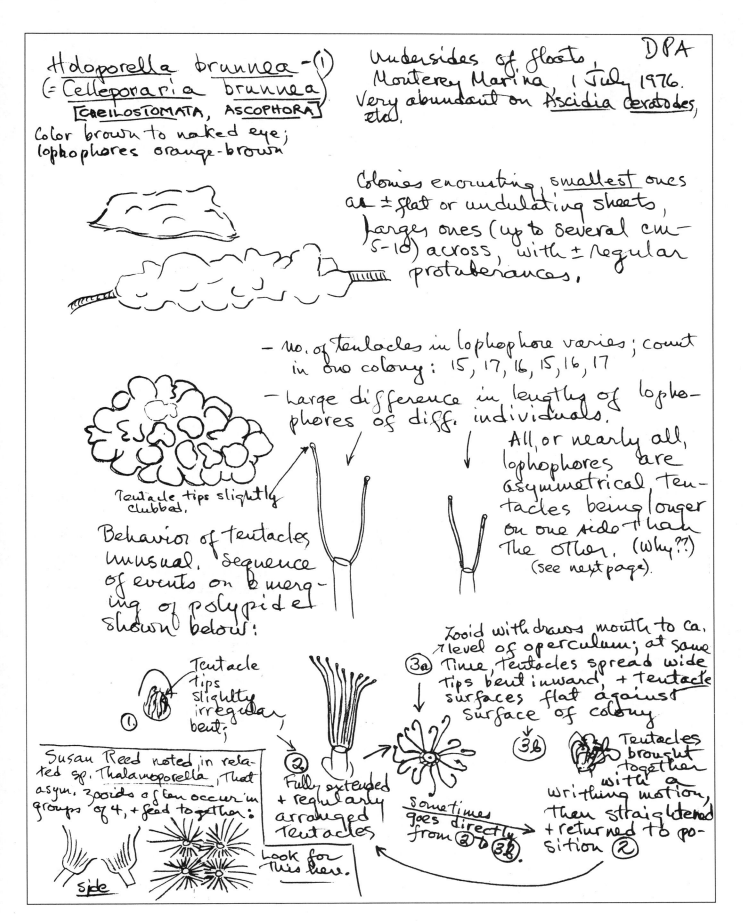

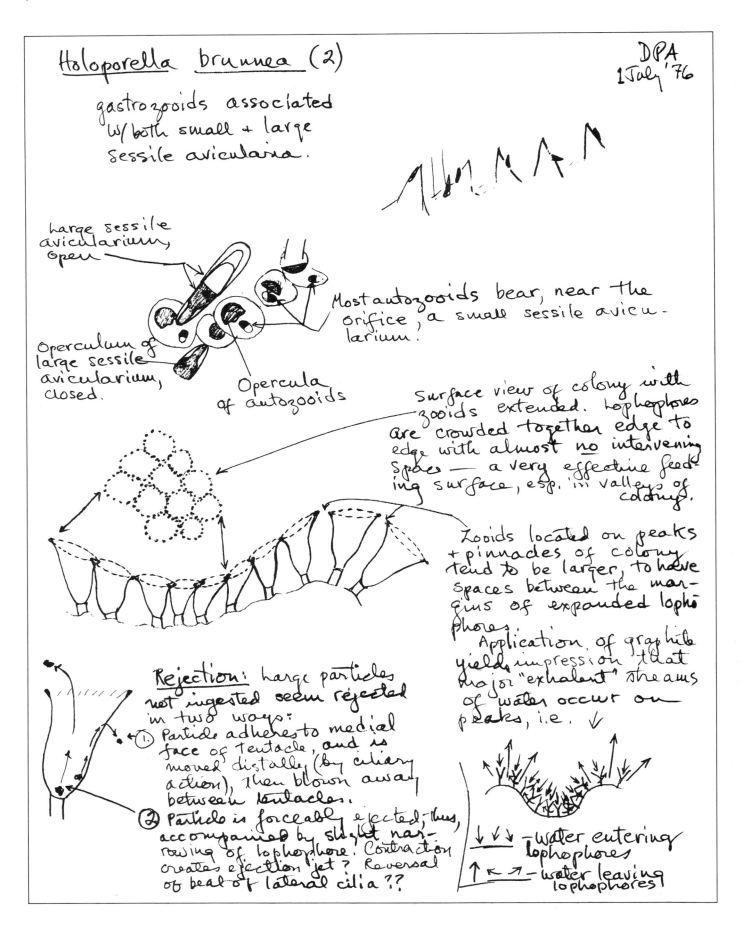

Ectoprocta/Cheilostomata/Ascophora

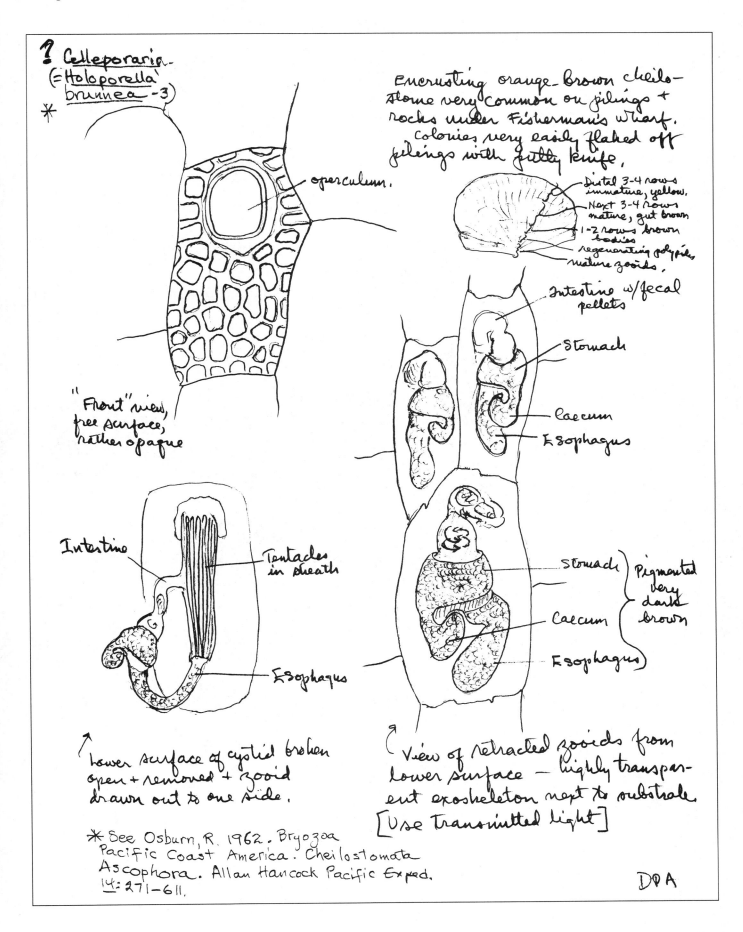

Crisia Pescadero Pt., 30 June 1961.

Ectoprocta
Cyclostomata or
Stenolaemata

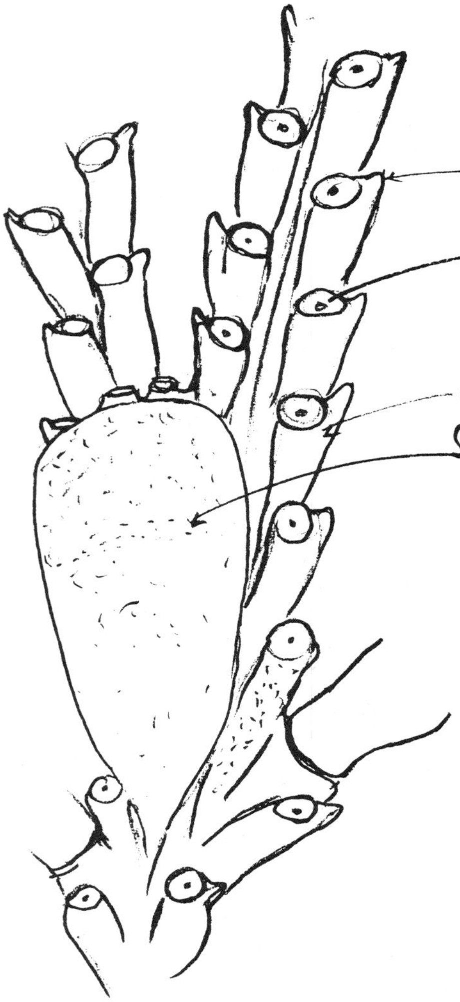

- spine on autozooid
- orifice
- autozooid
- gonozooid or ooecium — filled with orange tissue (body wall, tentacle sheath, remnant of polypide, + poss. nutritive cells in coelom, and numerous (ca. 20?) small yellowish early larvae — with yolky centers showing little beyond a mesentodermal mass, but with ciliated girdle well-developed, + capable of rapid, erratic swimming).

DPA

Ectoprocta/Ctenostomata

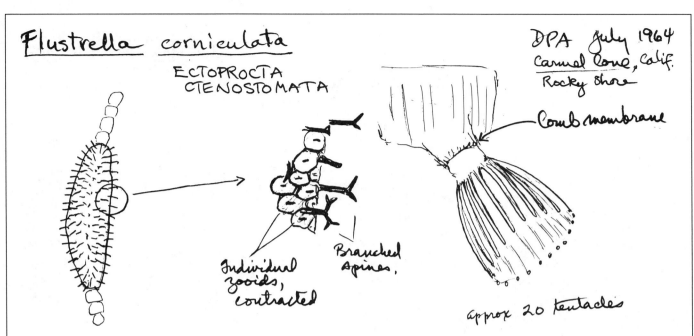

Flustrella corniculata
ECTOPROCTA
CTENOSTOMATA

DPA July 1964
Carmel Cove, Calif.
Rocky shore

Comb membrane

Individual zooids, contracted

Branched spines

approx 20 tentacles

Colony on coarse Coralline alga (*Bossea* or *Calliarthron*). Brownish, with dark brown spines

Expands easily — good for observations of feeding, since zooids can be viewed from all angles. Exoskeleton opaque; internal anatomy not visible from outside.

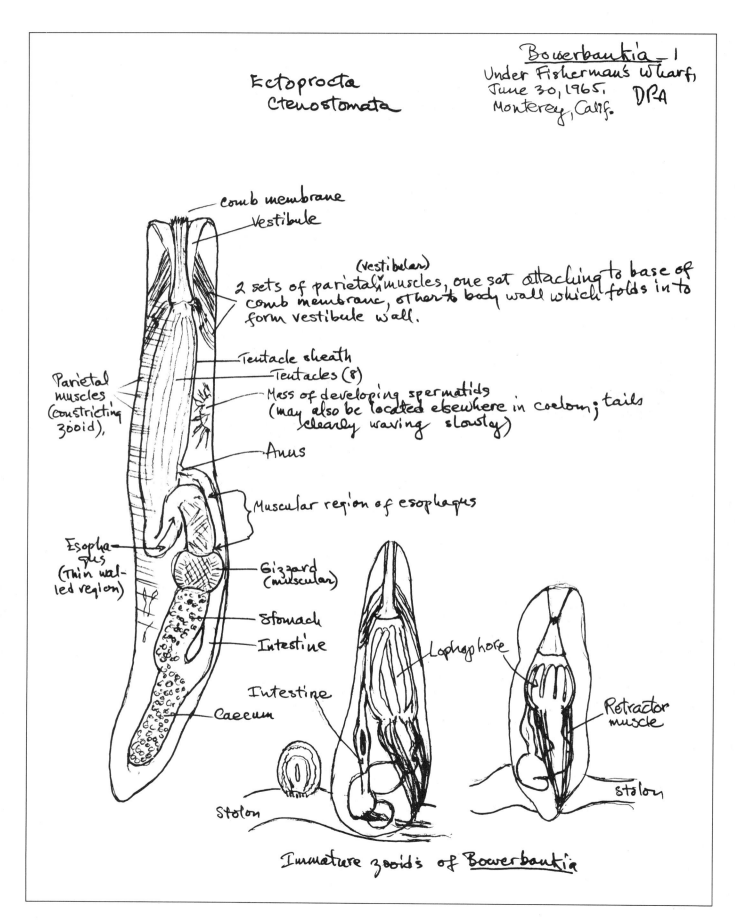

Ectoprocta/Ctenostomata

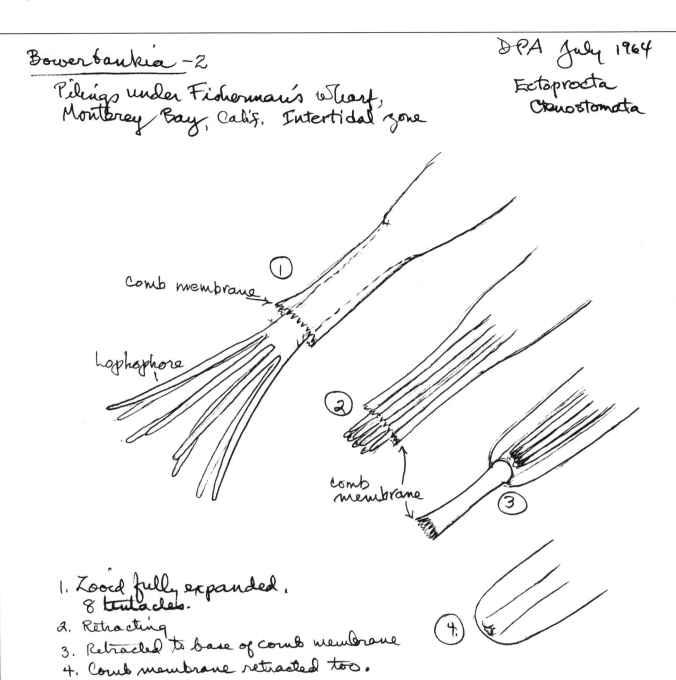

Bowerbankia - 2
 Pilings under Fisherman's Wharf,
 Monterey Bay, Calif. Intertidal zone

DPA July 1964
Ectoprocta
Ctenostomata

1. Zooid fully expanded. 8 tentacles.
2. Retracting
3. Retracted to base of comb membrane
4. Comb membrane retracted too.

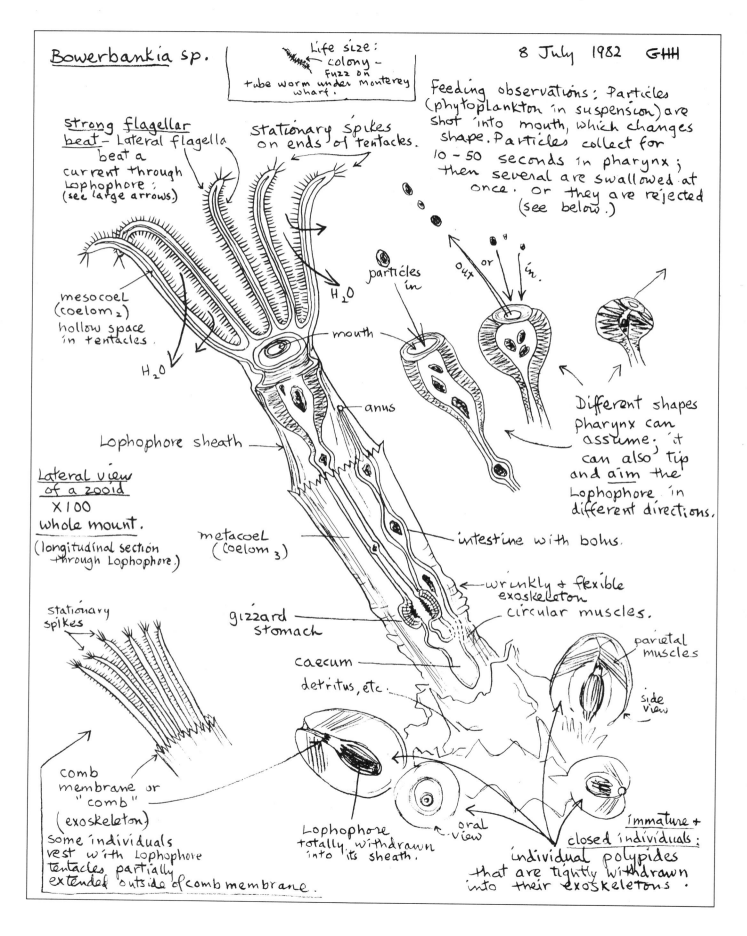

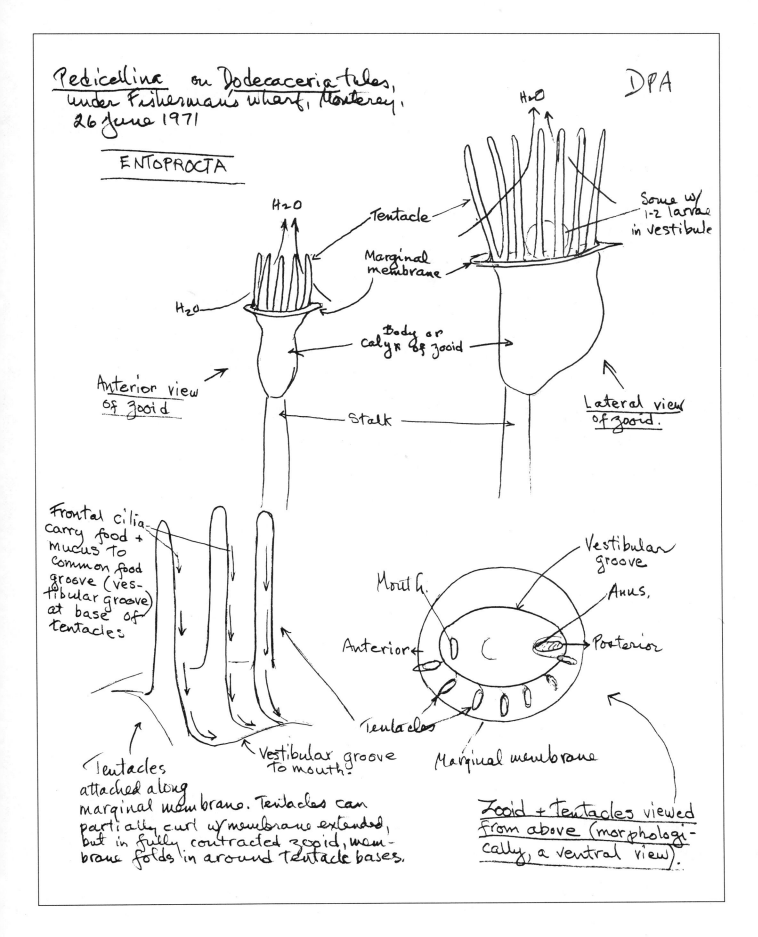

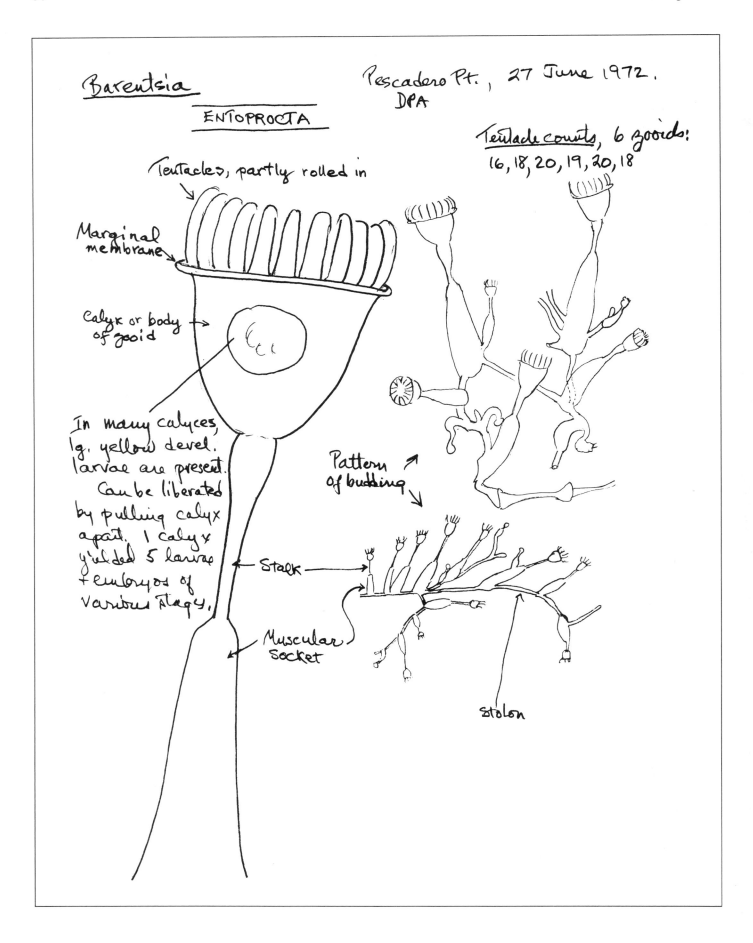

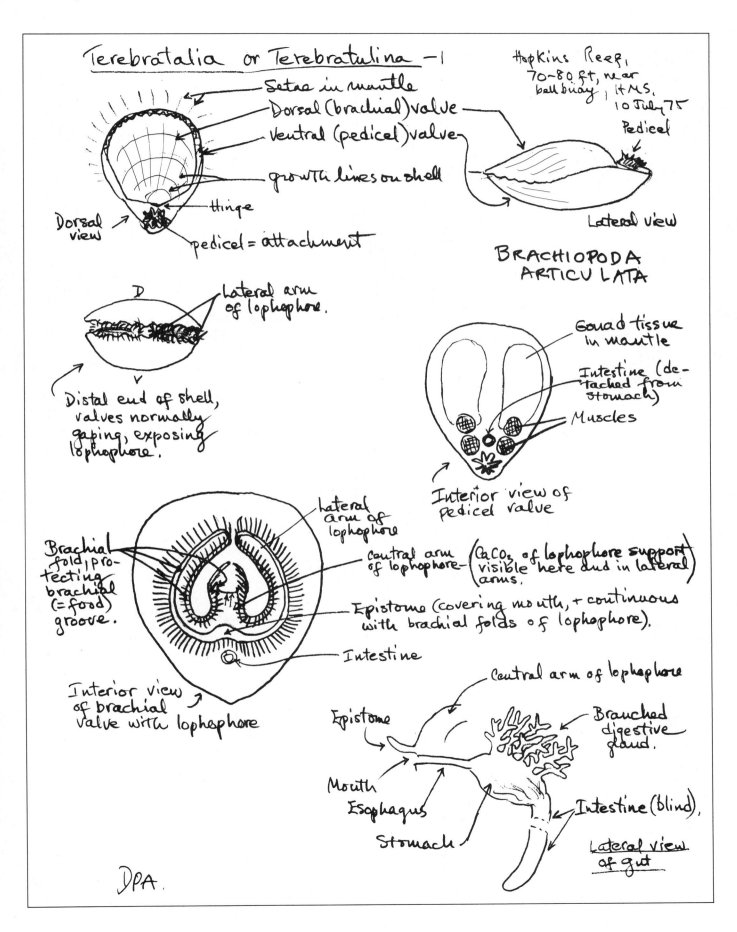

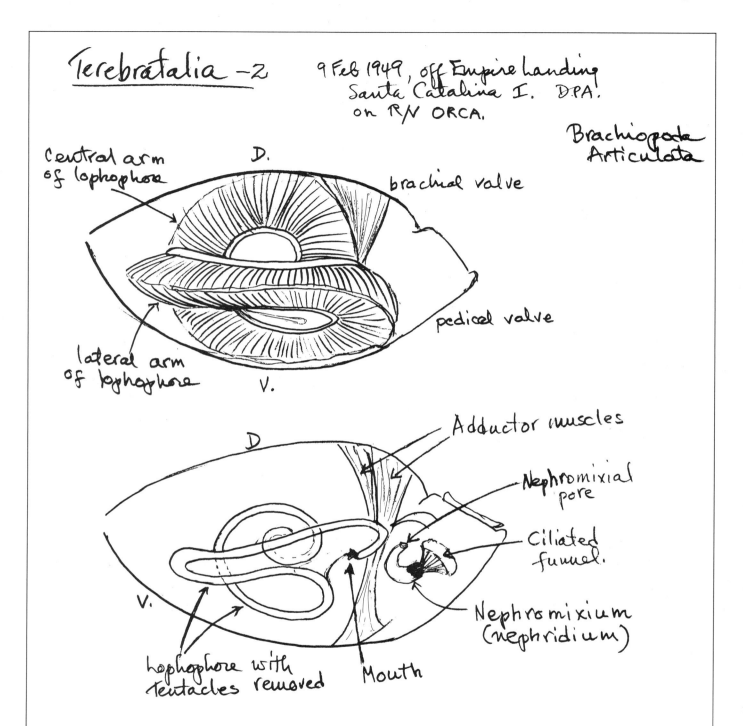

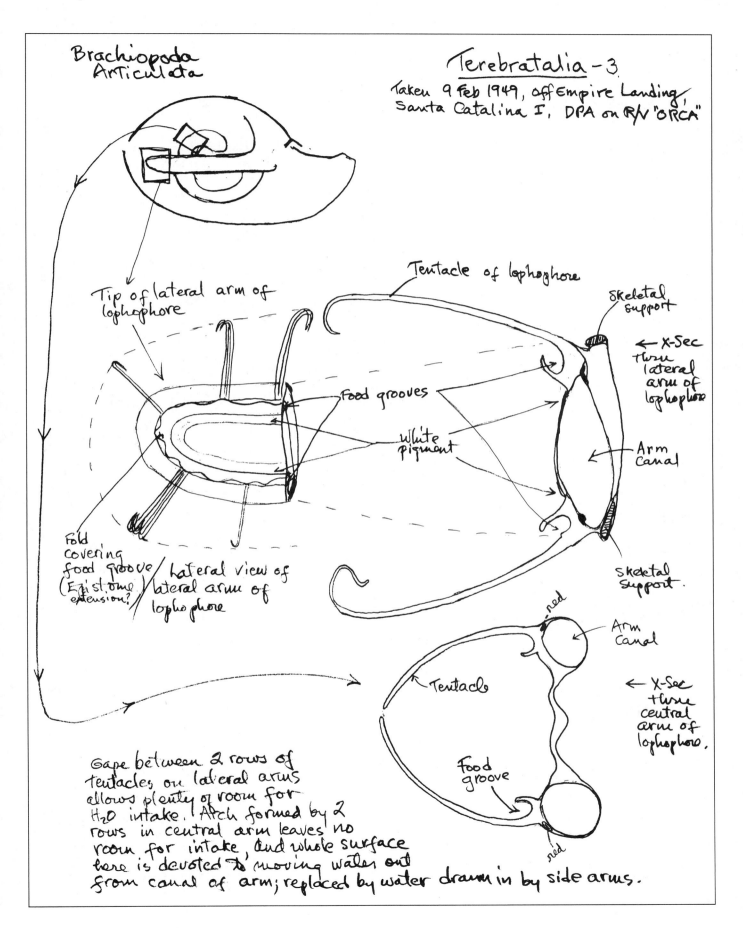

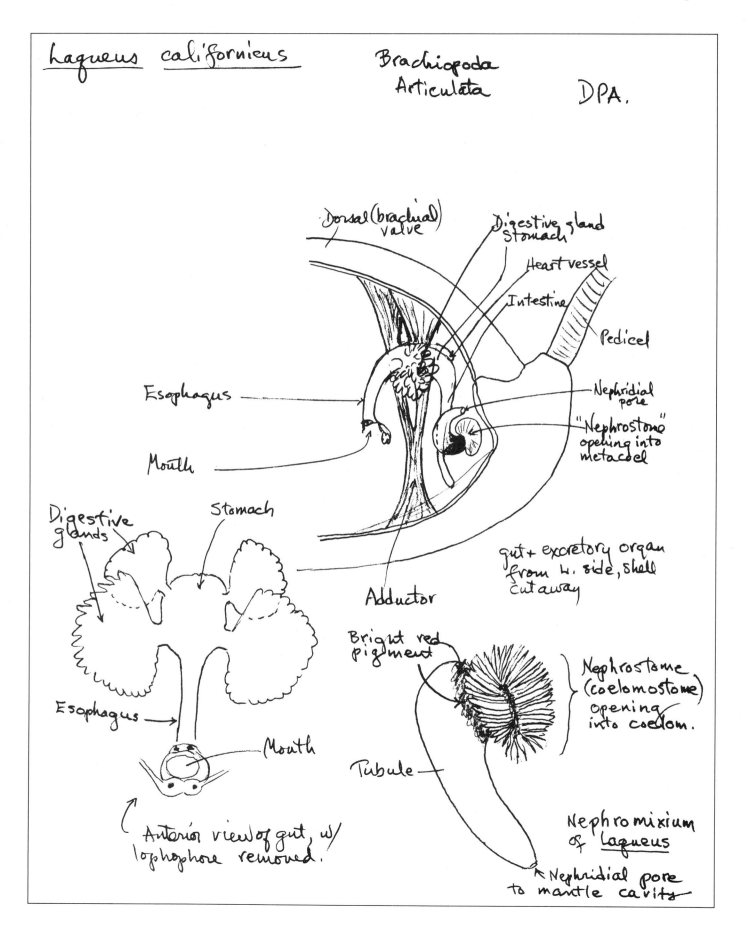

Phoronida

Phoronis sp. HMS Rocks, central island on ocean side of channel beyond drain outflow from Loeb Lab. 8 Aug 1962. Midtide zone, forming sandy mat at bases of algae, near upper limit of _Clavelina_.

DPA

Posterior view of lophophore

Almost Dorsal view.

dorsal vessel; some ebb + flow, but major flow is downward (ventral)

Ebb + flow circulation in spiral vessels in tentacles.

to dorsal vessel w/ posterior flow.

Posterior Right view.

Anterior flow — vessel appears contractile

Numerous specimens w/ short tentacles in this region — new tentacles being added here.

E. longate fecal pellets emerge here.

Nephridium with brownish matter in tubule lumen.

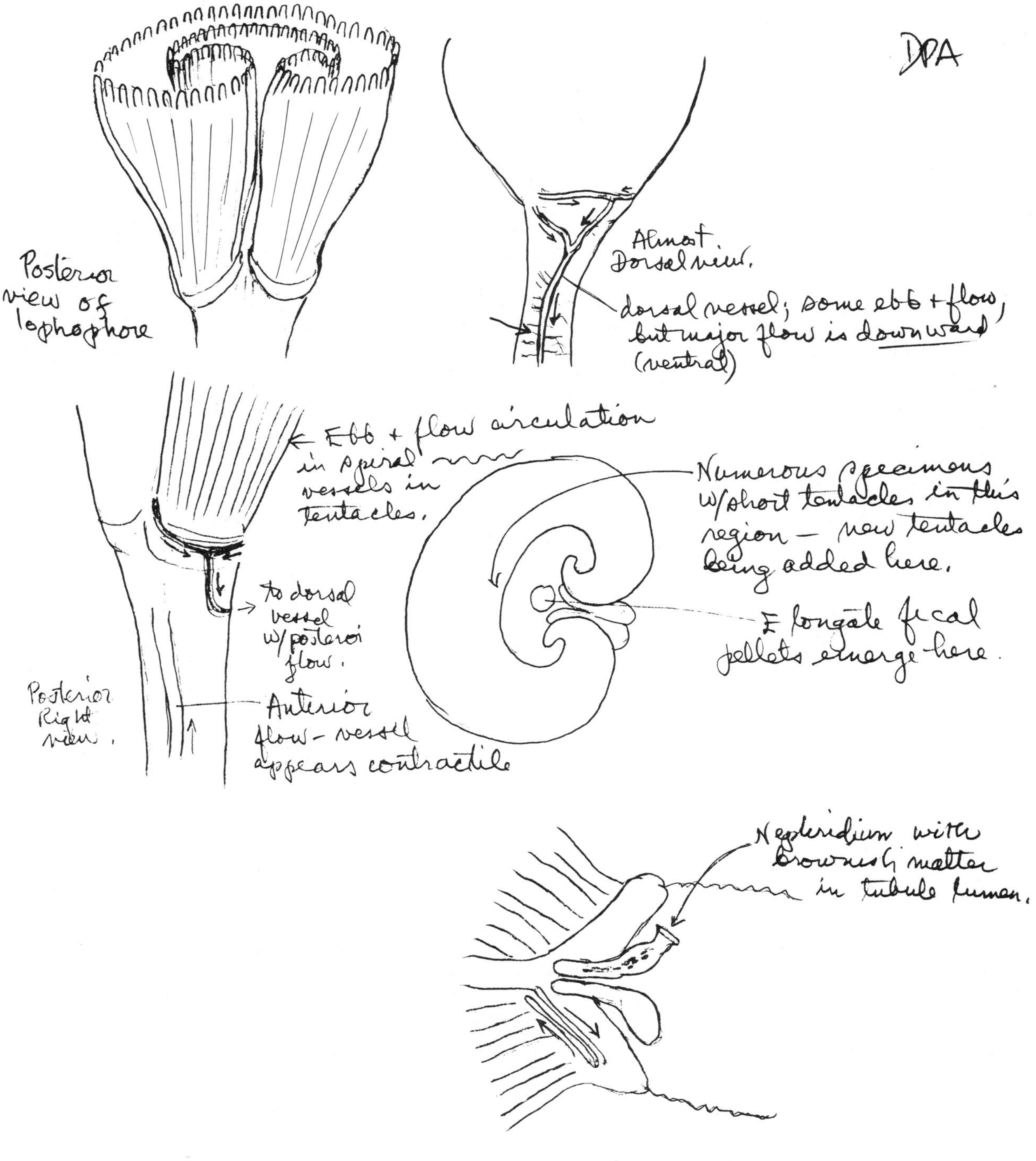

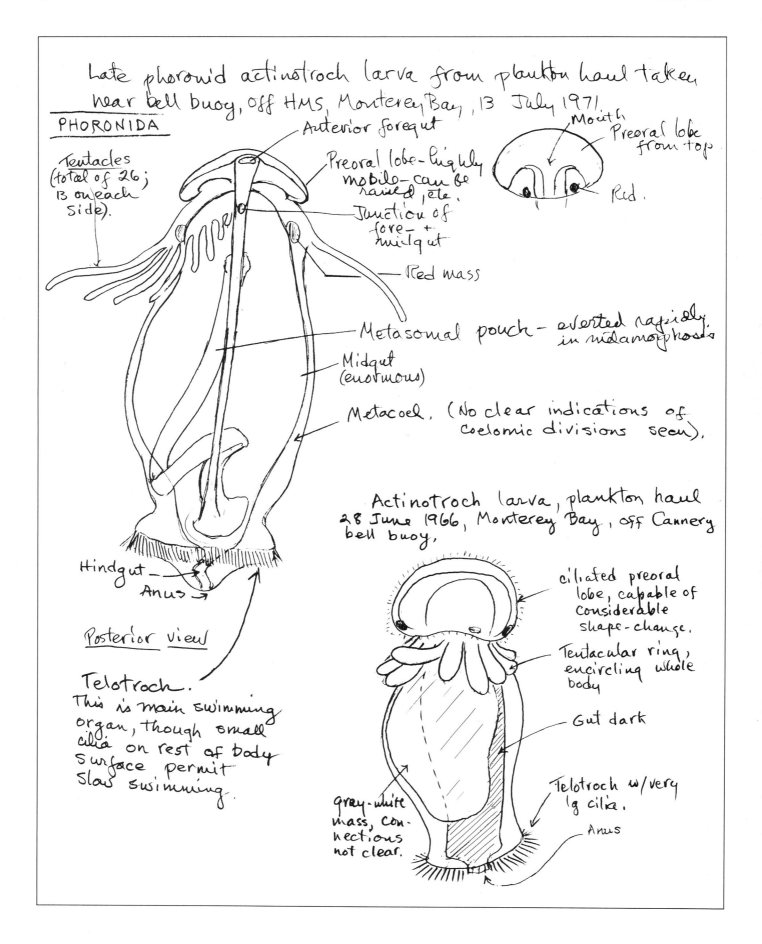

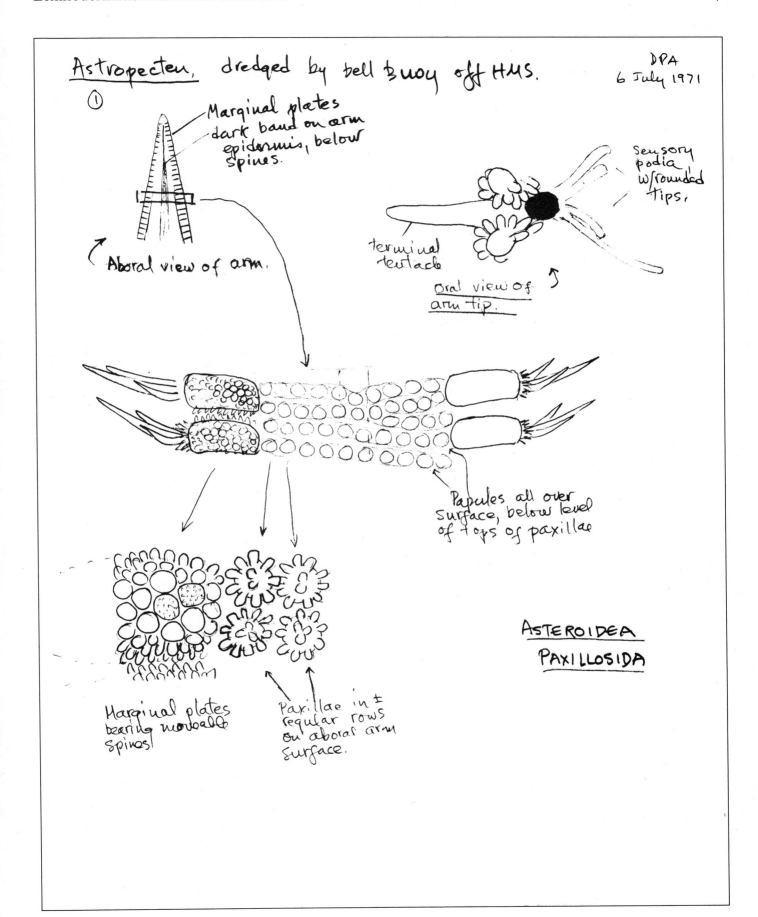

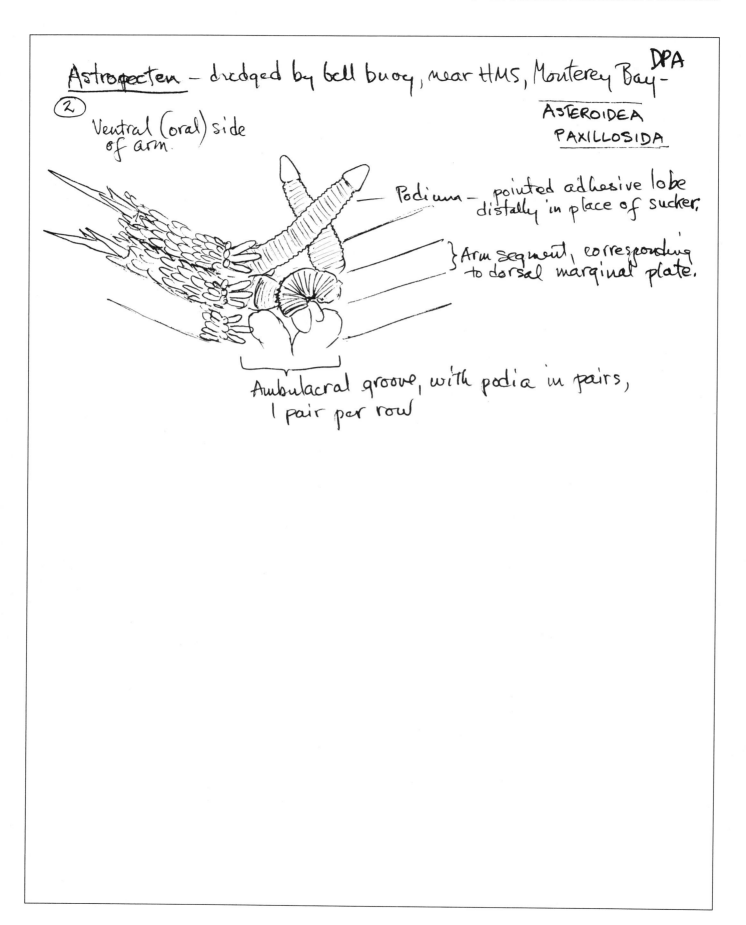

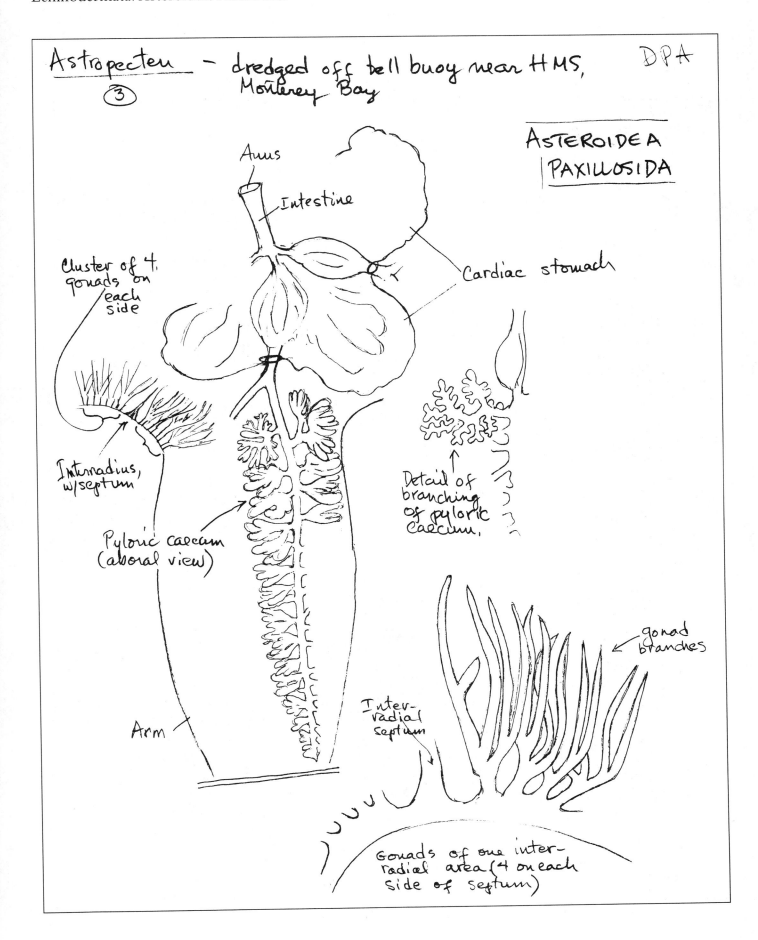

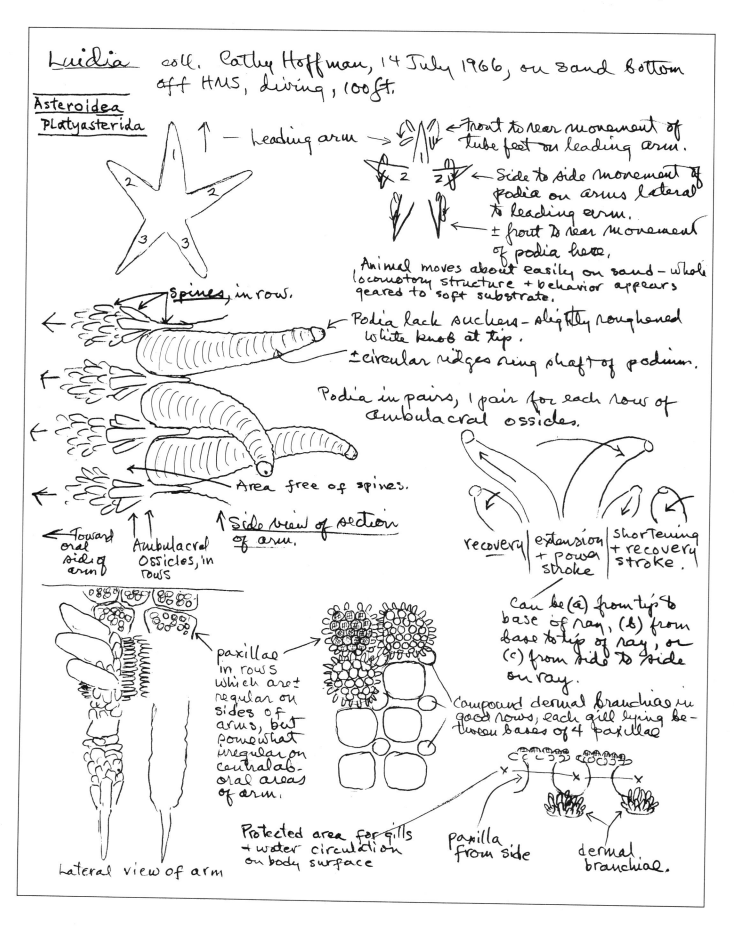

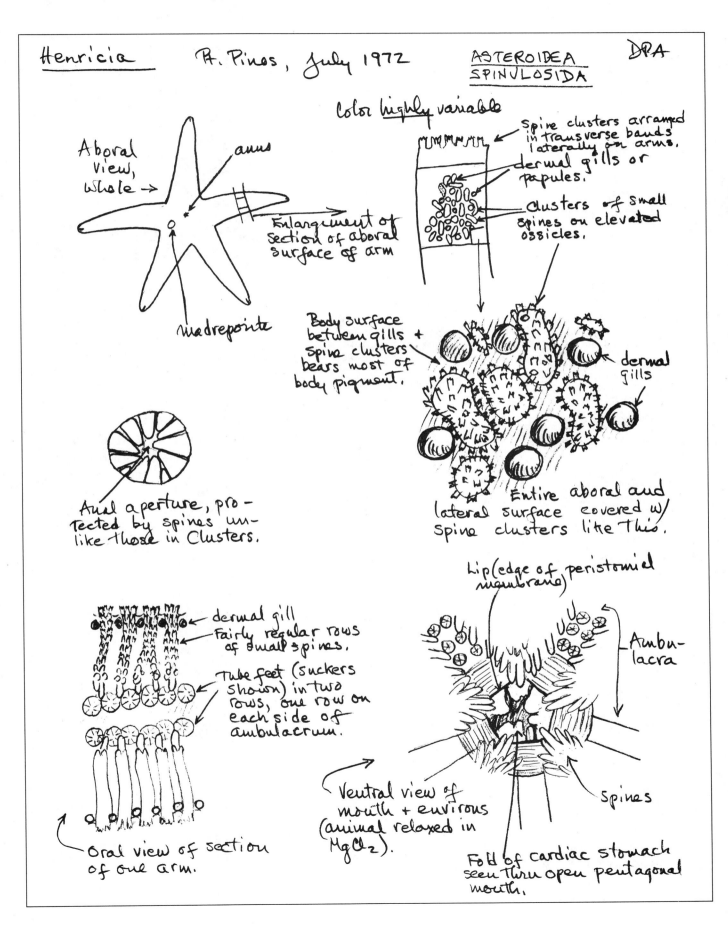

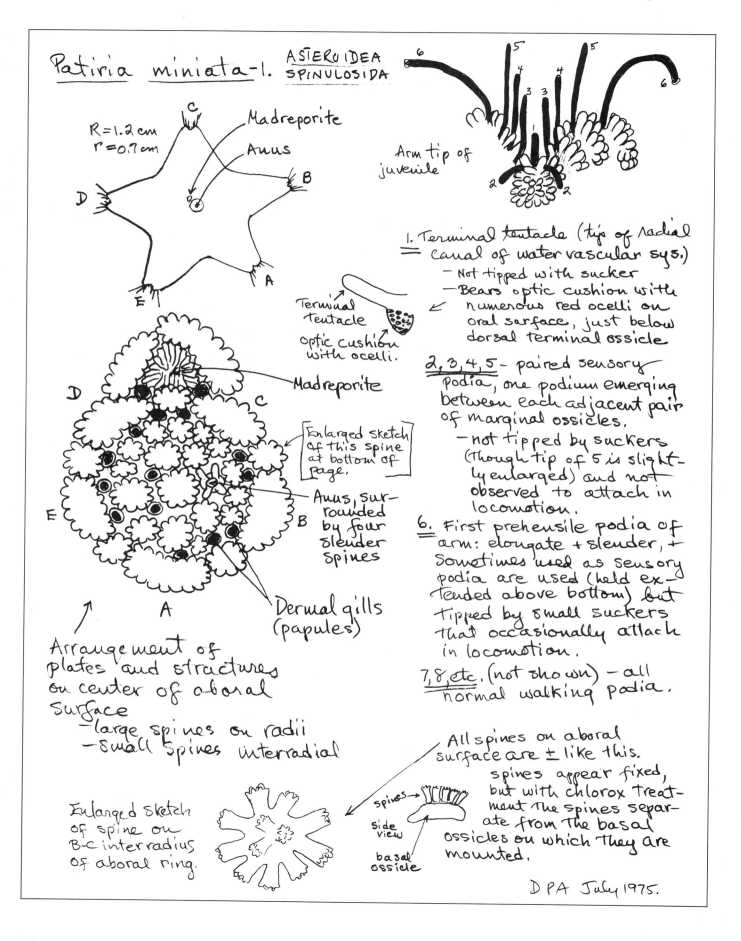

Echinodermata/Asteroidea/Spinulosida

Patiria miniata - 2.
ASTEROIDEA
SPINULOSIDA

DPA
July, 1975

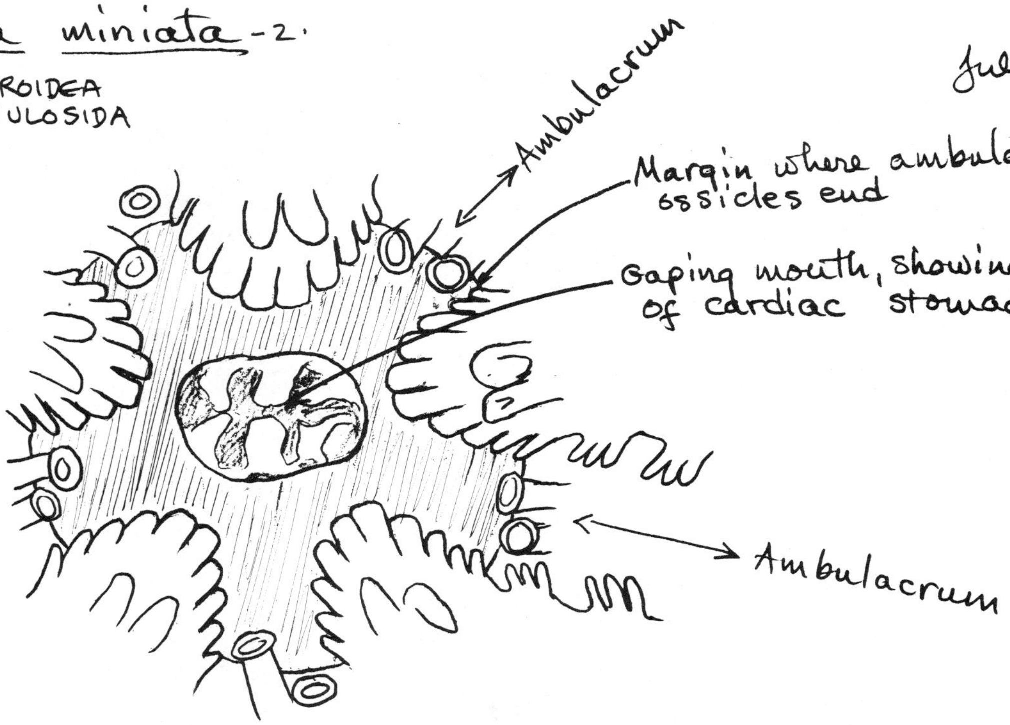

- Ambulacrum
- Margin where ambulacral ossicles end
- Gaping mouth, showing folds of cardiac stomach.
- Ambulacrum

Oral view of mouth

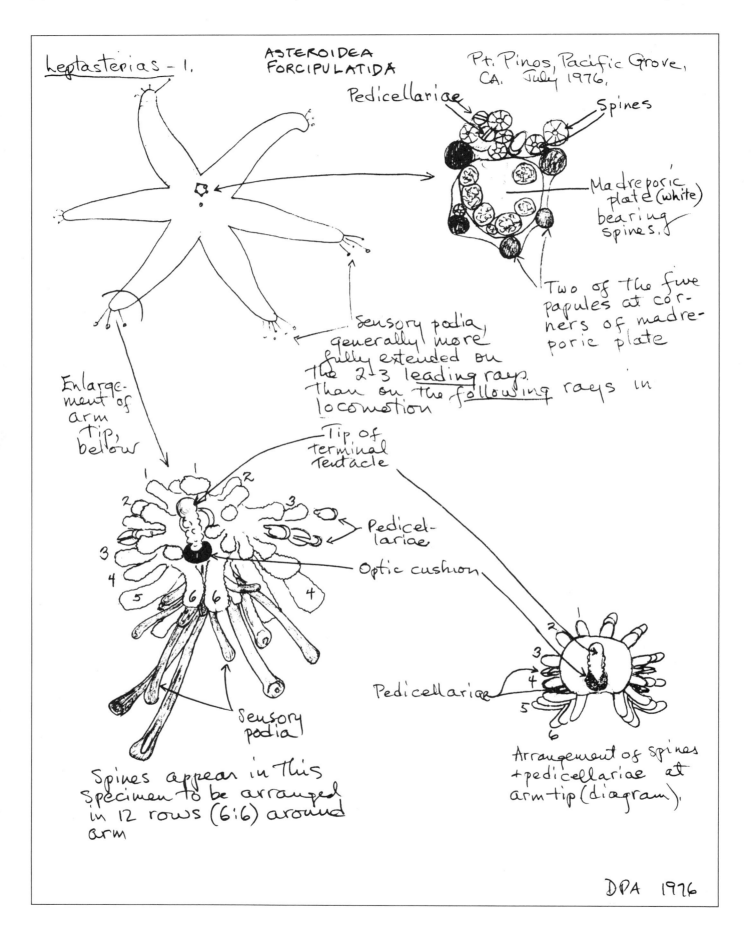

Echinodermata/Asteroidea/Forcipulatida

Leptasterias - 2.

ASTEROIDEA
FORCIPULATIDA

DPA
July, 1976

spines and external tissues removed with chlorox, showing supporting ossicles of arm skeleton.

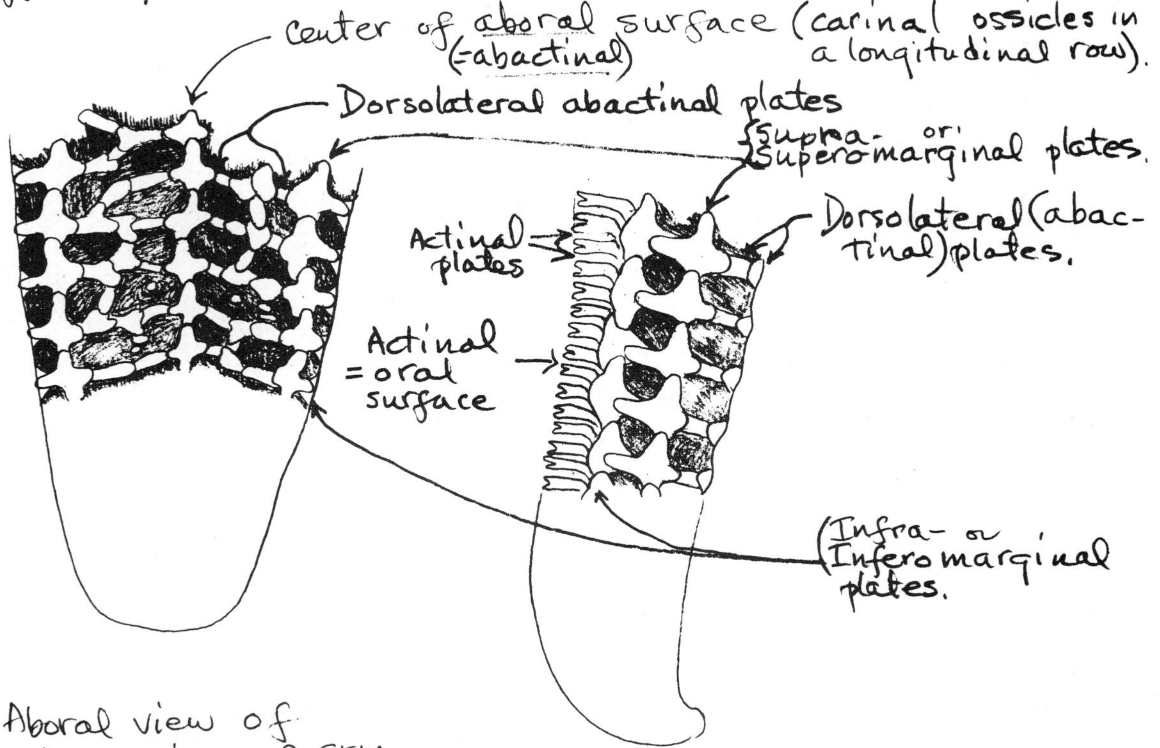

- Center of aboral surface (carinal ossicles in a longitudinal row). (=abactinal)
- Dorsolateral abactinal plates
- Supra- or Superomarginal plates.
- Actinal plates
- Dorsolateral (abactinal) plates.
- Actinal = oral surface
- Infra- or Inferomarginal plates.

Aboral view of distal region of arm (ossicles _white_; connective tissue + muscle _dark_)

Lateral view of arm

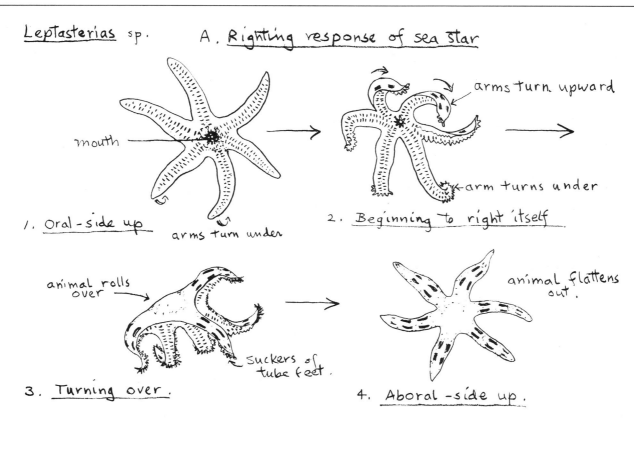
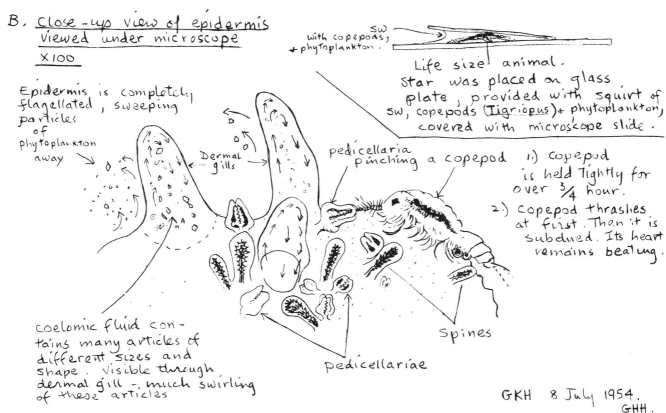

GKH 8 July 1954.
GHH.

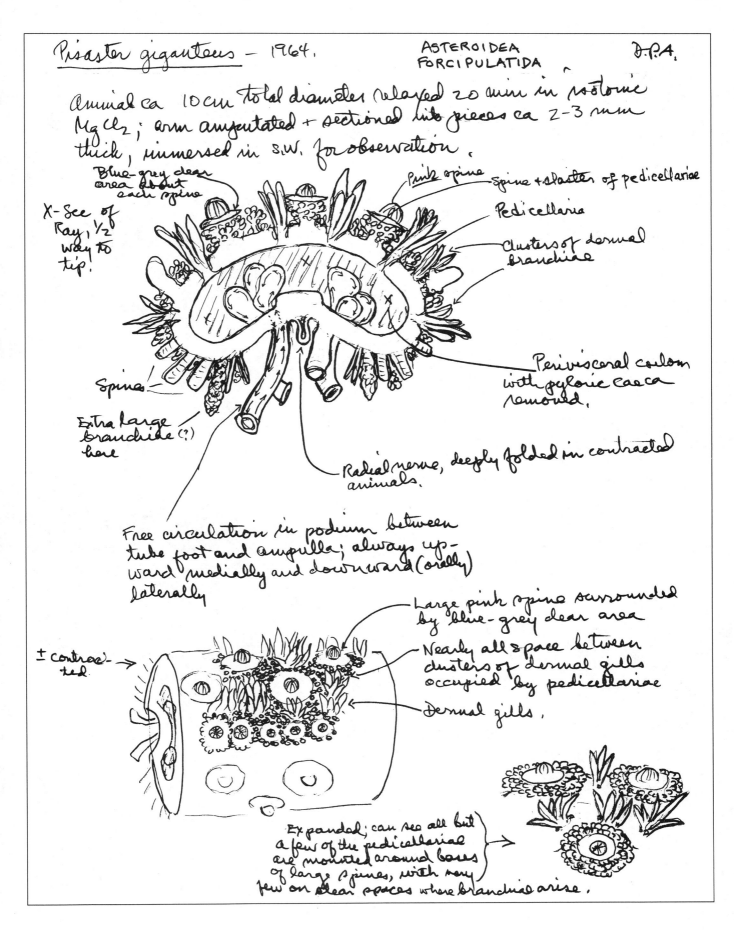

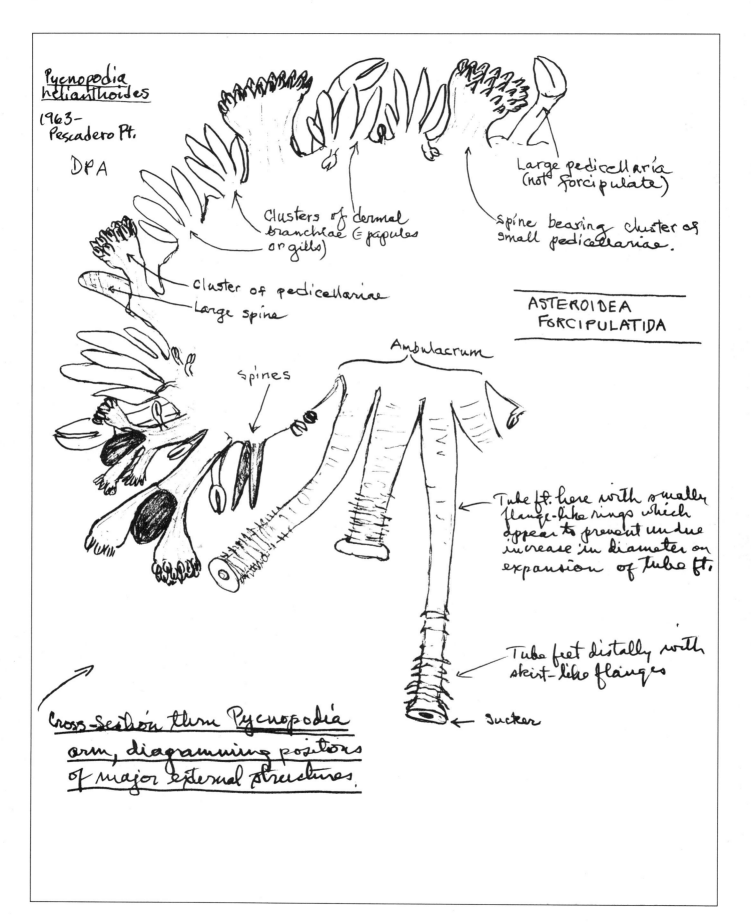

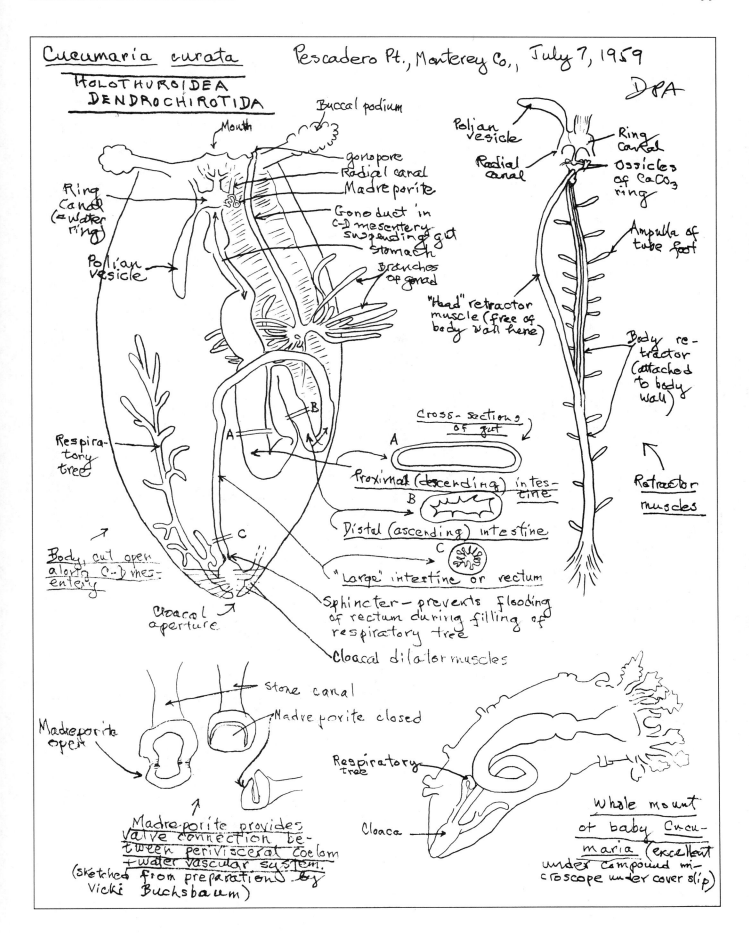

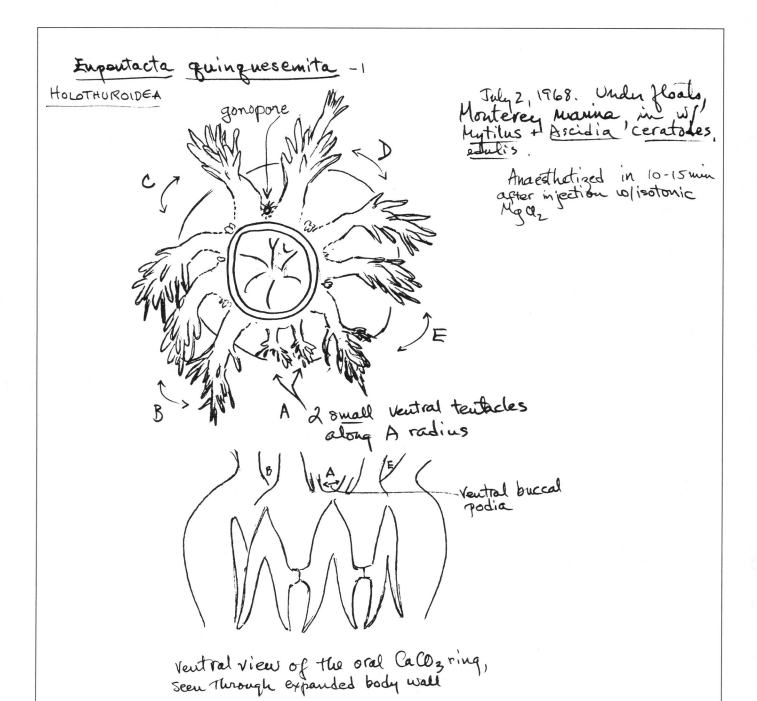

Eupentacta quinquesemita – 1
HOLOTHUROIDEA

July 2, 1968. Under floats, Monterey marina, in w/ *Mytilus edulis* + *Ascidia ceratodes*.

Anaesthetized in 10–15 min after injection w/ isotonic $MgCl_2$

gonopore

A 2 small ventral tentacles along A radius

ventral buccal podia

Ventral view of the oral $CaCO_3$ ring, seen through expanded body wall

Echinodermata/Holothuroidea

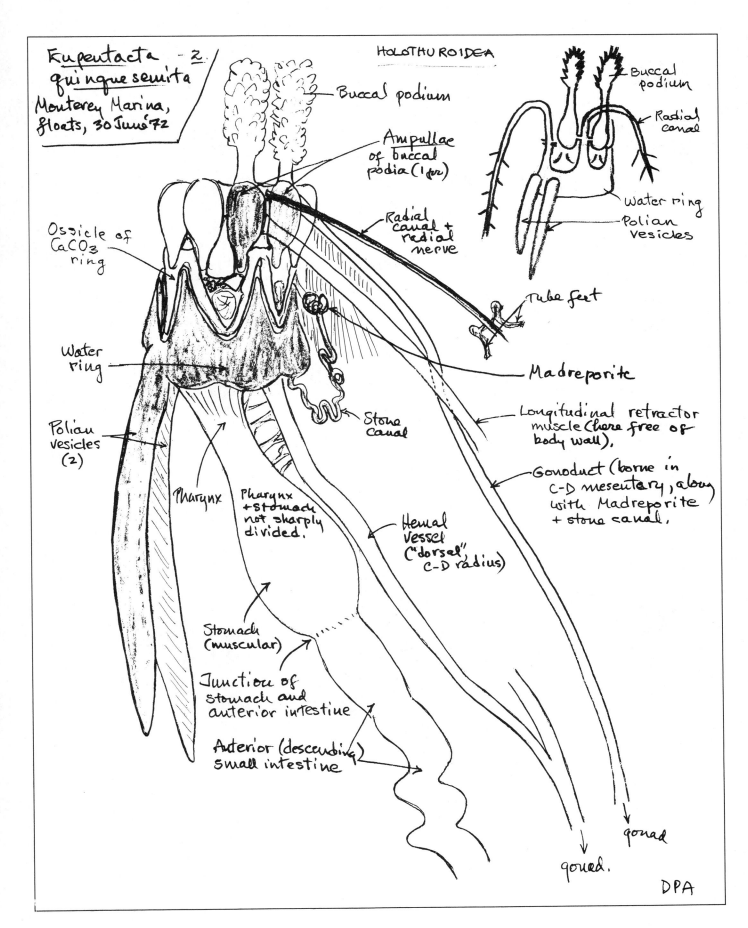

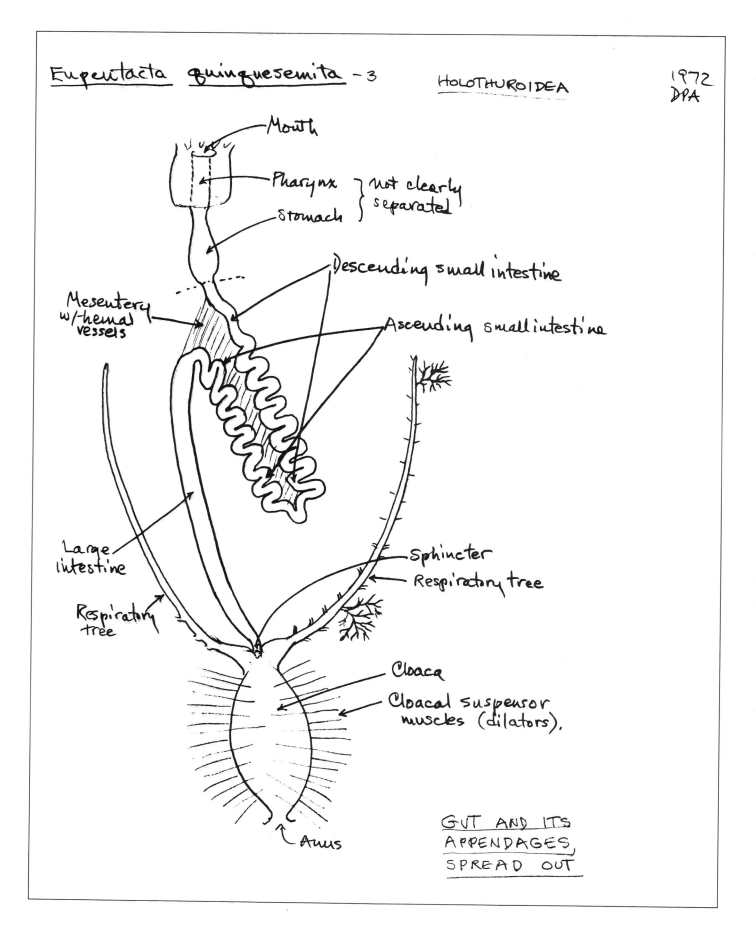

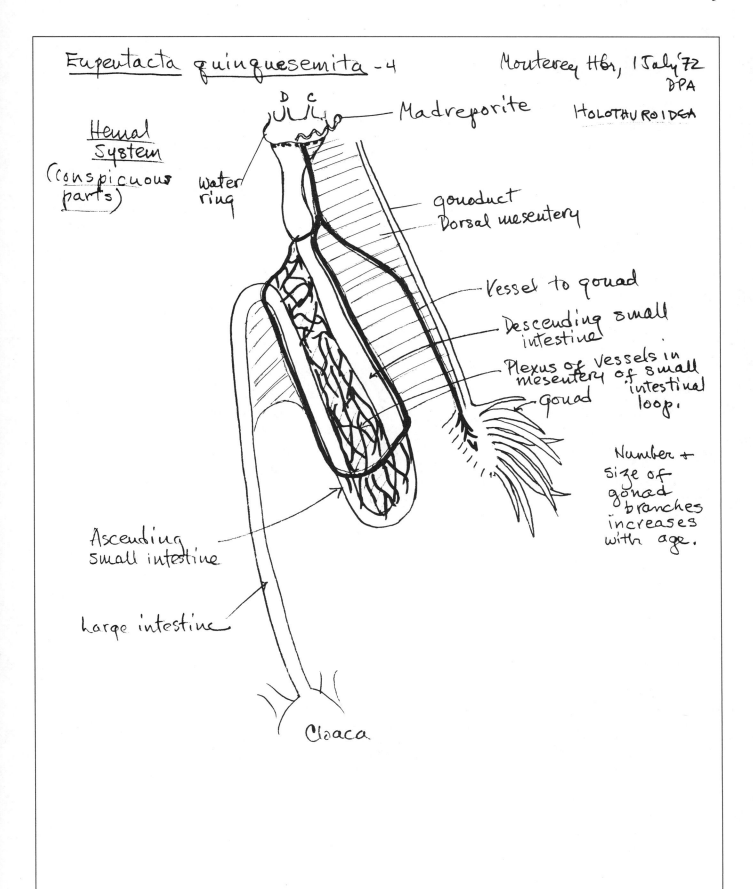

Leptosynapta (Pt Pinos, July 5, 1966).
①

HOLOTHUROIDEA
APODIDA
DPA

Locomotion on glass dish bottom, submerged.
— Pinnate tentacles sticky; alternately extended on bottom & retracted & brushed across mouth — extension & retraction anchors then pulls body forward at rate of ca. 3 cm./min. Forward motion accompanied by rear-to-front peristaltic waves along body, which appear to aid in extending ant. end & pulling rear end forward.

$MgCl_2$ 7.3% — KO's animal completely in 5 min. or less at room temp.

Ant. end — 12 tentacles, all pinnate & ± same size.

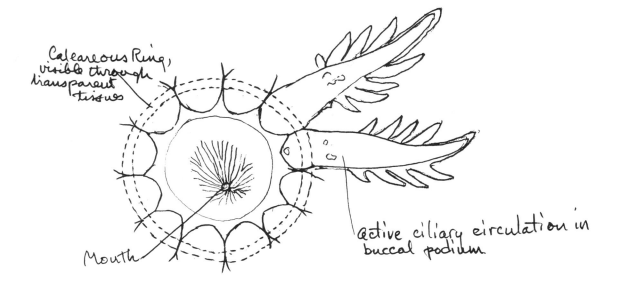

Calcareous Ring, visible through transparent tissues

Mouth

Active ciliary circulation in buccal podium.

WHOLE MOUNT OF TIP OF BUCCAL PODIUM UNDER COMP. SCOPE →

spicules, arranged along walls of canals, possibly supporting canal.
Epidermis

Water vascular canal of buccal podium

Echinodermata/Holothuroidea 105

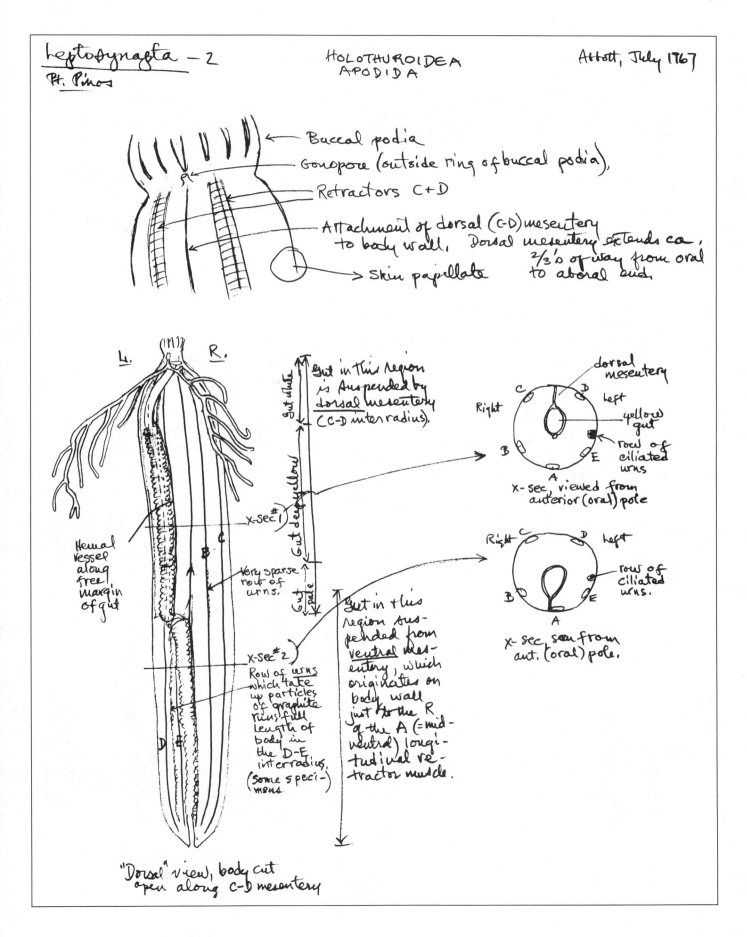

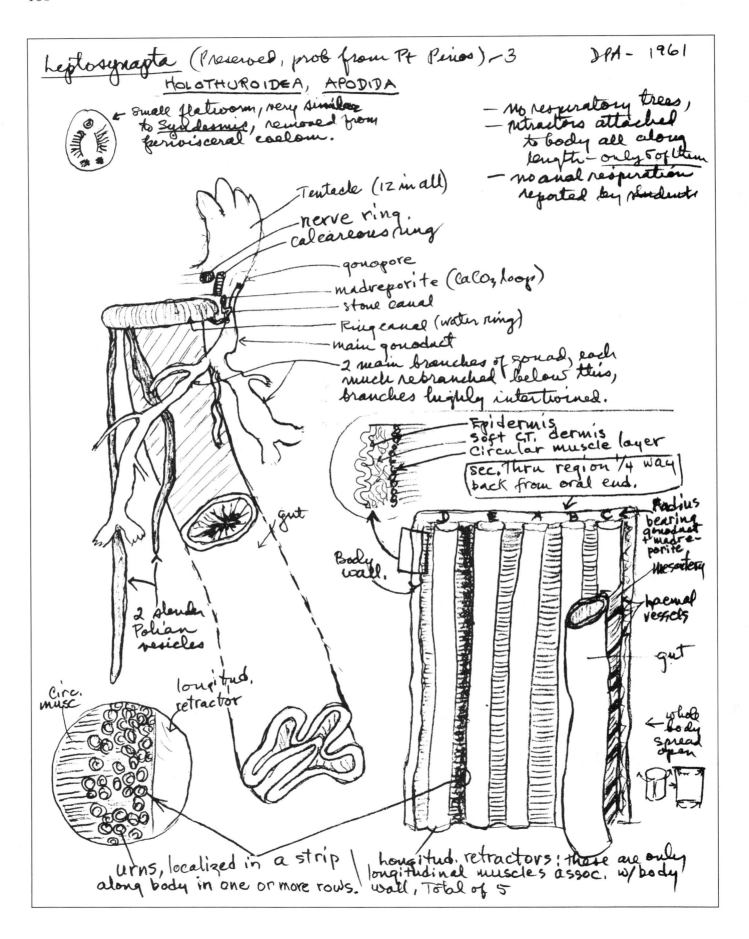

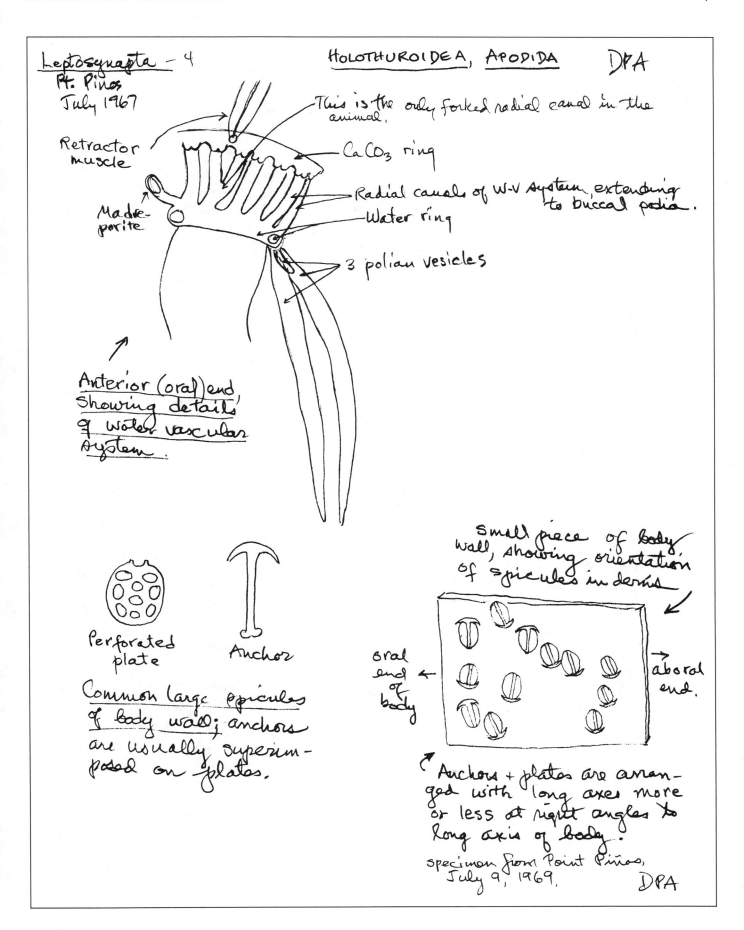

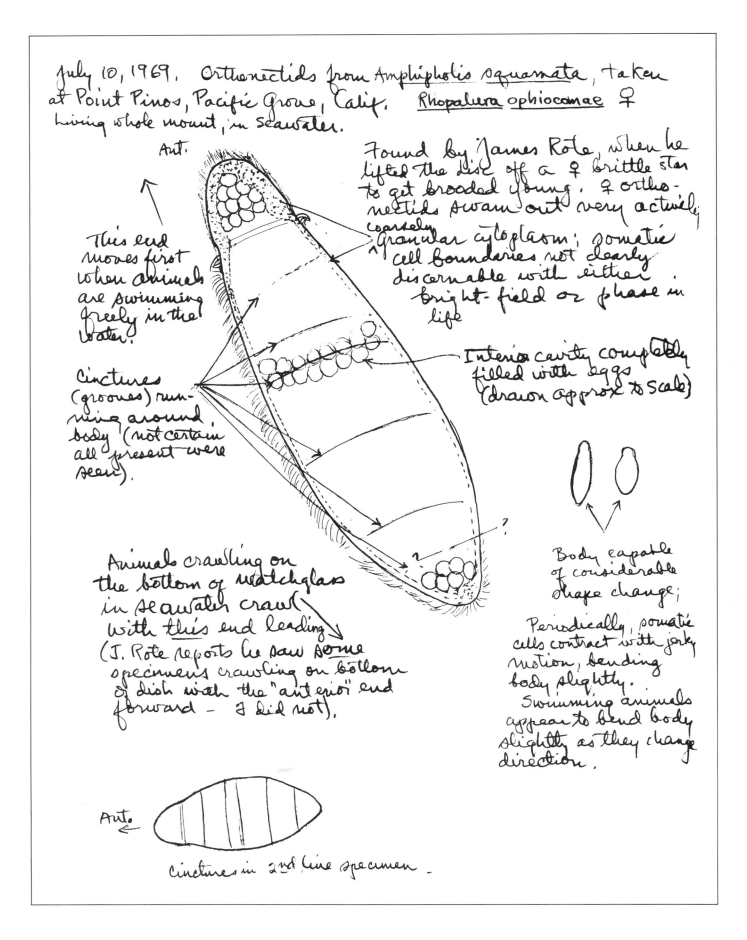

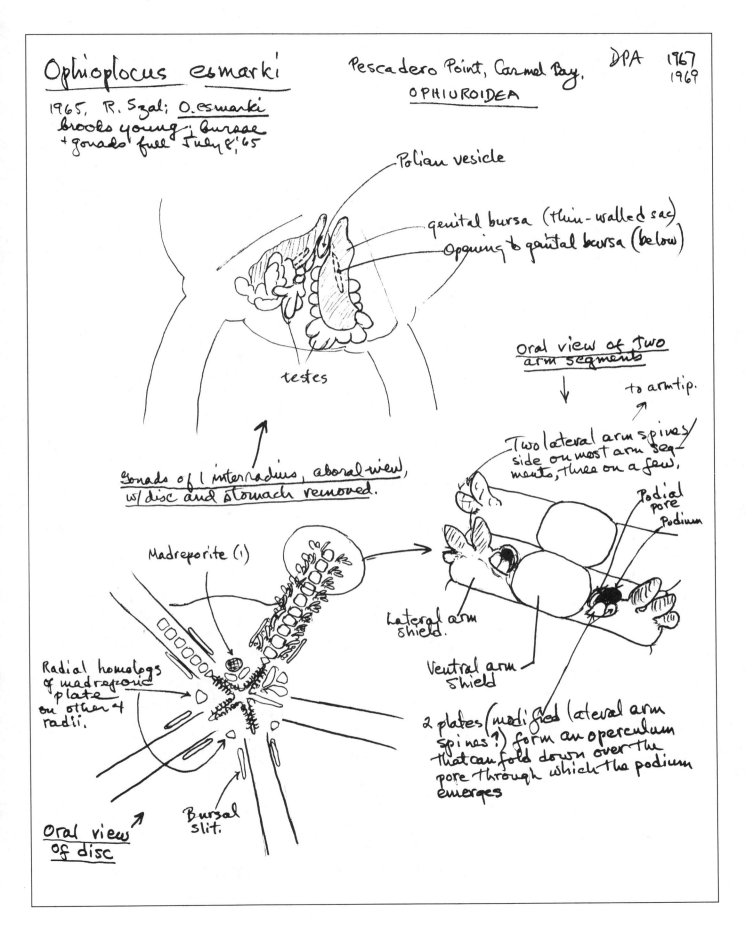

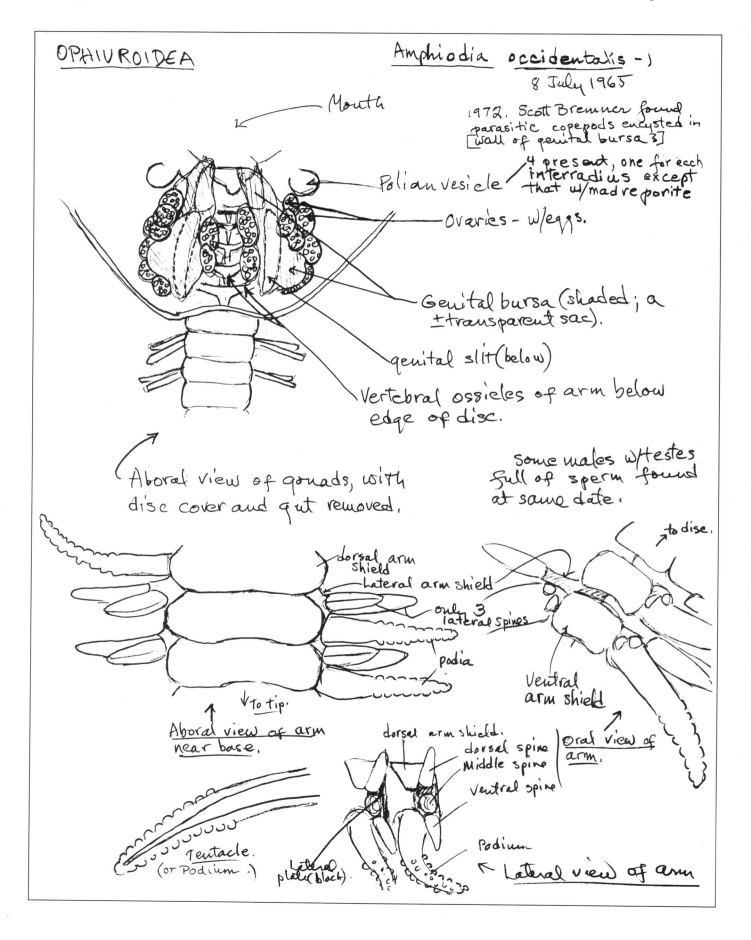

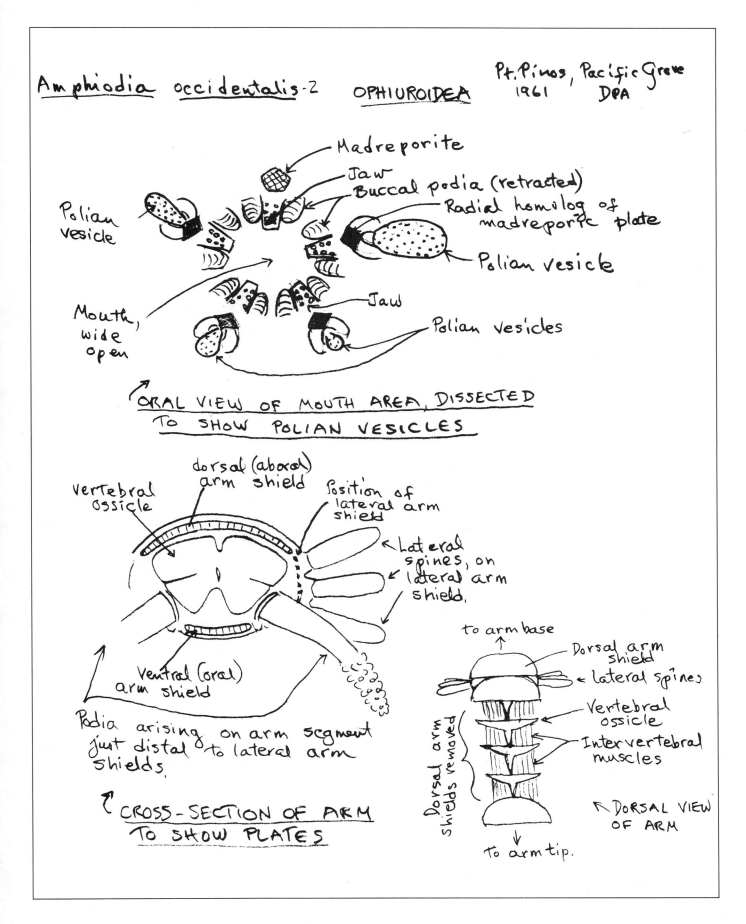

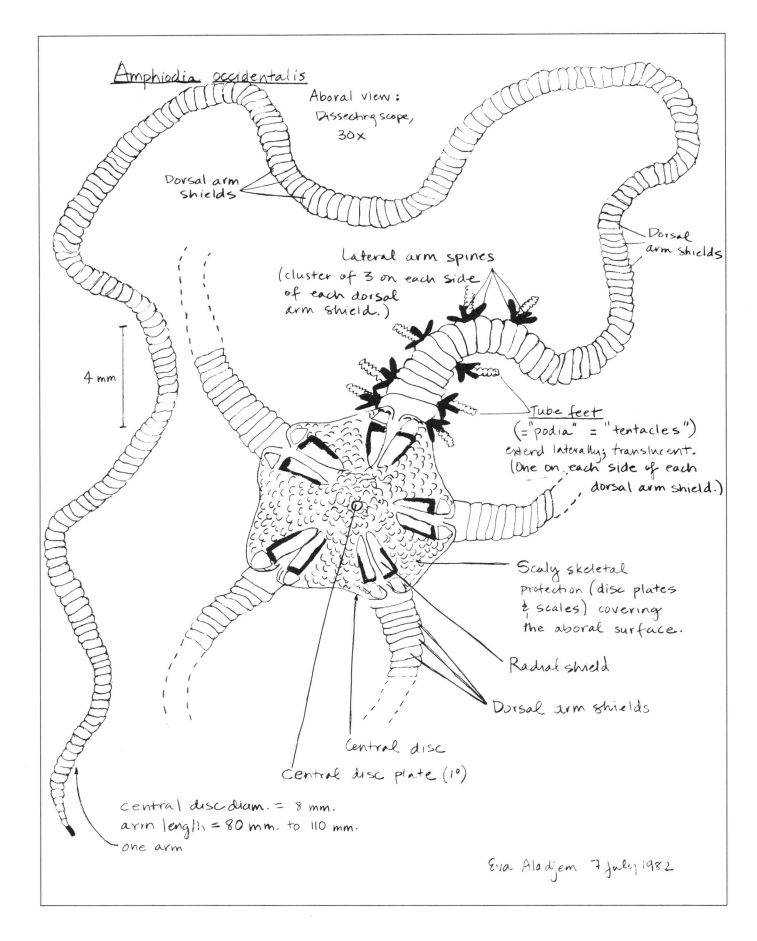

Amphiodia occidentalis

LOCOMOTION: (in glass dish)

Amphiodia ambulates by overall arm movement. The arms are highly flexible, they twist & curl; they are rarely straight. Usually, not all of an arm is in contact with the substrate.

Podia provide the force to move arms. There are two rows of podia on each arm, one row projecting from each side. These contact the substrate — not by using sucker action — but by use of adhesive papillae.

There is no single leading arm. One "random" arm will begin a sidewinding movement, encompassing a 2-3 cm. section of the arm, ① generally at the body end of the arm. This sidewinding and coiling is then straightened out ② partially, pulling the Amphiodia behind it. Several other, or all other arms also sidewind toward the body. Thus there is a sidewinding, snake-like movement.

Amphiodia move in a variety of ways. Another method is for one arm to coil up ③ and push the body ahead of it: ④

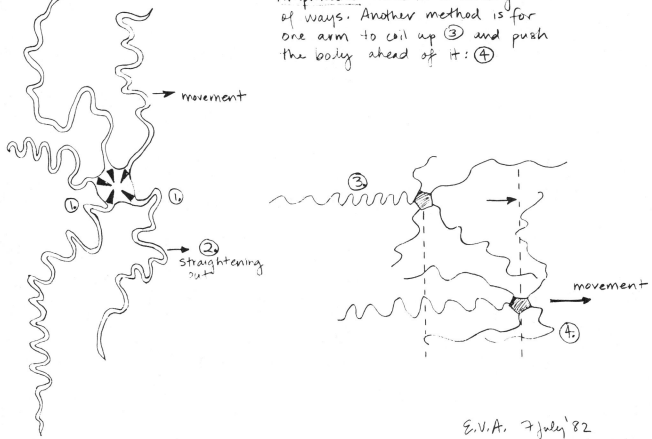

E.V.A. 7 July '82

RIGHTING RESPONSE:

One method:

Amphiodia first raises its body about 5 mm off the substrate using all five arms. Two of the arms are preferred, and these flip the animal. The arms land directly on the other side. The remaining three arms orient themselves in a new "normal" position:

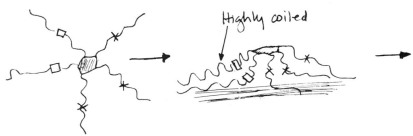

Flat on substrate, oral side up. Raises body

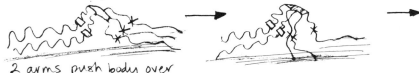

2 arms push body over

Other 3 arms orient themselves for the flip

Flip!

Land. Start walking. Oral side down.

Of the 6-7 righting responses observed, there was no repetition of method.

The righting response just described also can be accomplished using straight arms rather than coiled arms:

oral side up   oral side down

E.V.A. 7 July 82

BURROWING:

In about 1 cm of sand, covered with several cm of sea water.

Amphiodia began burrowing by quickly flapping its podia, so that sand flipped on top of the podia. In this manner, the animal covered almost all of the five arms. Podia all over the arms flipped sand at once, there was no pattern or synchronization to the burrowing. Sometimes tips of the arms remained unburied.

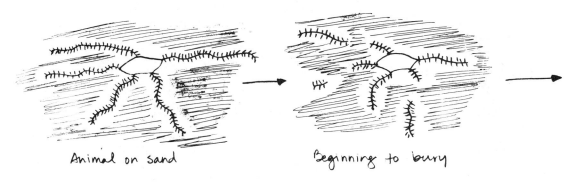

Animal on sand → Beginning to bury →

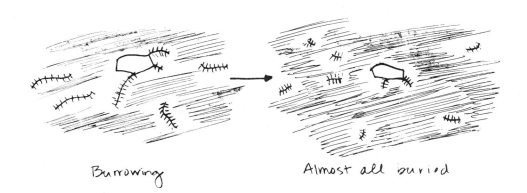

Burrowing → Almost all buried

Amphiodia may begin to walk via arm movements, so that it buries its body, and is completely hidden from view.

E.V.A. 7 July 1982

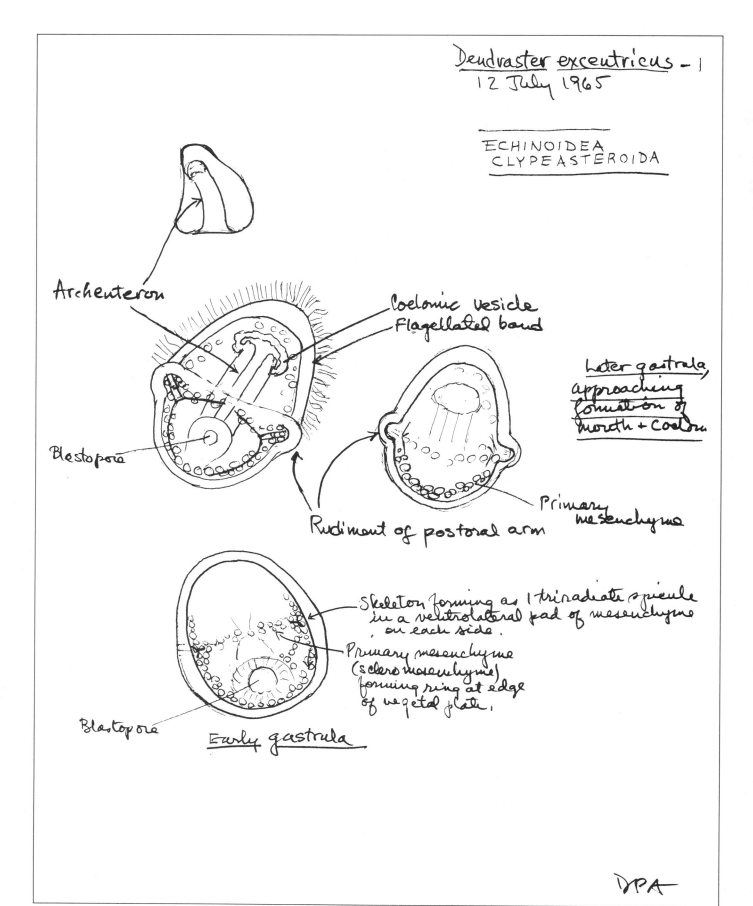

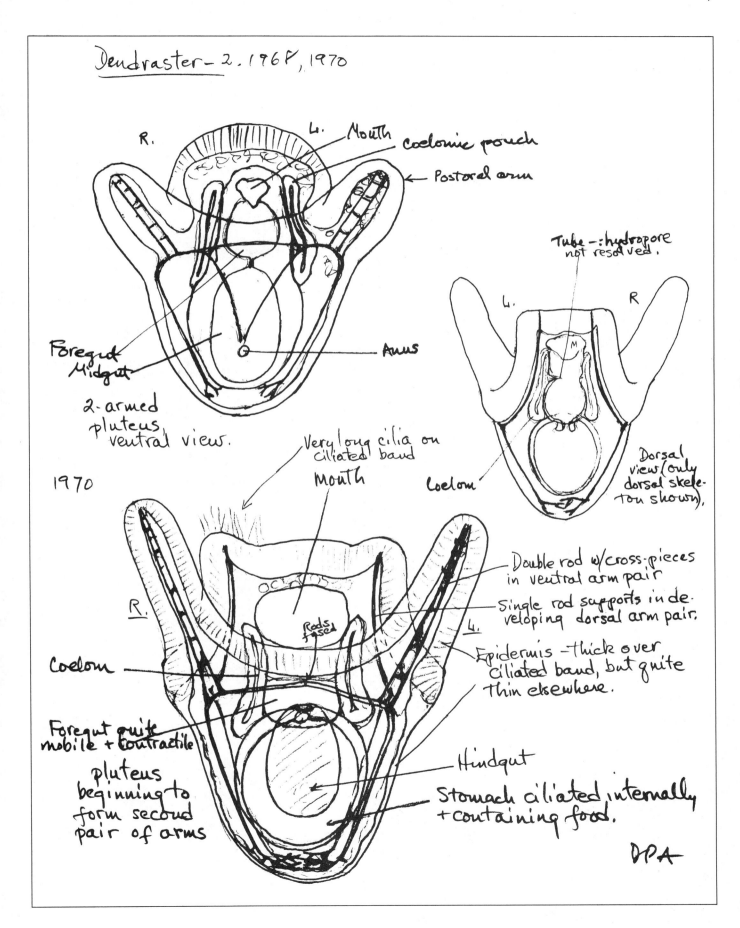

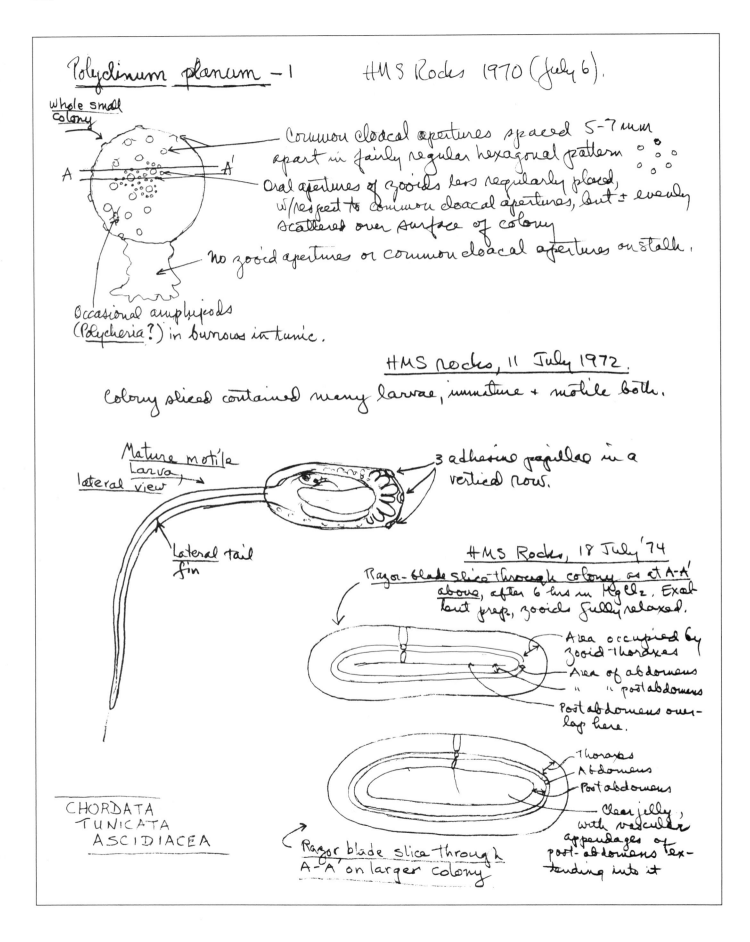

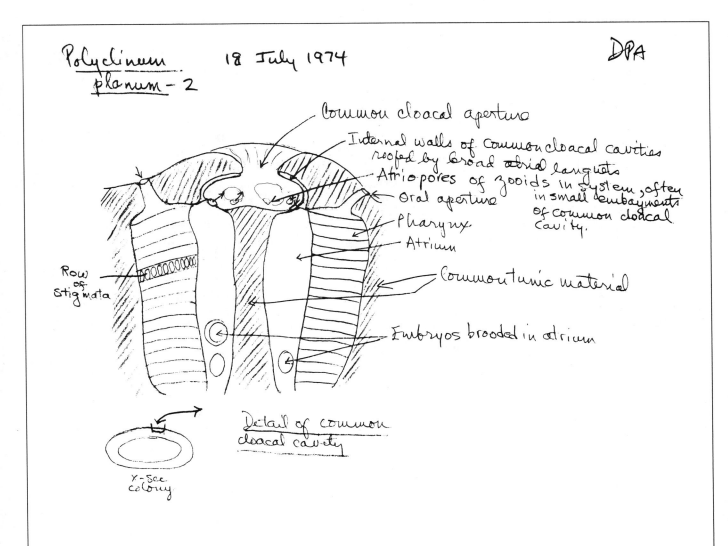

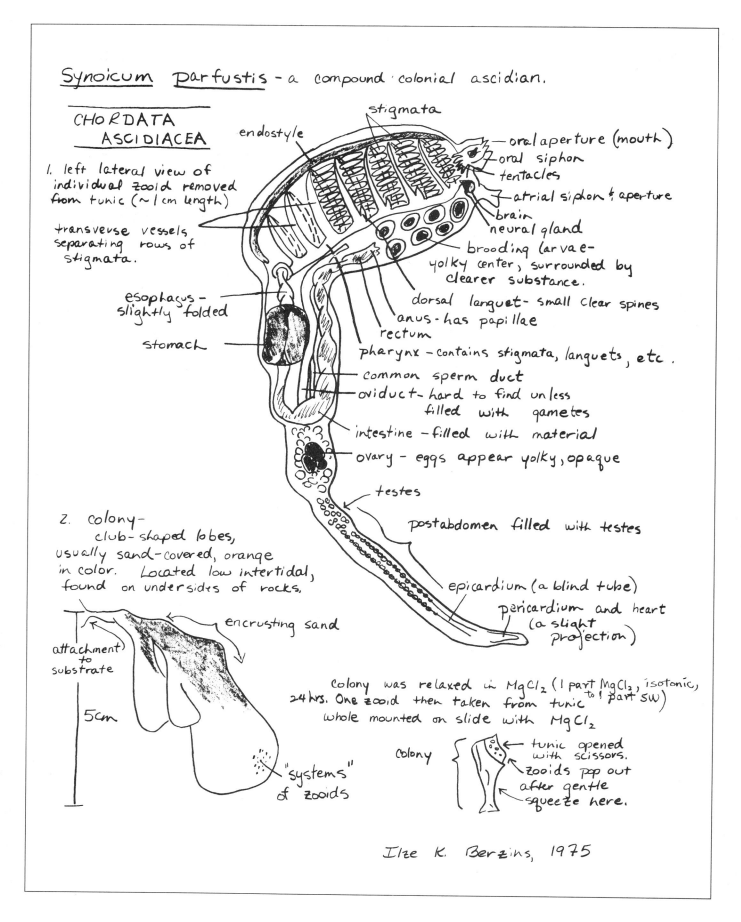

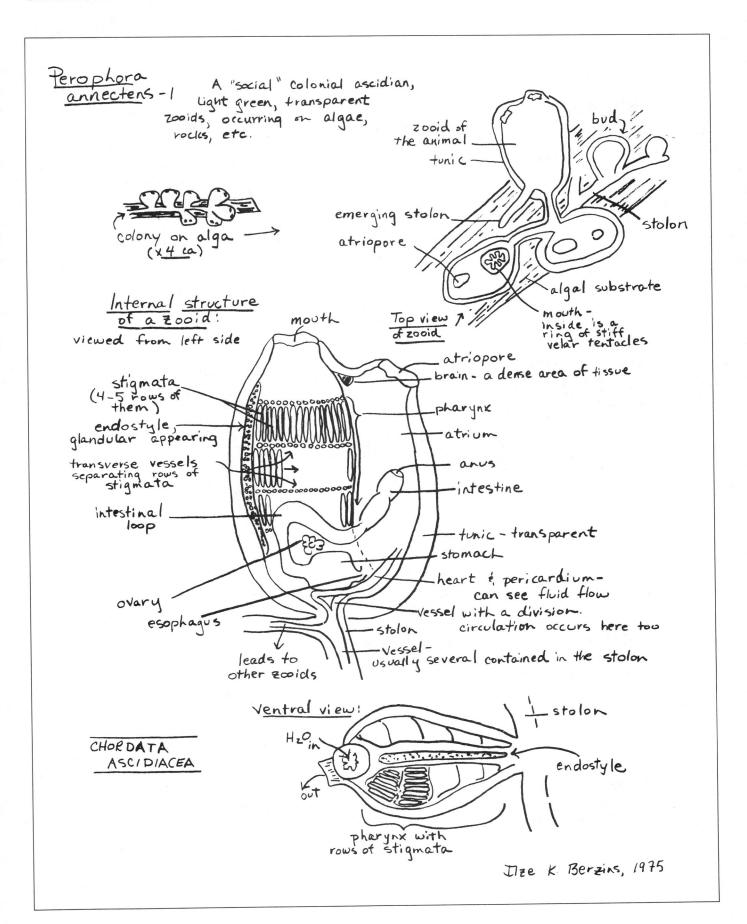

Perophora annectens - 2

BLOOD FLOW OBSERVATIONS

The heart is a clear, tubular structure located below or near the stomach. In the atrium, a fine net of blood passages can be seen. Blood flow can be observed in both the passages and the heart.

The flow of blood alternates direction. I counted about 50 beats in one direction before flow reversed and started moving in the opposite direction. When the flow of blood reverses in the heart, flow reverses in the vessel net.

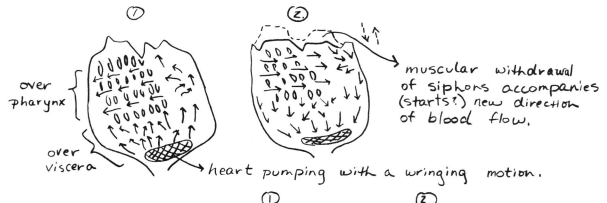

over pharynx / over viscera

→ muscular withdrawal of siphons accompanies (starts?) new direction of blood flow.

→ heart pumping with a wringing motion.

Flow in vessel net:

Before reversal of flow there is always a short time when the blood is pumped first in one direction, then in the other direction (alternating)

1 → 2 → 3 → 4 → 5
6 →
← 7
8 ←
← 9
10 →
← 11 ← 12 ← 13

Blood flow in the connecting stolons between individual zooids also reverses direction when blood flow reverses in the heart.

Ilze K. Berzins, 1975

Chordata/Tunicata/Ascidiacea

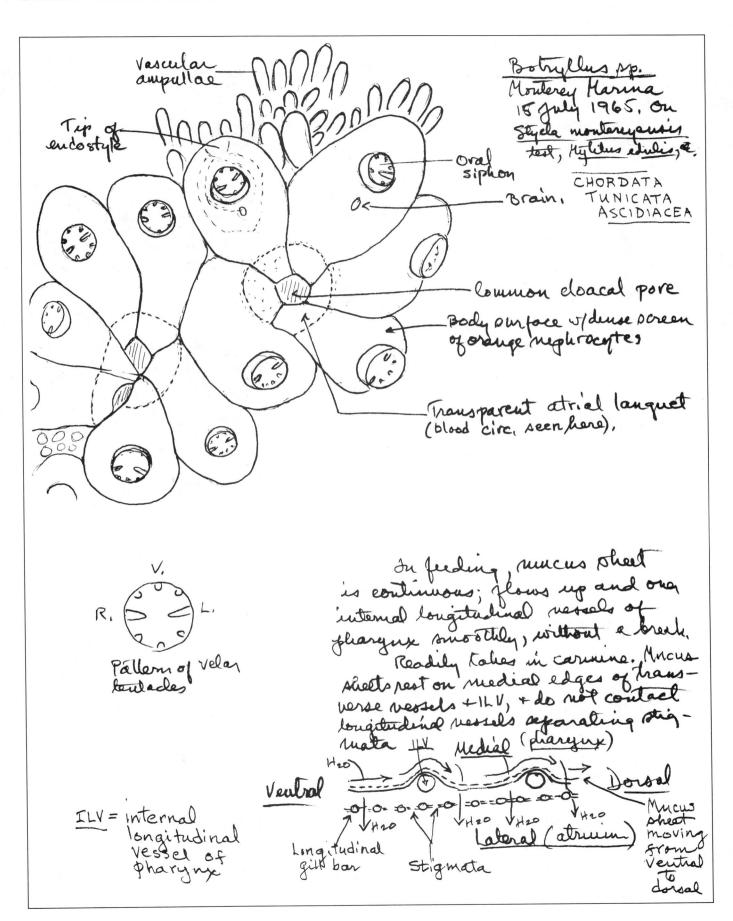

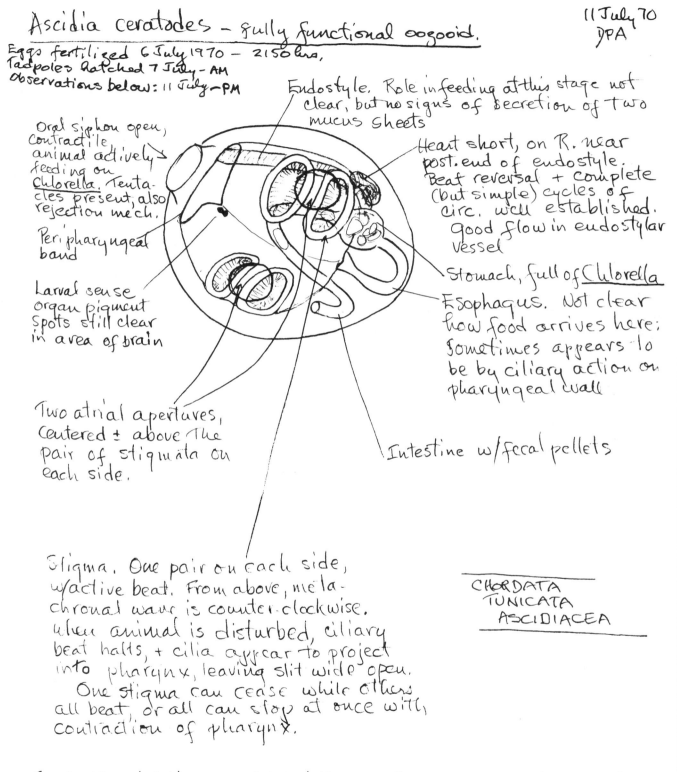

Ascidia ceratodes – fully functional oozooid.   11 July 70 DPA

Eggs fertilized 6 July 1970 – 2150 hrs.
Tadpoles hatched 7 July – AM
Observations below: 11 July – PM

Oral siphon open, contractile animal actively feeding on *Chlorella*. Tentacles present, also rejection mech.

Peripharyngeal band

Larval sense organ pigment spots still clear in area of brain

Endostyle. Role in feeding at this stage not clear, but no signs of secretion of two mucus sheets

Heart short, on R. near post. end of endostyle. Beat reversal + complete (but simple) cycles of circ. well established. Good flow in endostylar vessel

Stomach, full of *Chlorella*

Esophagus. Not clear how food arrives here; sometimes appears to be by ciliary action on pharyngeal wall

Intestine w/ fecal pellets

Two atrial apertures, centered ± above the pair of stigmata on each side.

Stigma. One pair on each side, w/ active beat. From above, metachronal wave is counter-clockwise. When animal is disturbed, ciliary beat halts, + cilia appear to project into pharynx, leaving slit wide open. One stigma can cease while others all beat, or all can stop at once with contraction of pharynx.

CHORDATA
TUNICATA
ASCIDIACEA

For further details see Allen, Neil, 1971. Morphogenesis + behavior during metamorphosis + post-metamorphic growth in... *Ascidia ceratodes*. 26 pp. Unpubl. MS on file at HMS Library.

Chordata/Tunicata/Ascidiacea

Ascidia ceratodes, fert. 6 July, raised in fingerbowl-running water since 6 Sept. Drawn 3 Oct 1970.

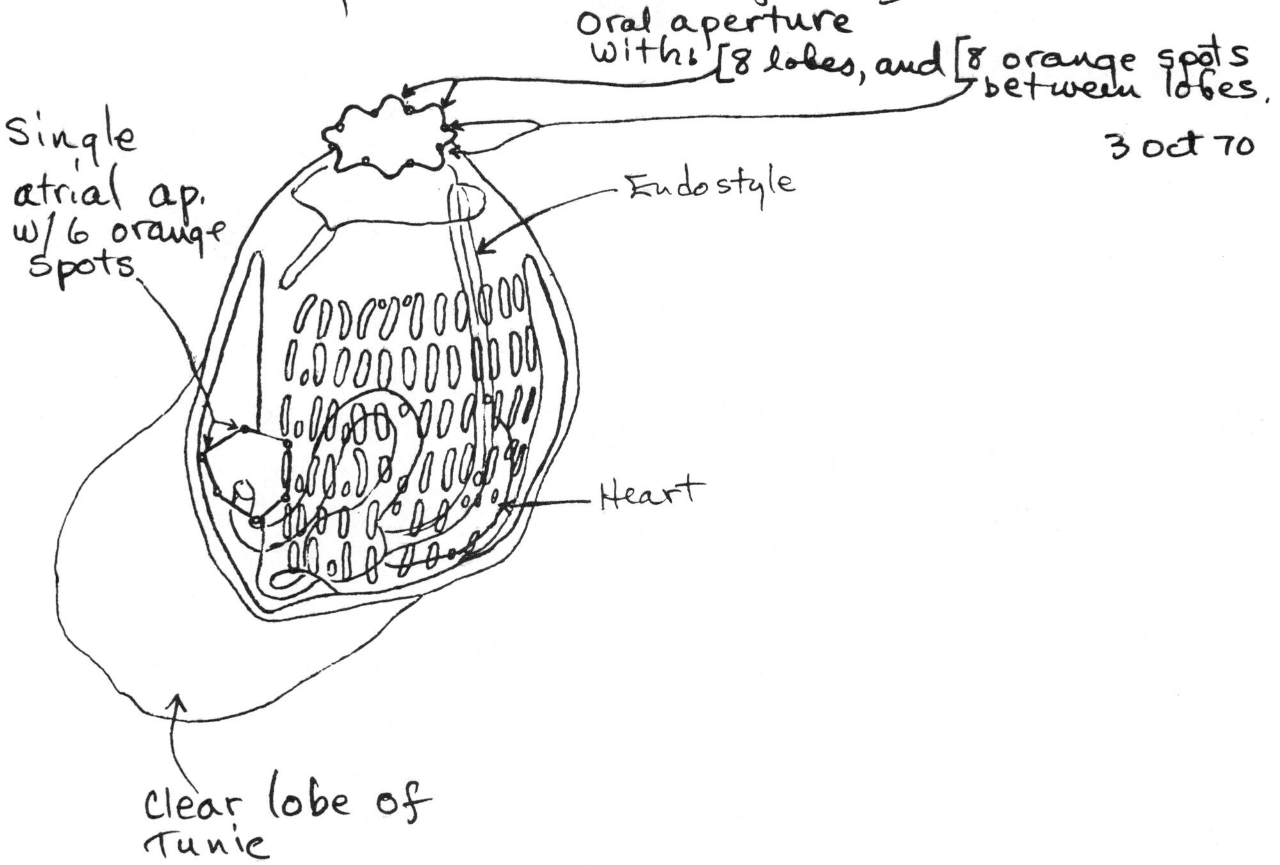

Oral aperture with: [8 lobes, and [8 orange spots between lobes.

3 Oct 70

Single atrial ap. w/ 6 orange spots

Endostyle

Heart

Clear lobe of Tunic

tunic opened to show
6 rows of stigmata
single atrial aperture

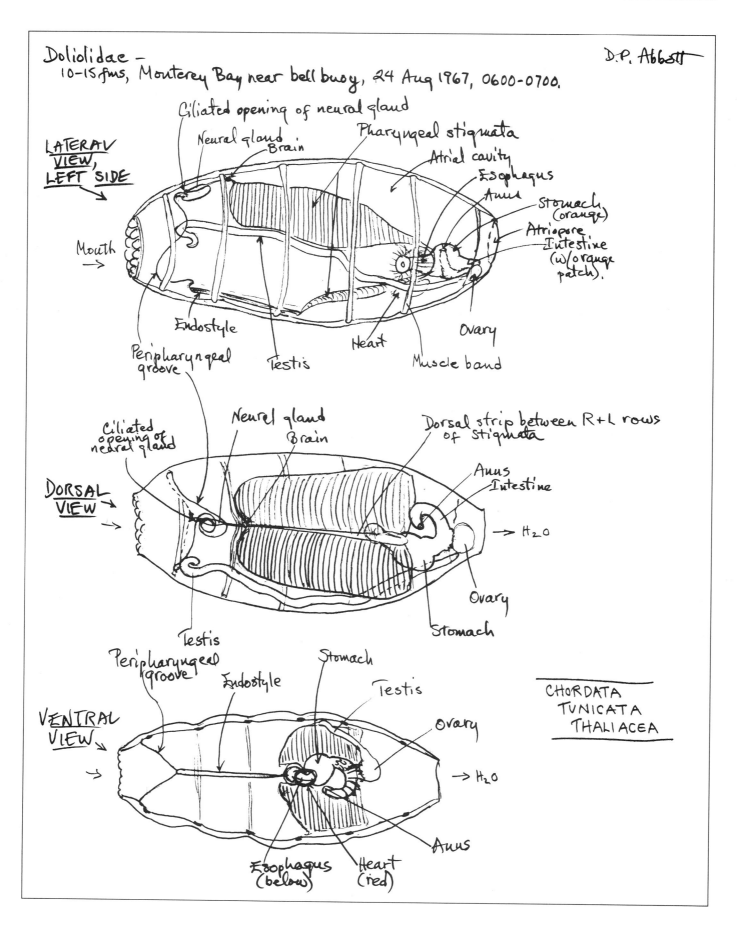

Mollusca/Gastropoda/Prosobranchia/Archaeogastropoda

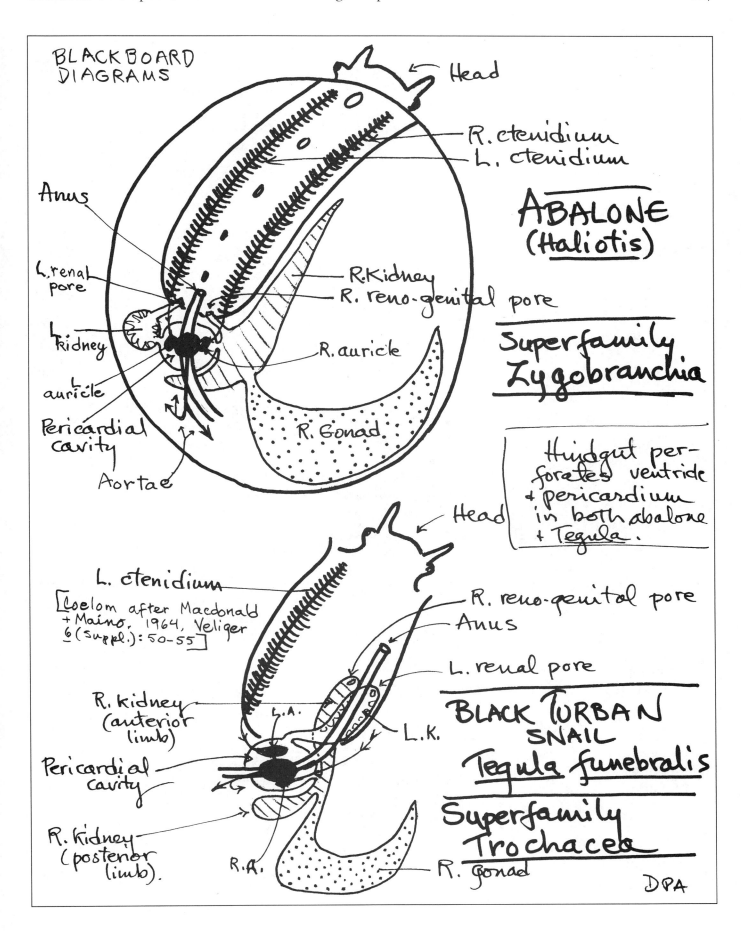

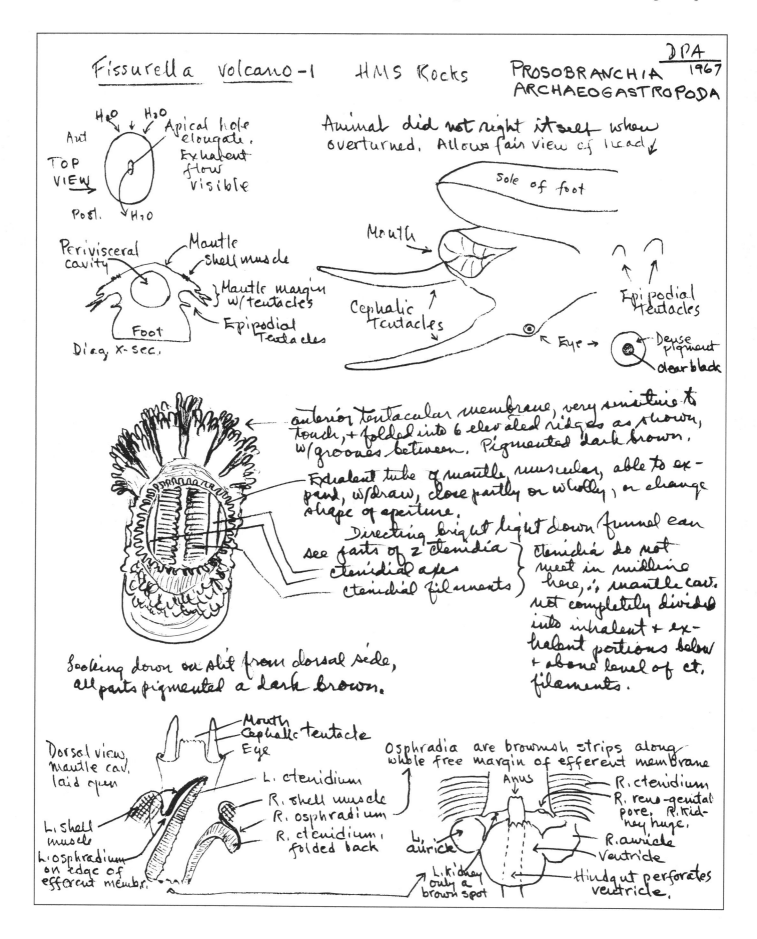

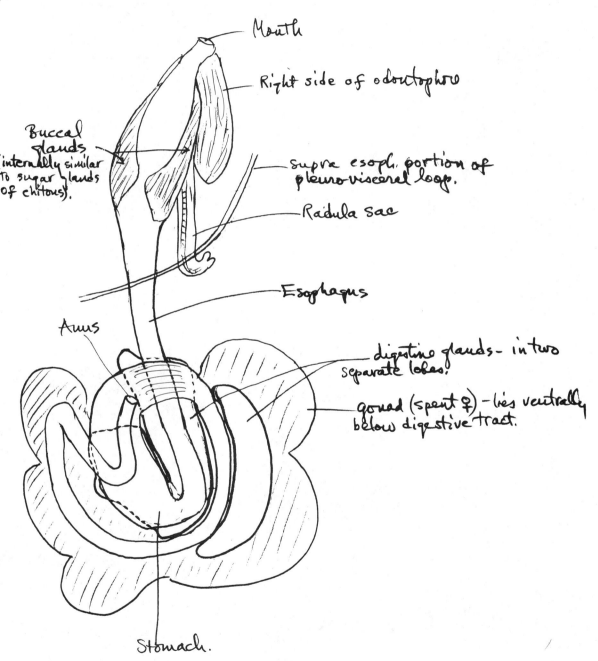

Fissurella volcano – 2
1967.

Dorsal view of gut + gonad

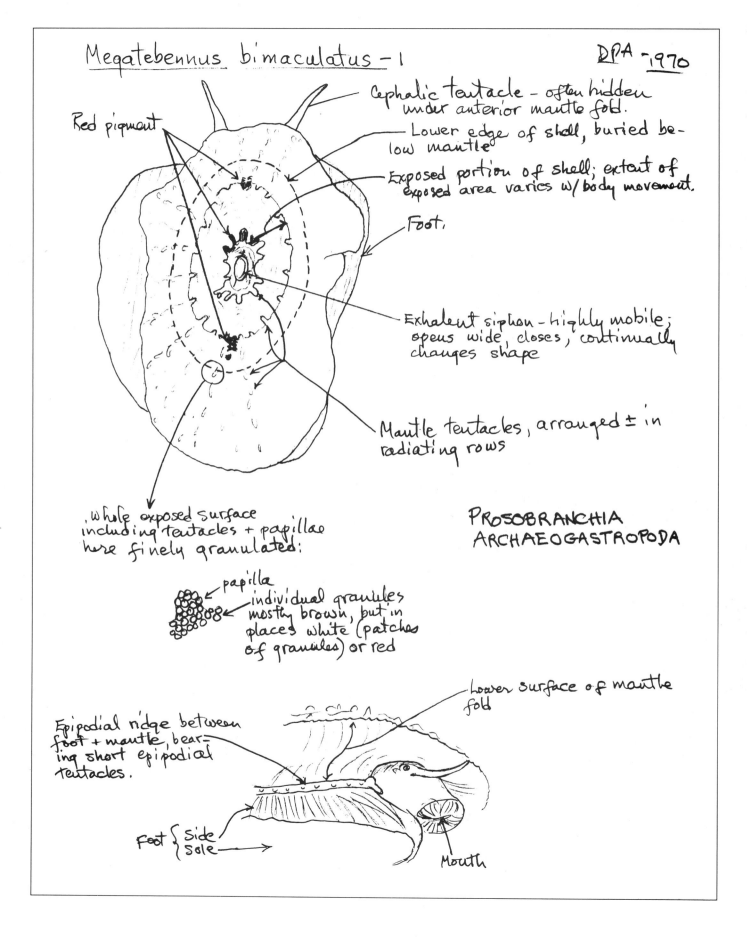

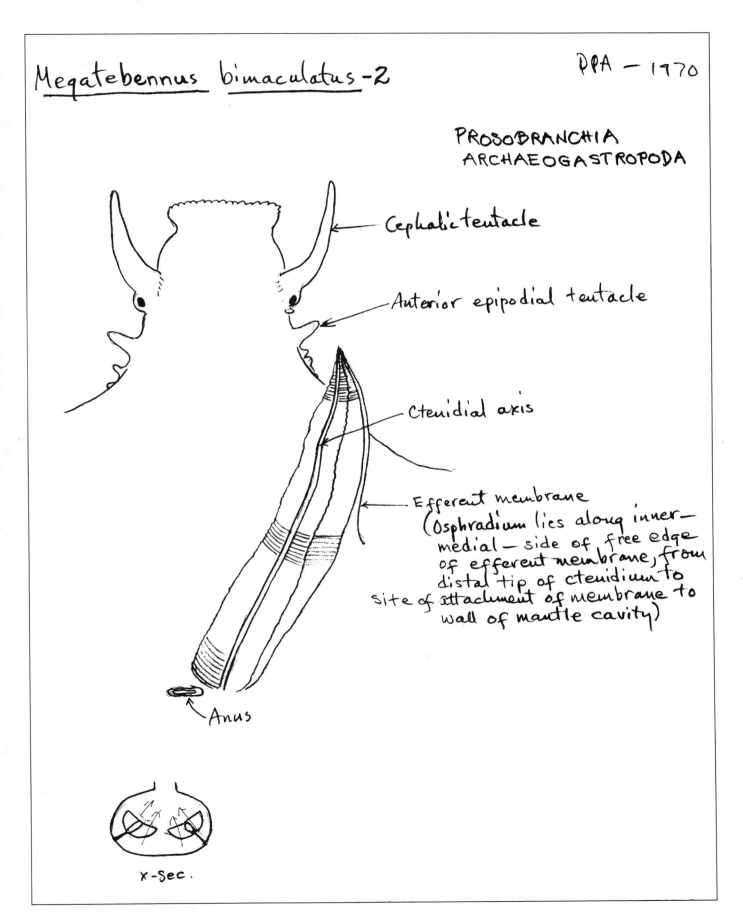

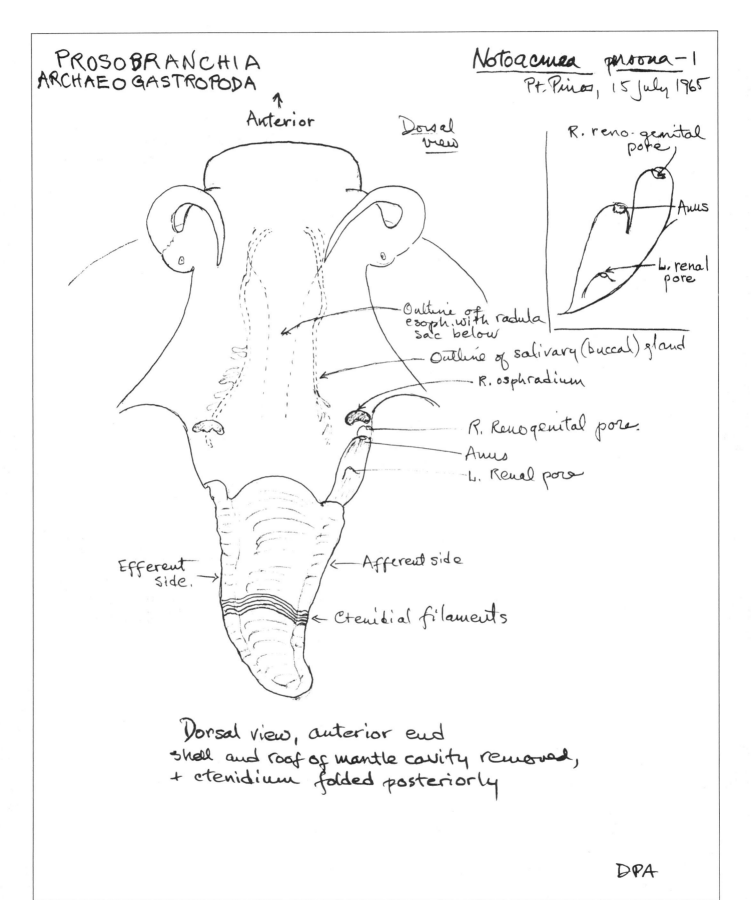

Mollusca/Gastropoda/Prosobranchia/Archaeogastropoda

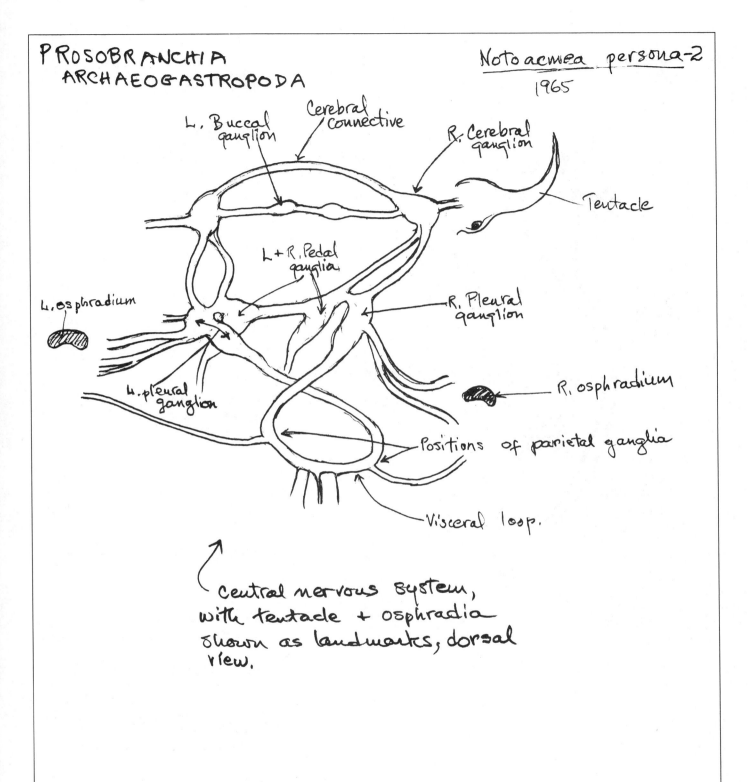

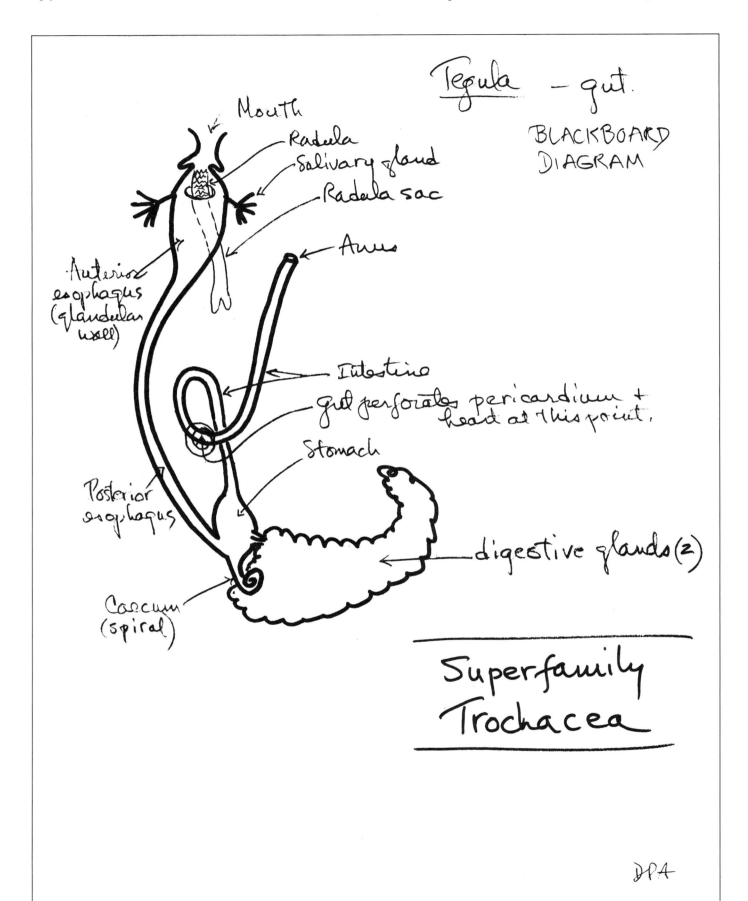

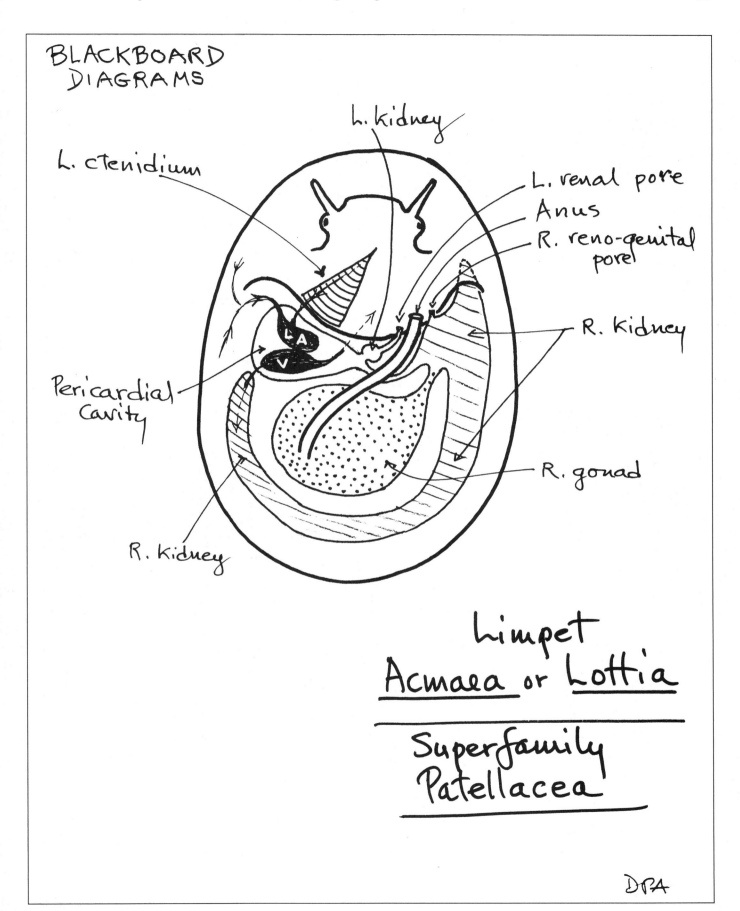

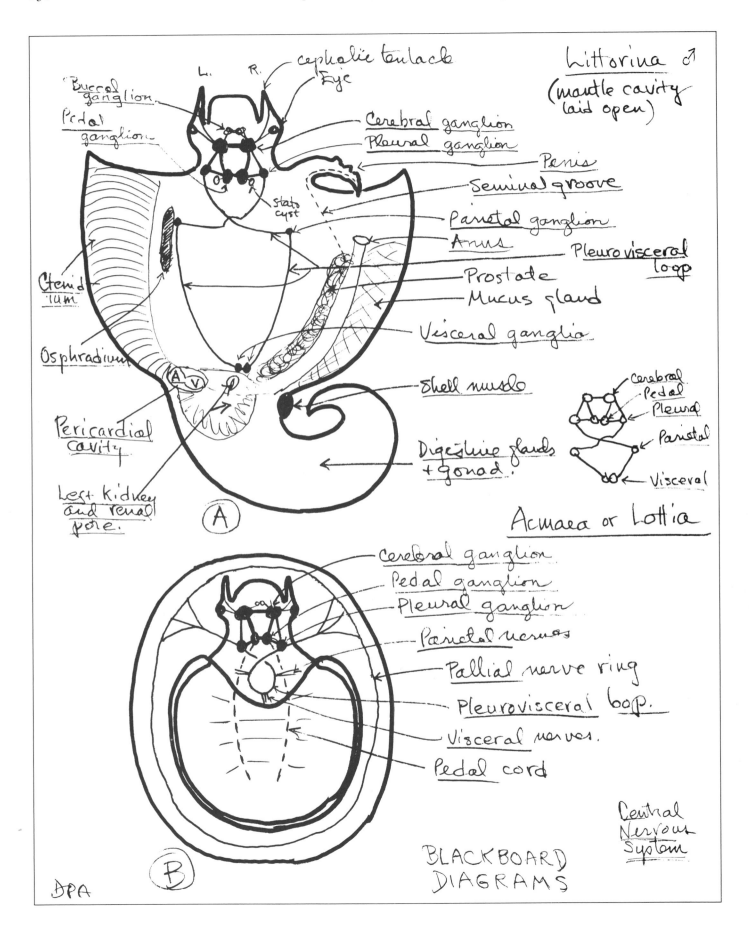

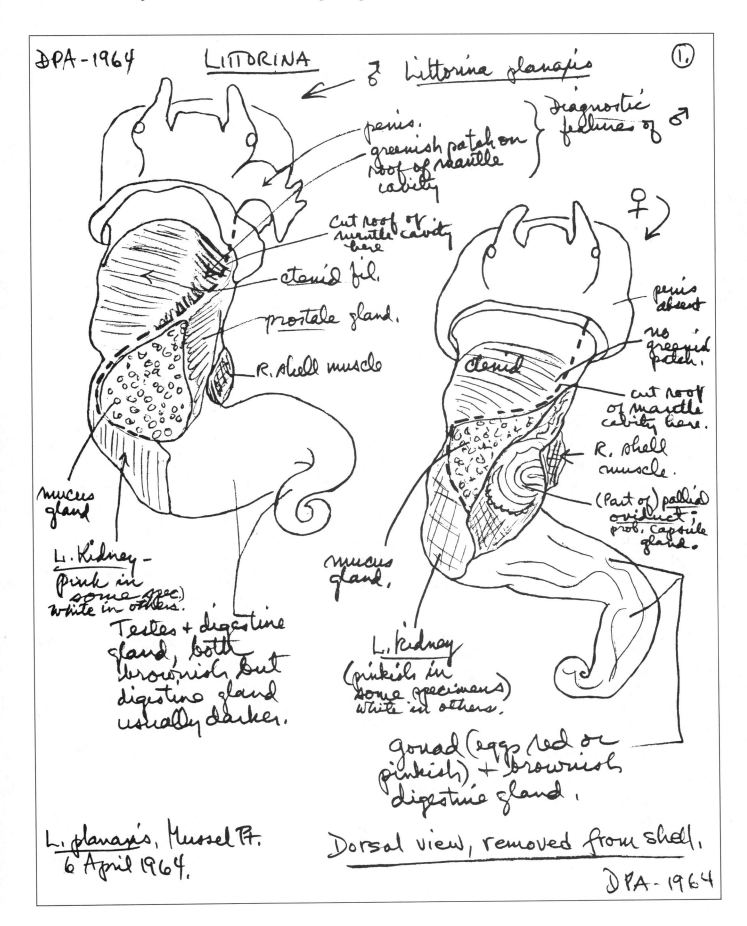

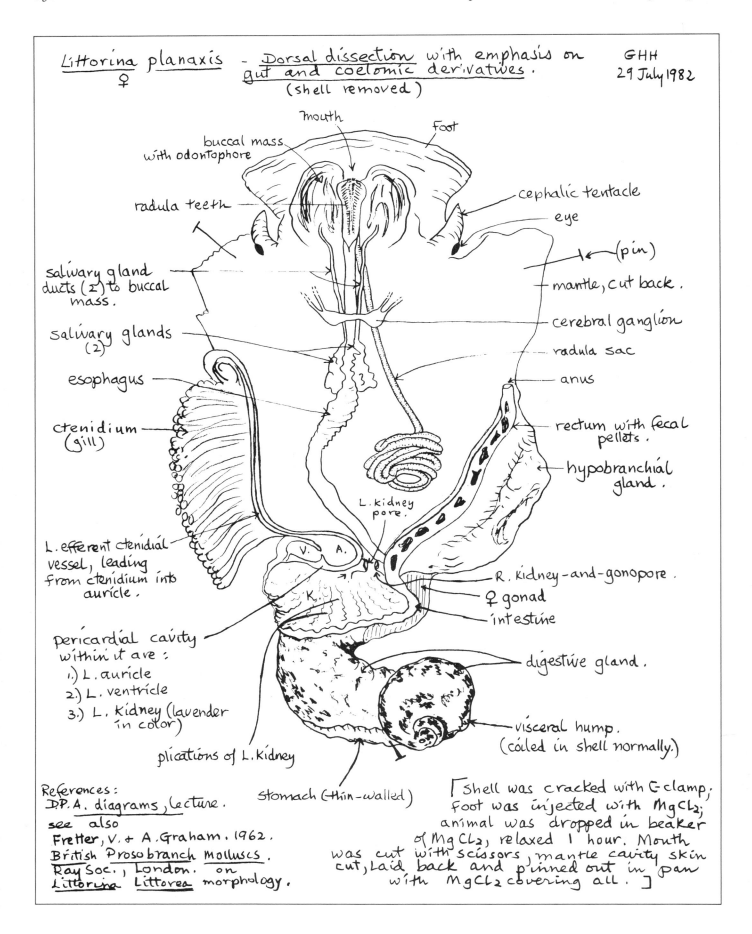

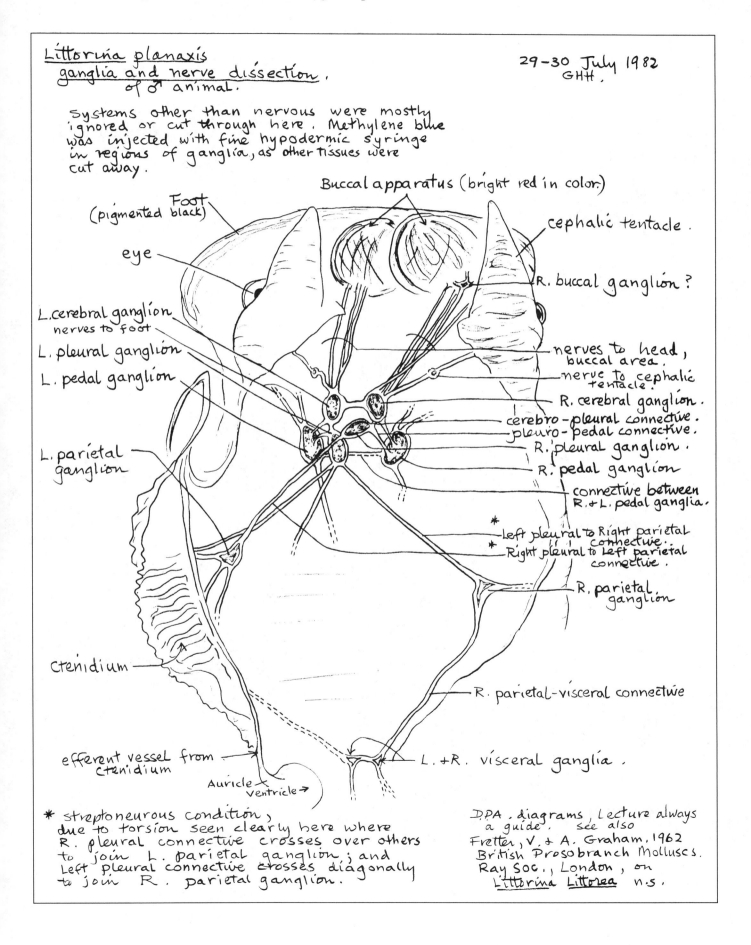

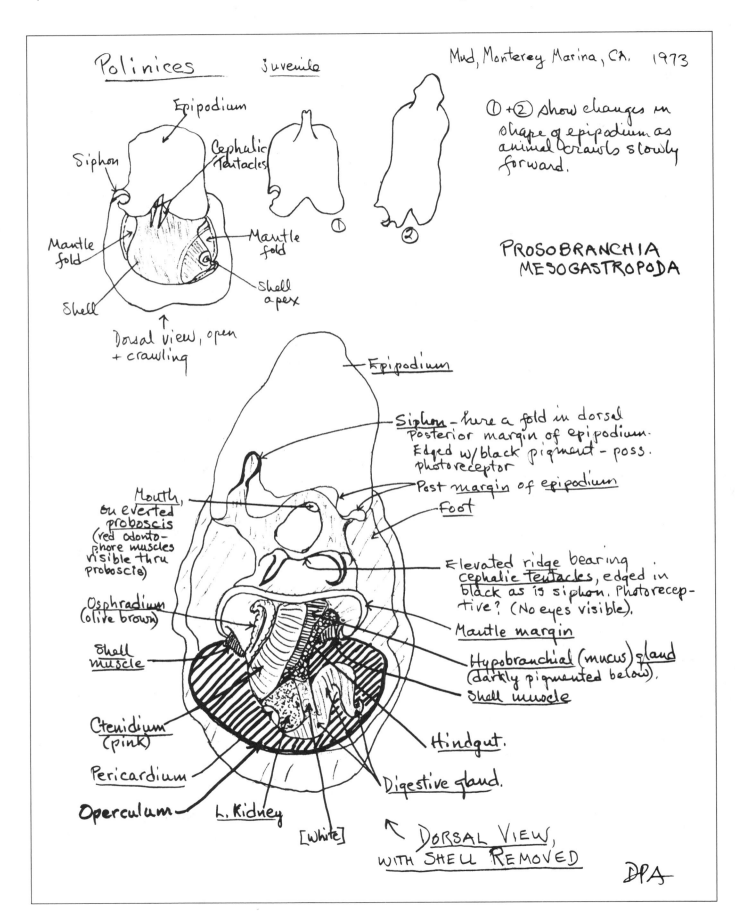

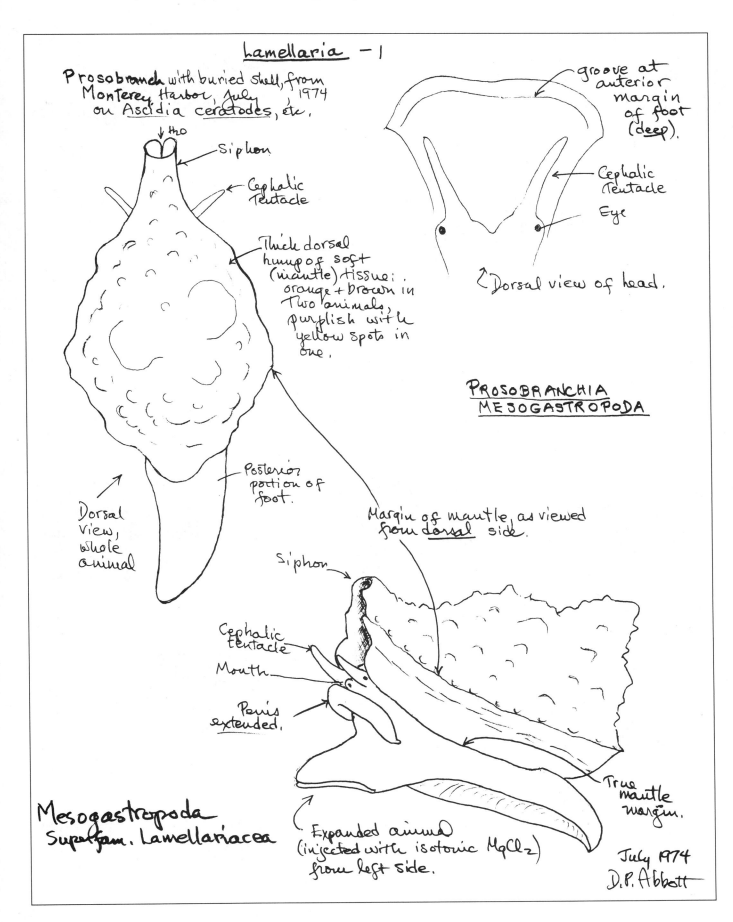

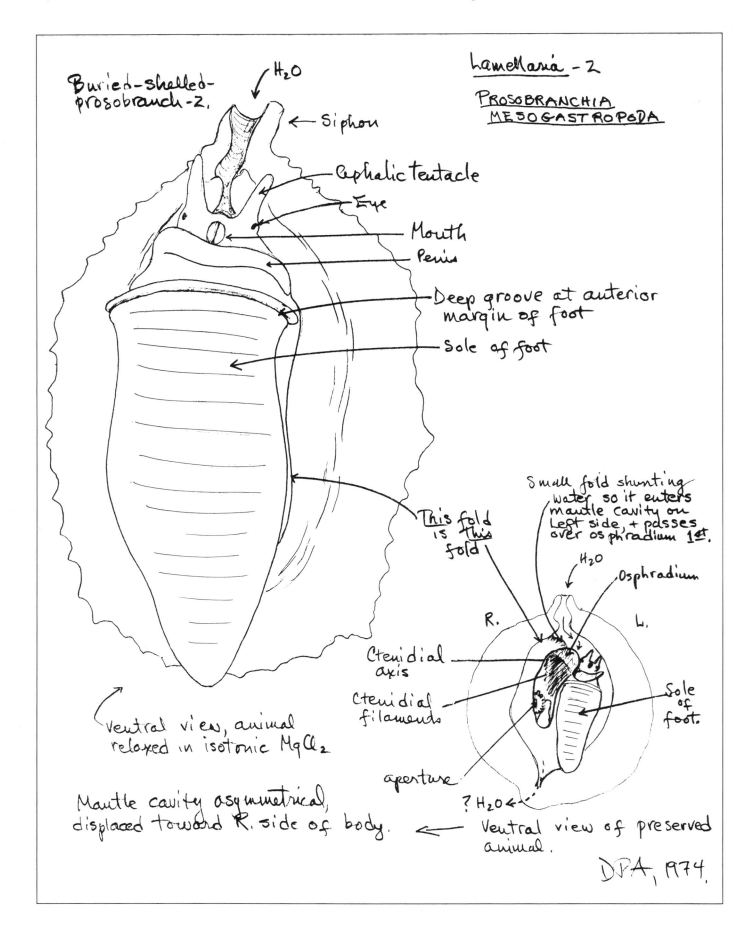

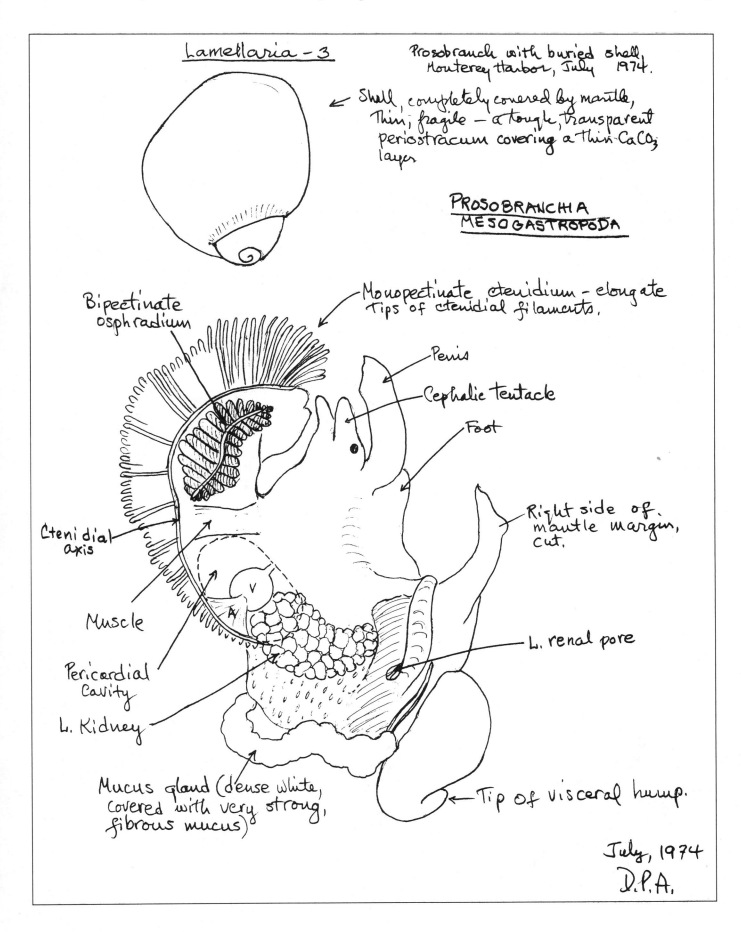

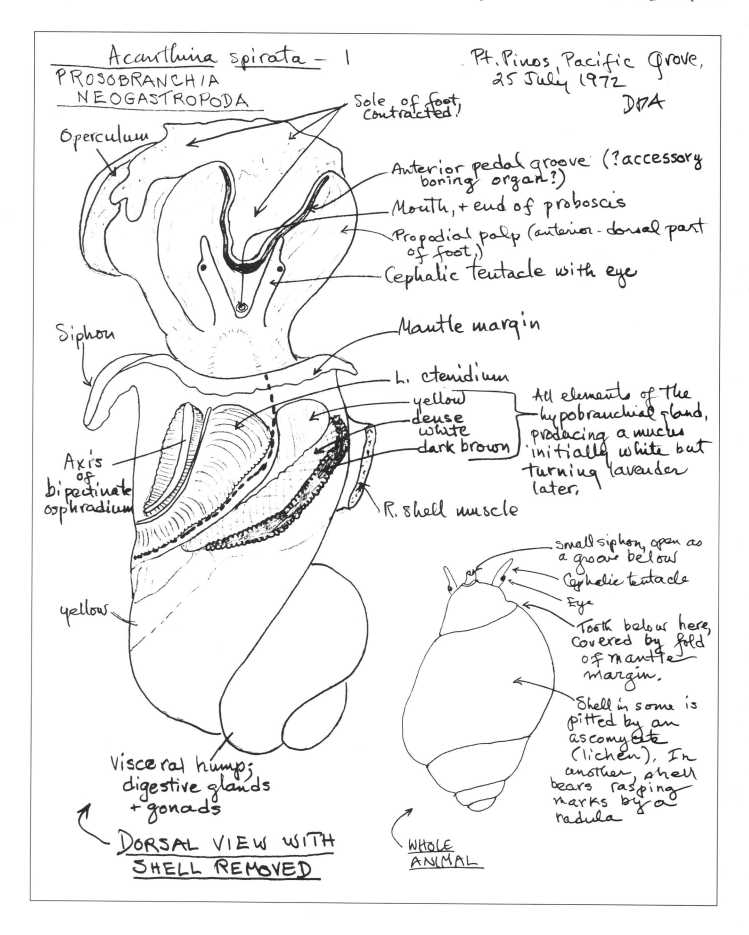

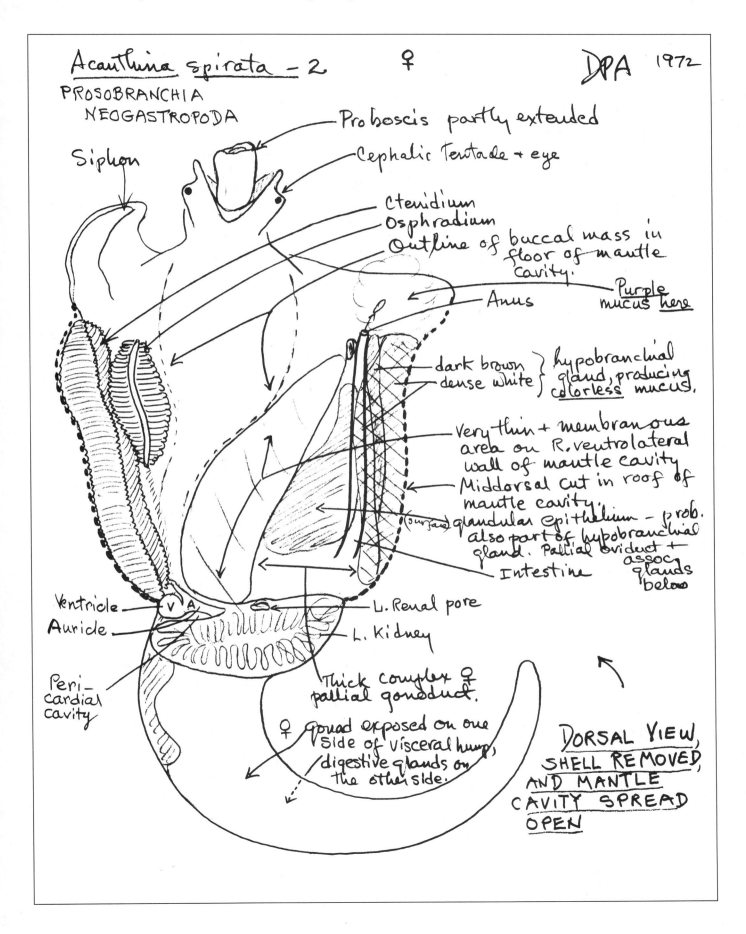

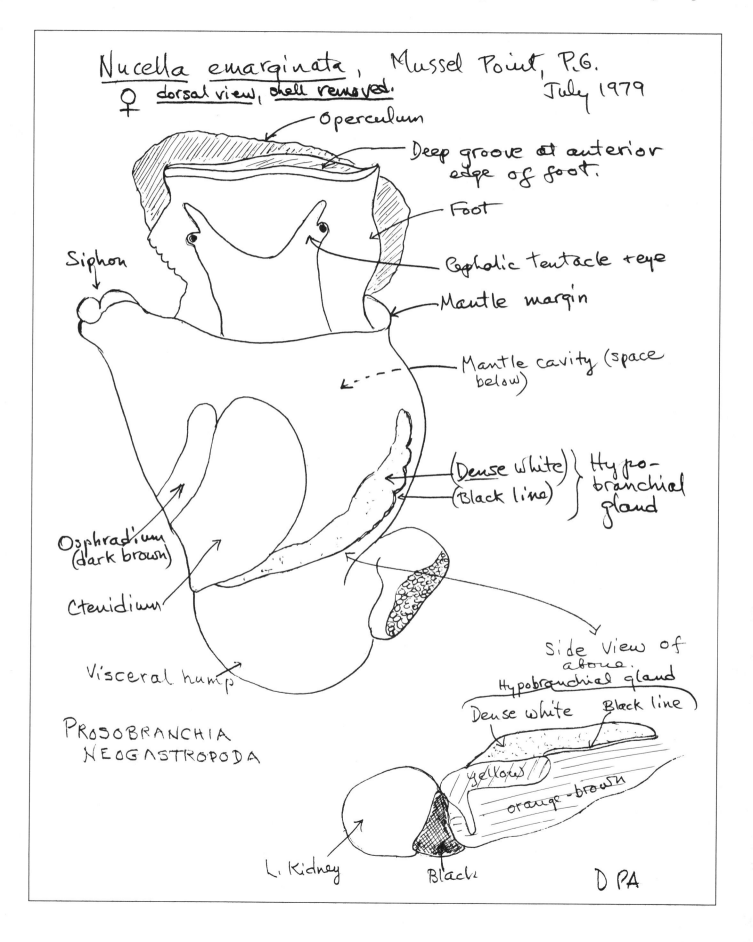

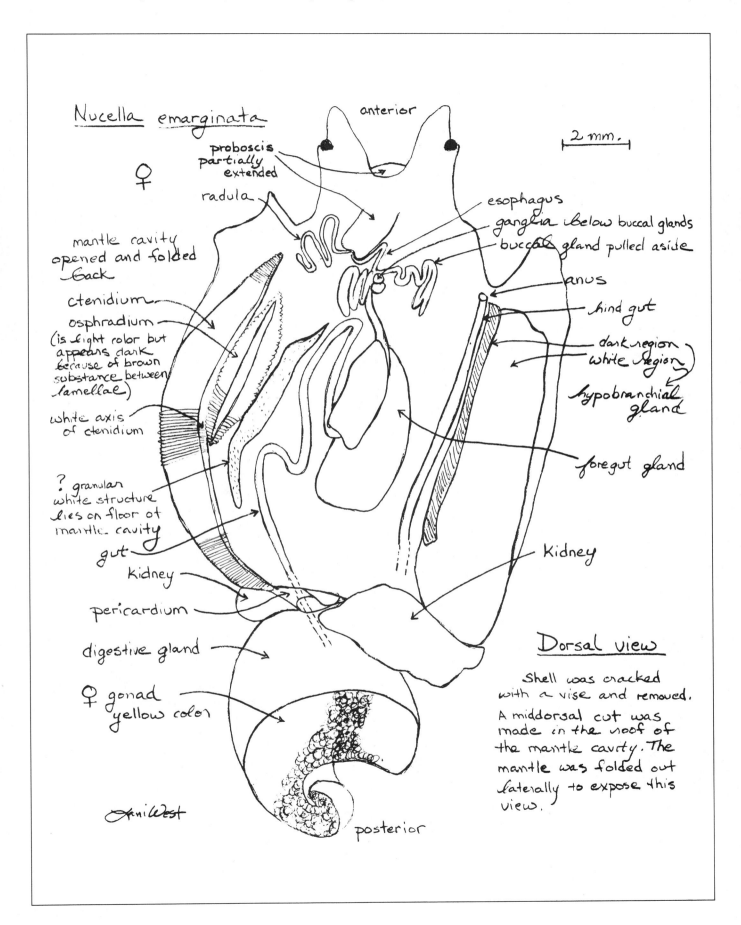

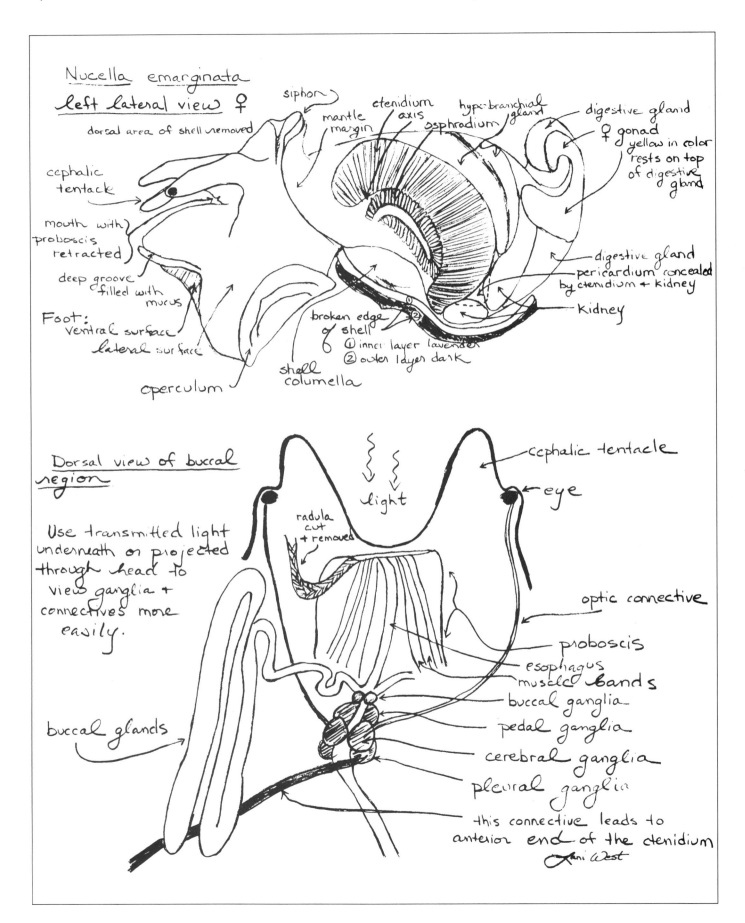

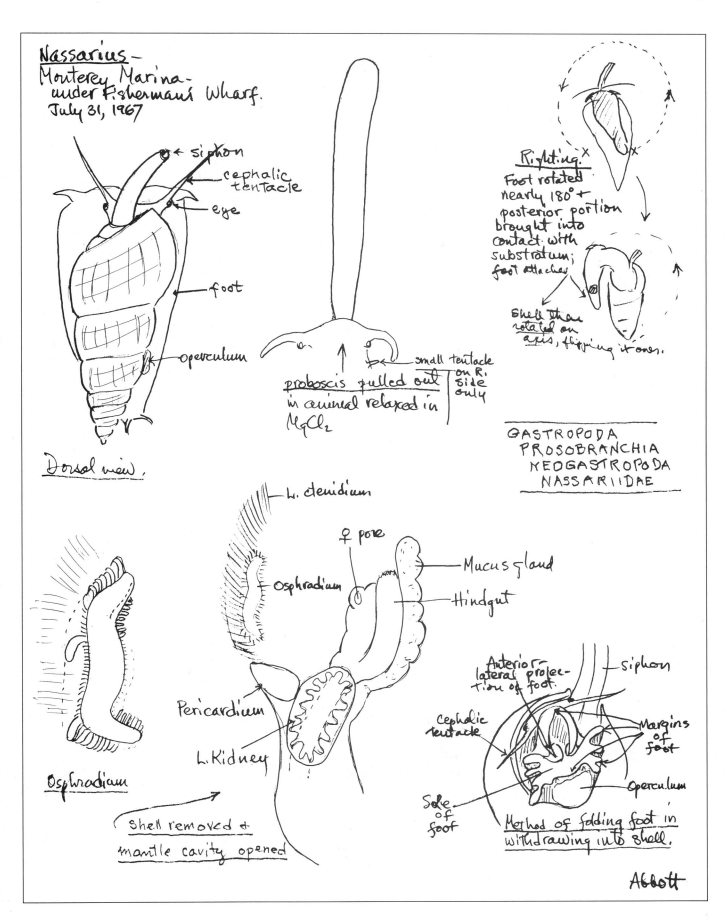

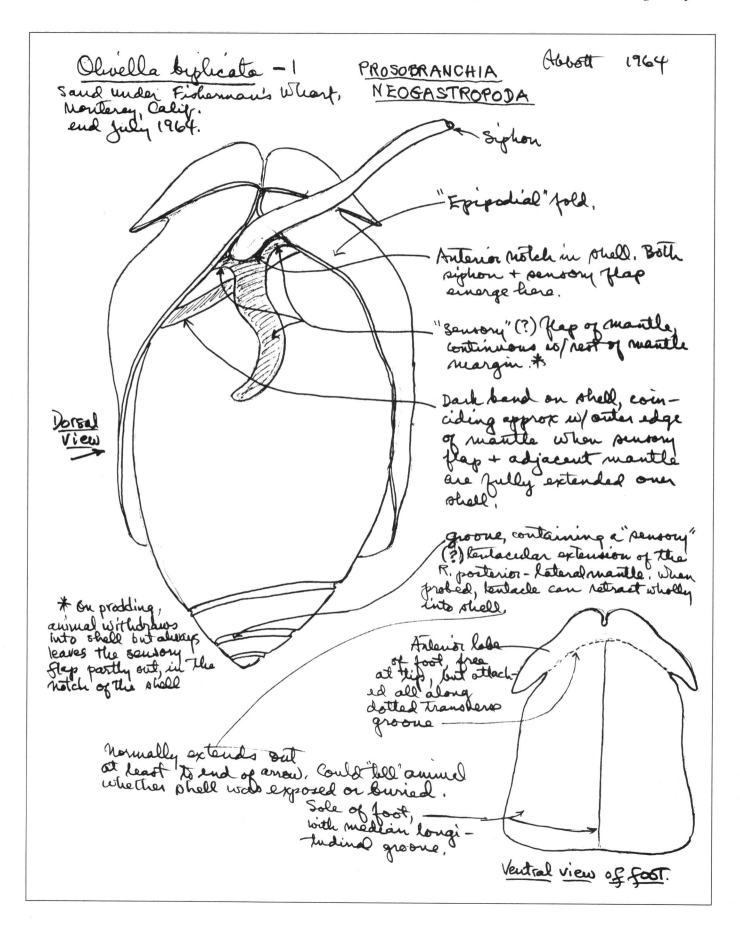

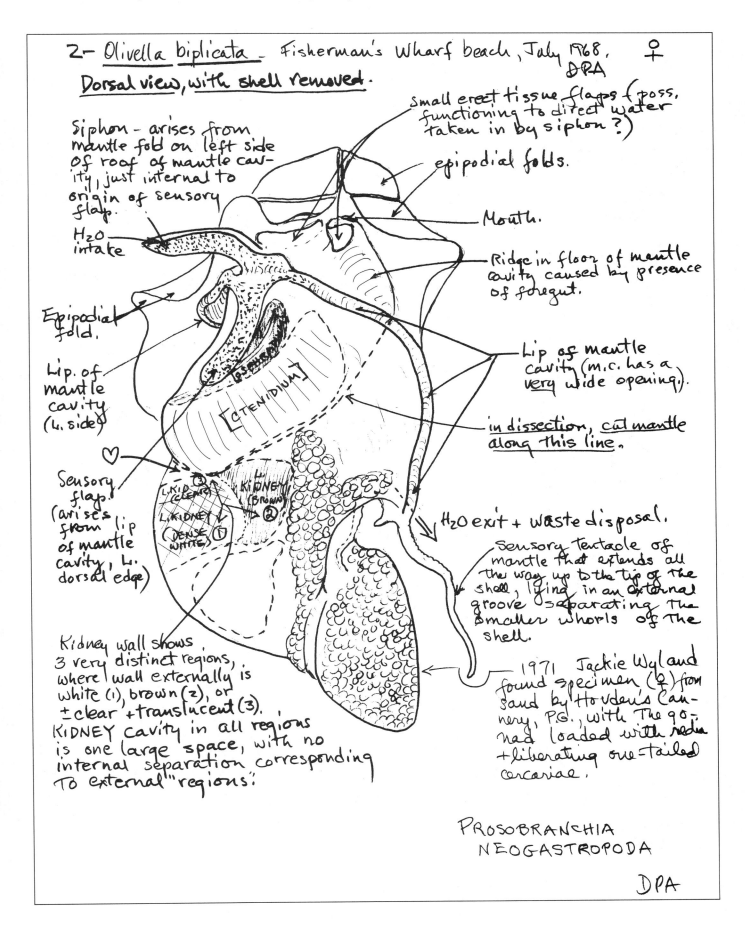

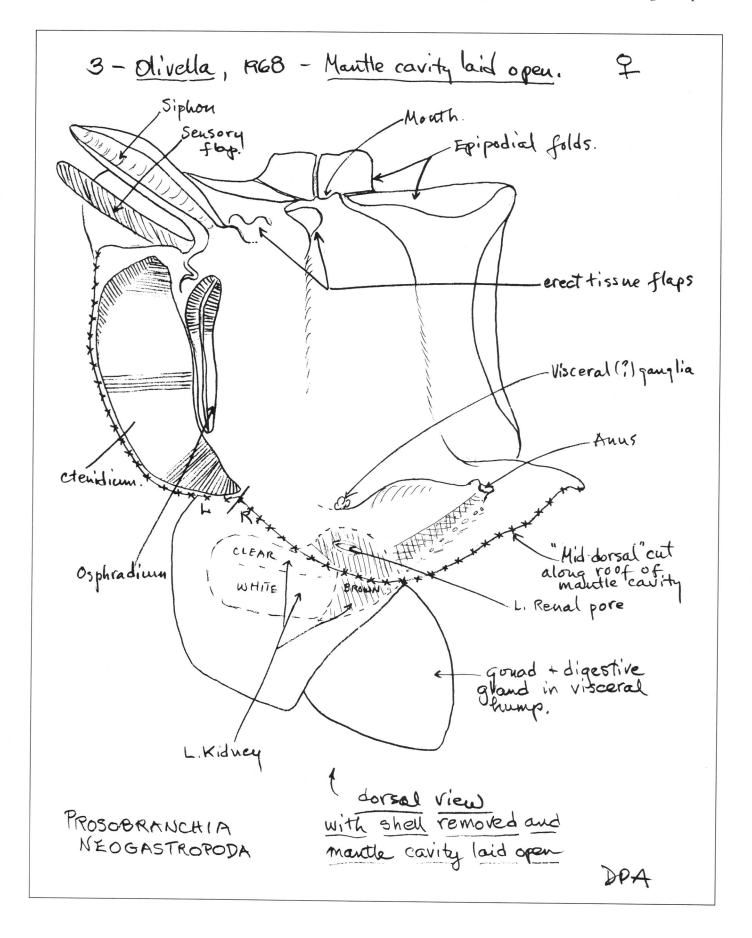

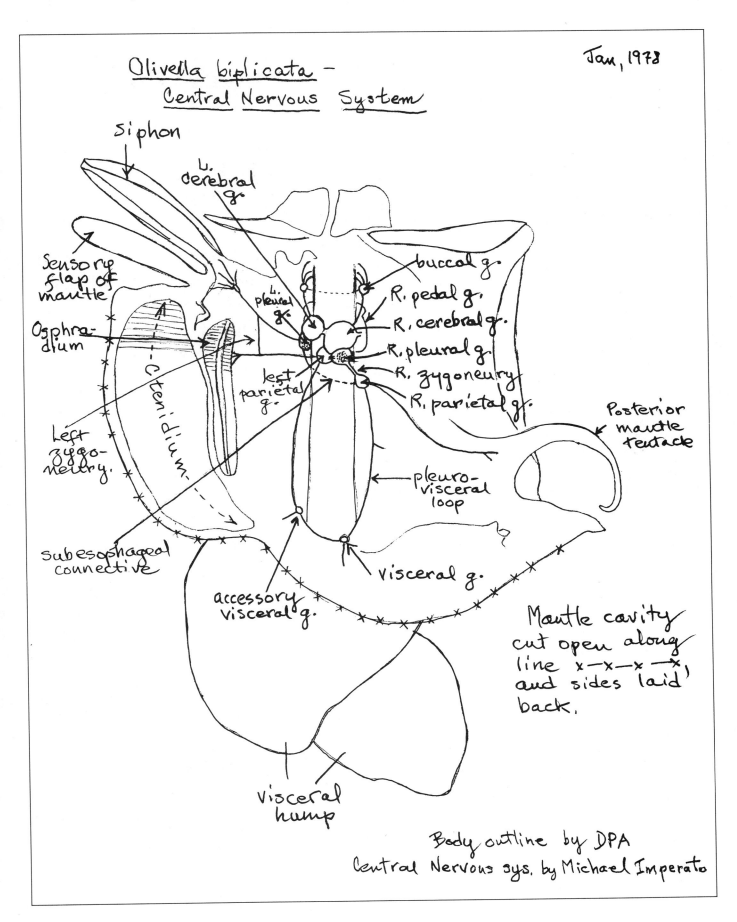

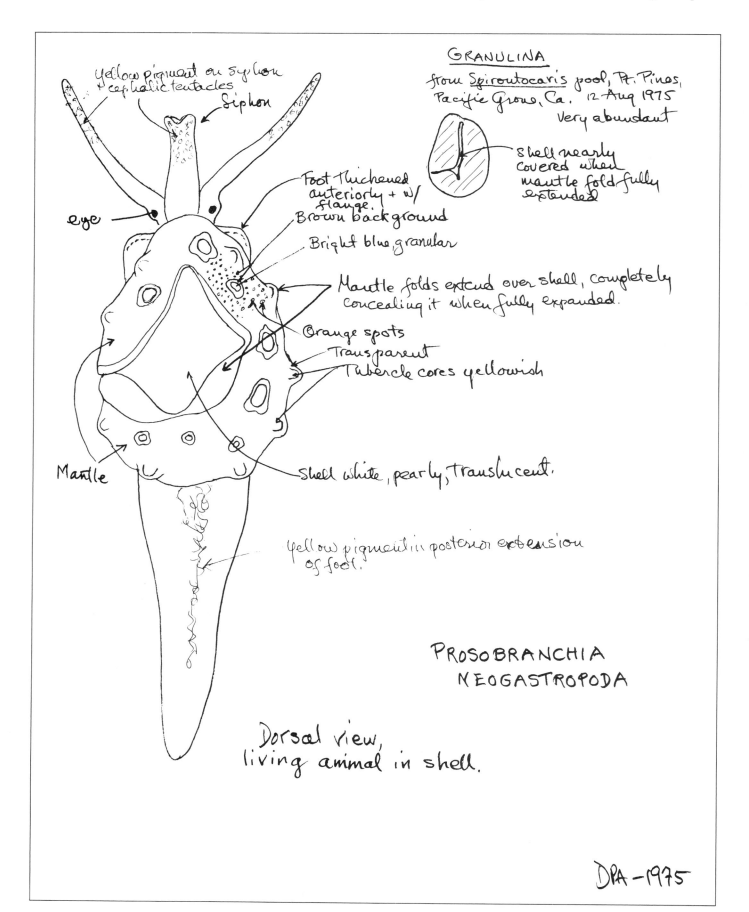

Mollusca/Gastropoda/Prosobranchia/Neogastropoda

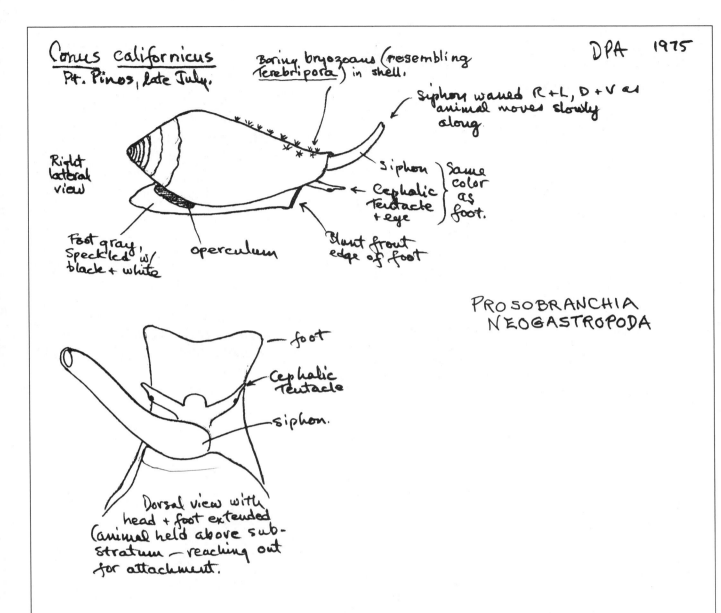

Aug, 1963. Monterey Harbor, shallow sand burrower.

DPA

__Rictaxis__ (= __Actaeon__) __punctocaelatus__ – 1

OPISTHOBRANCHIA
CEPHALASPIDEA

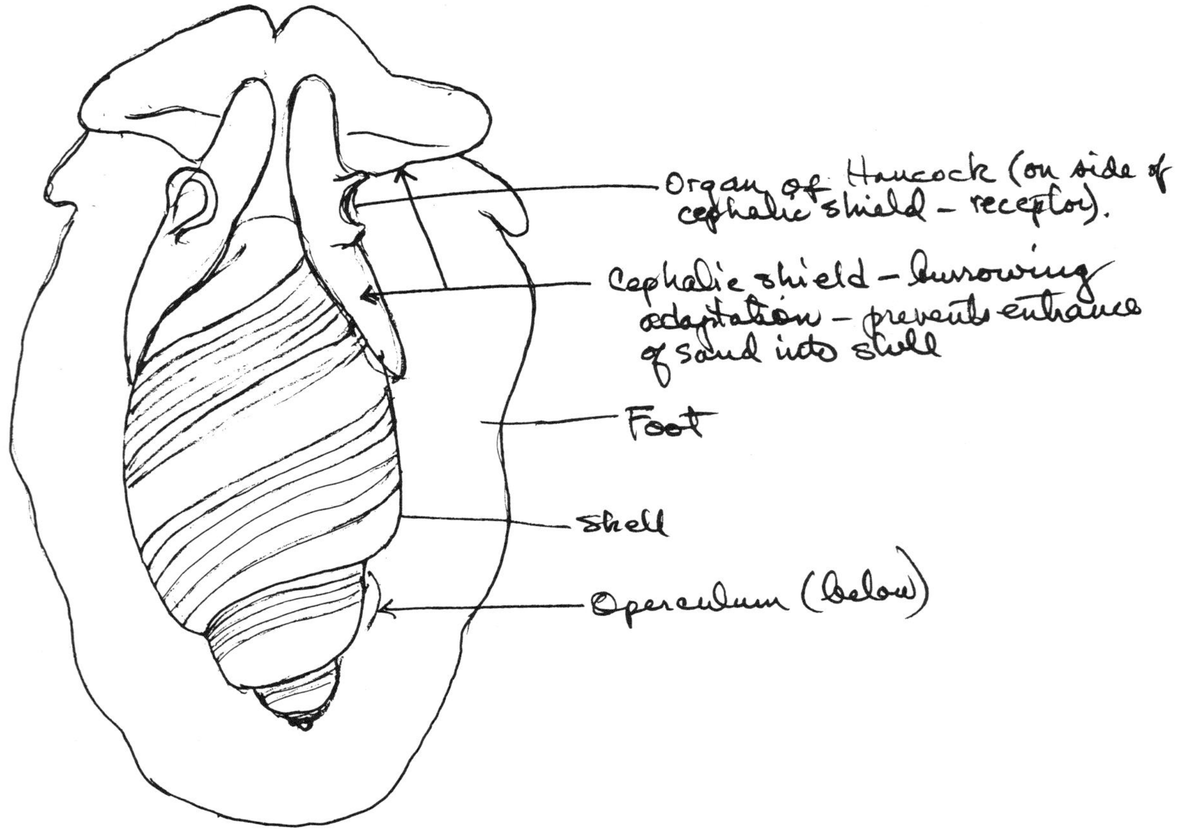

- Organ of Hancock (on side of cephalic shield – receptor).
- Cephalic shield – burrowing adaptation – prevents entrance of sand into shell
- Foot
- Shell
- Operculum (below)

CNS still slightly streptoneurous
Ctenidium + mantle cavity still present
Body can withdraw wholly into shell.

Mollusca/Gastropoda/Opisthobranchia/Cephalaspidea

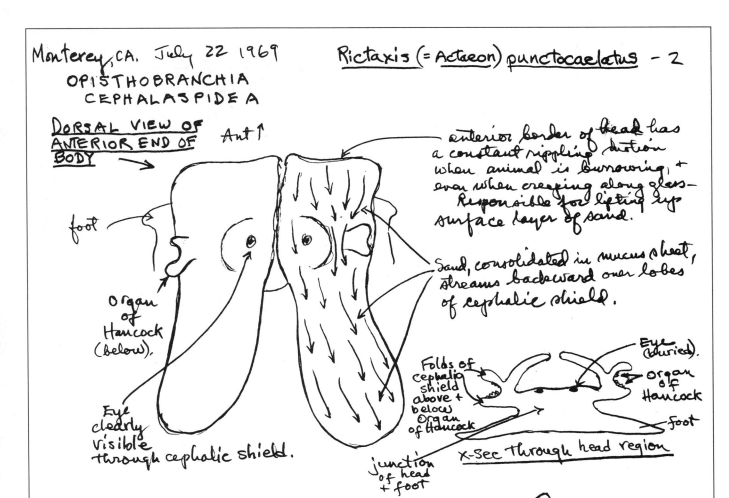

Monterey, CA. July 22 1969
OPISTHOBRANCHIA
CEPHALASPIDEA

Rictaxis (= Actaeon) punctocaelatus - 2

DORSAL VIEW OF ANTERIOR END OF BODY

- Ant↑
- foot
- Organ of Hancock (below).
- Eye clearly visible through cephalic shield.
- anterior border of head has a constant rippling motion when animal is burrowing, + even when creeping along glass — Responsible for lifting up surface layer of sand.
- Sand, consolidated in mucus sheet, streams backward over lobes of cephalic shield.
- Folds of cephalic shield above + below Organ of Hancock
- junction of head + foot

X-sec through head region
- Eye (buried)
- Organ of Hancock
- foot

In shallow burrowing, animal ploughs only the surface, passing over the cephalic shield a sand + mucus layer only about one sand grain thick.

In deep burrowing, ant. lobe of foot is directed downward, animal goes in w/ shell at 45° X or greater. Once ½ buried, foot is extended forward in powerful surges (hydrostatic expansion of foot ± as in bivalves), + shell is subsequently pulled down in jerks, with foot serving as subsurface anchor.

DPA - 1969

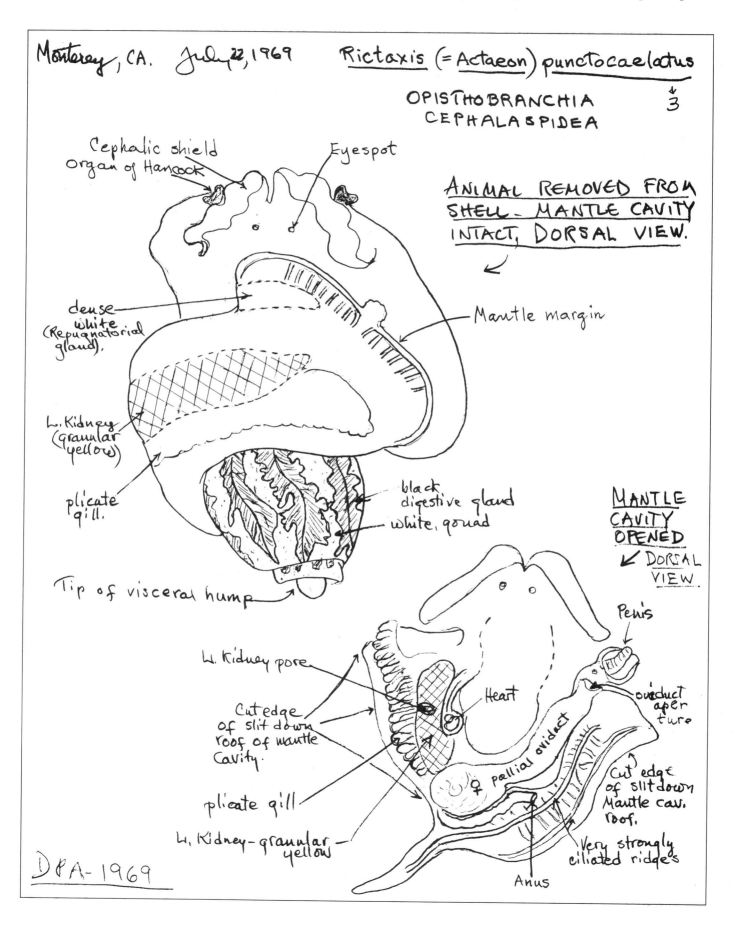

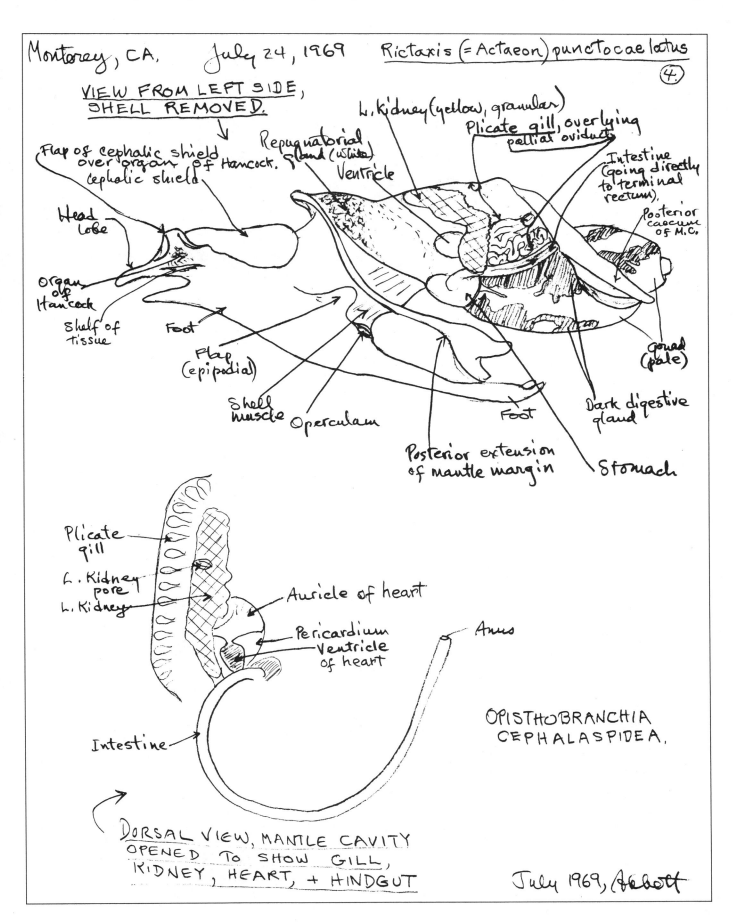

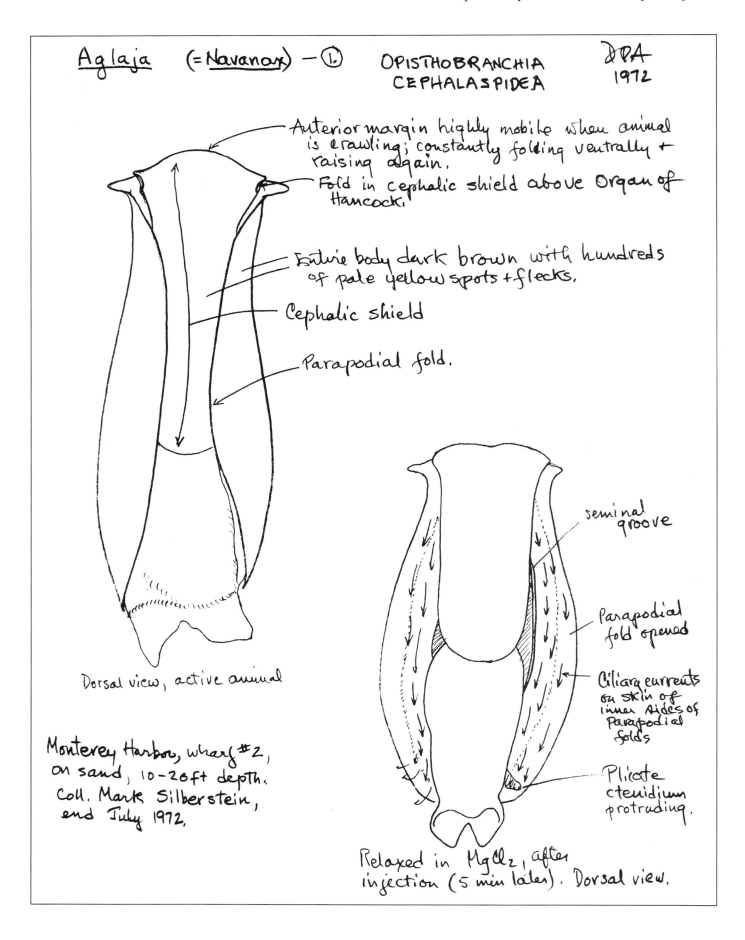

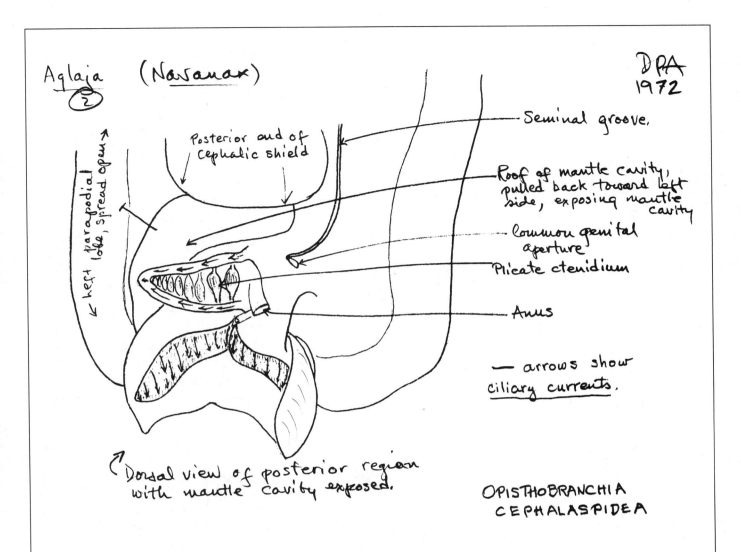

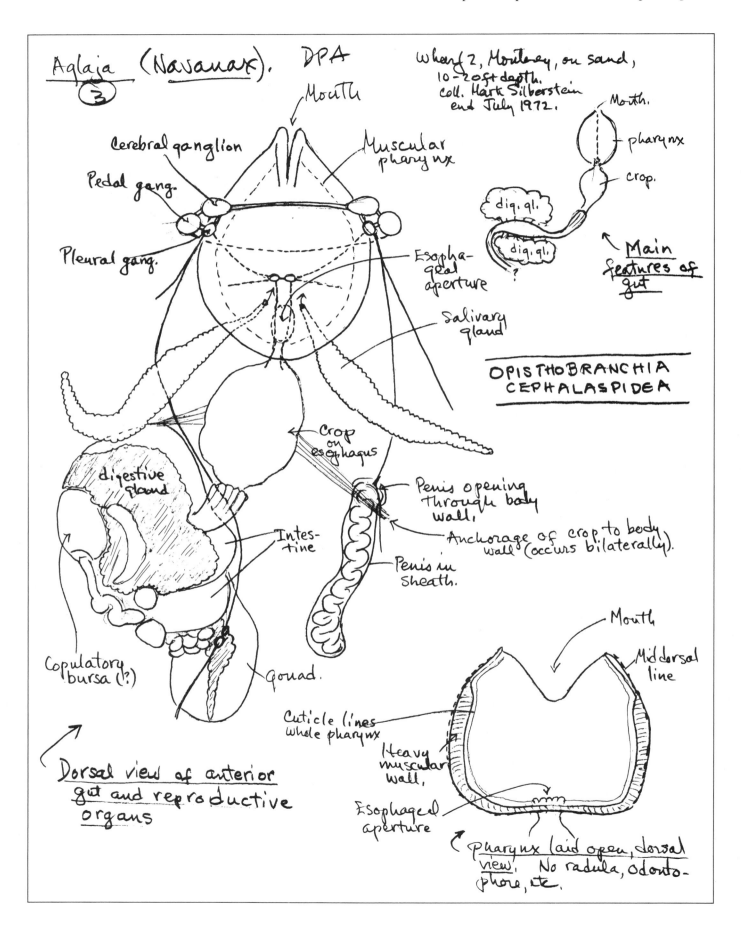

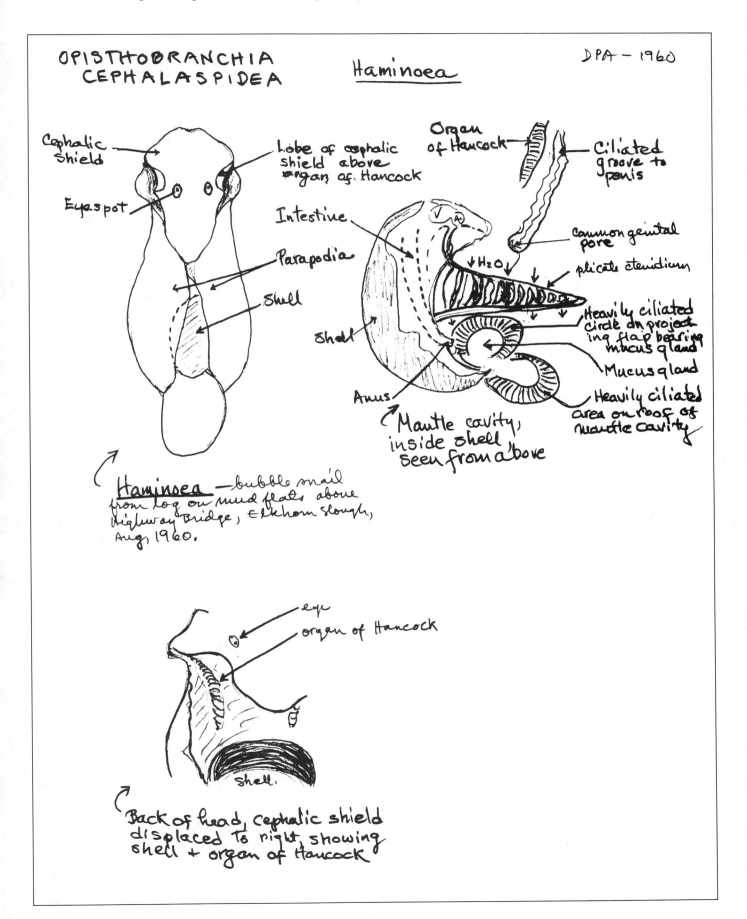

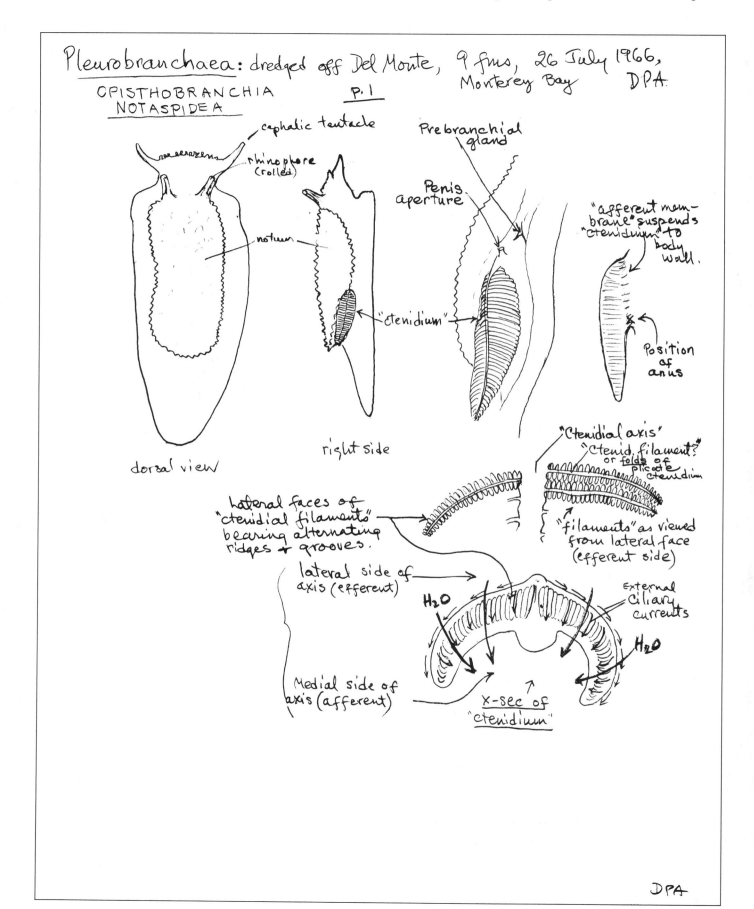

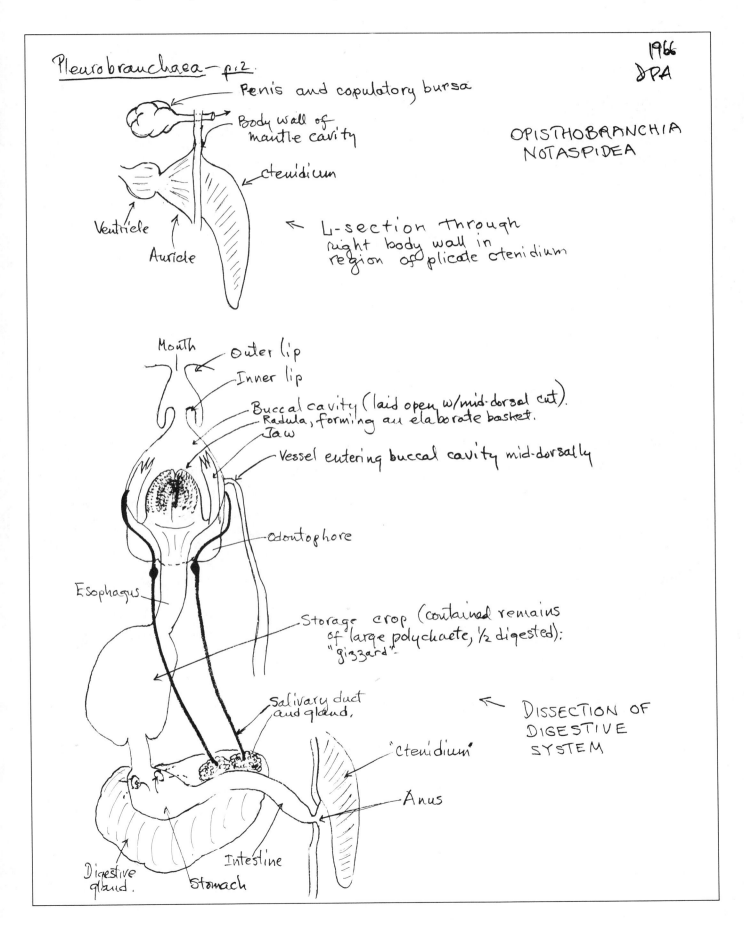

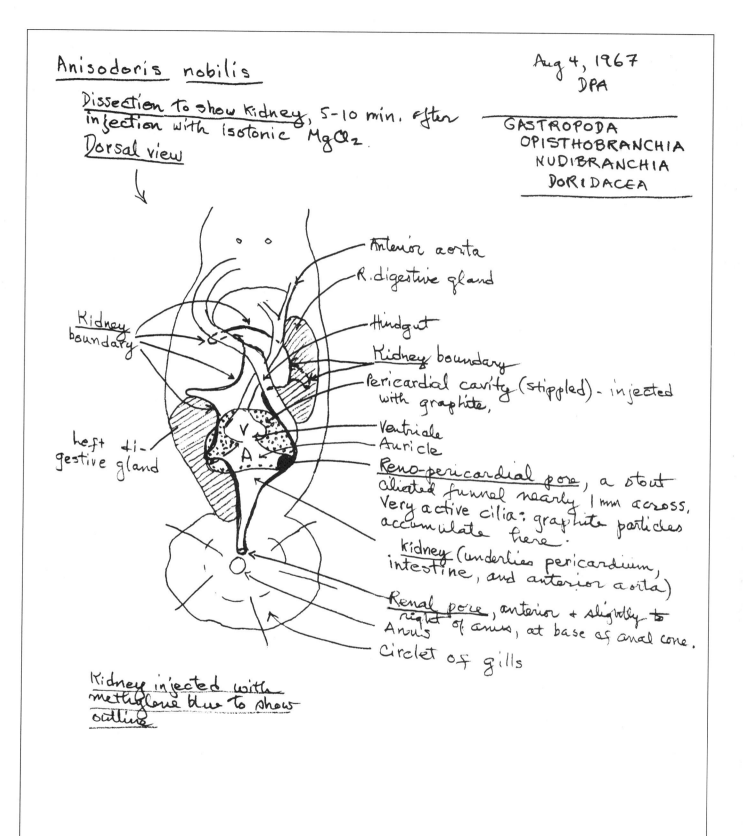

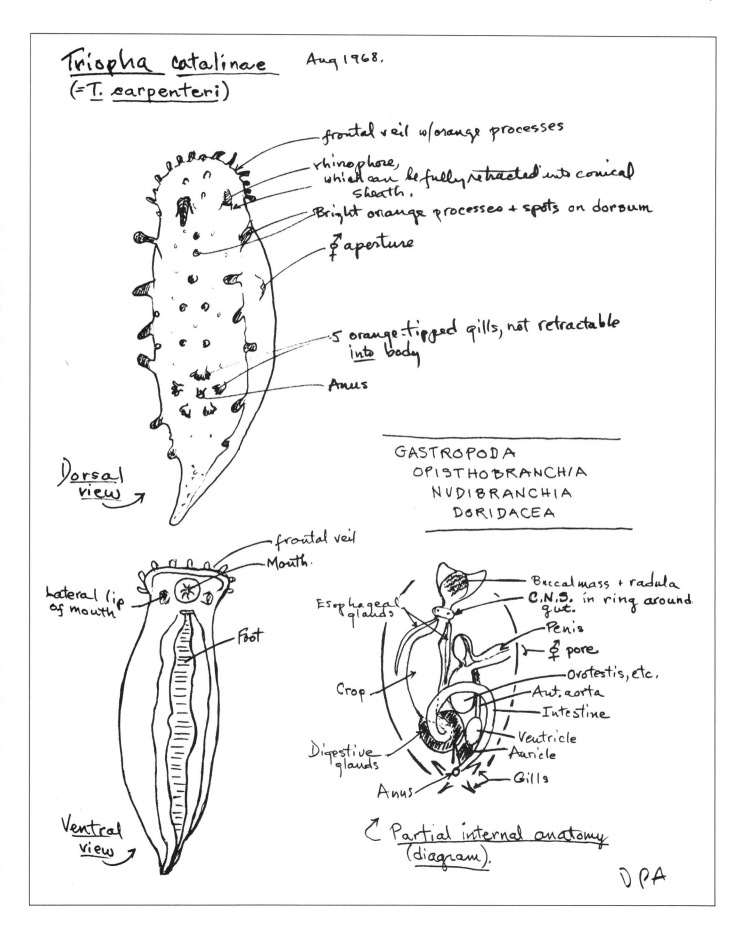

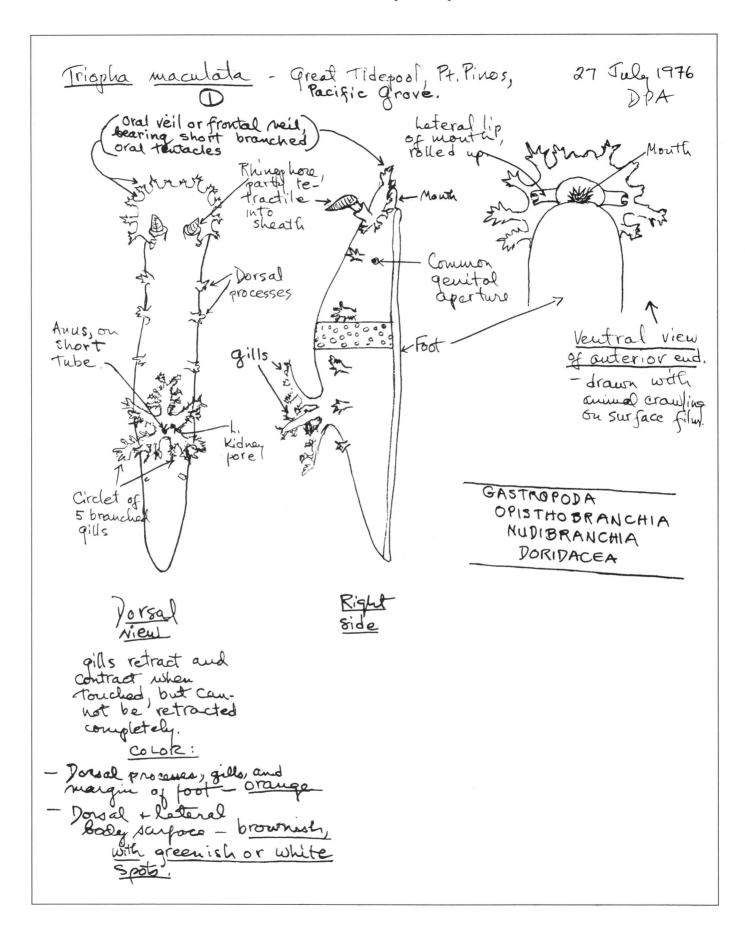

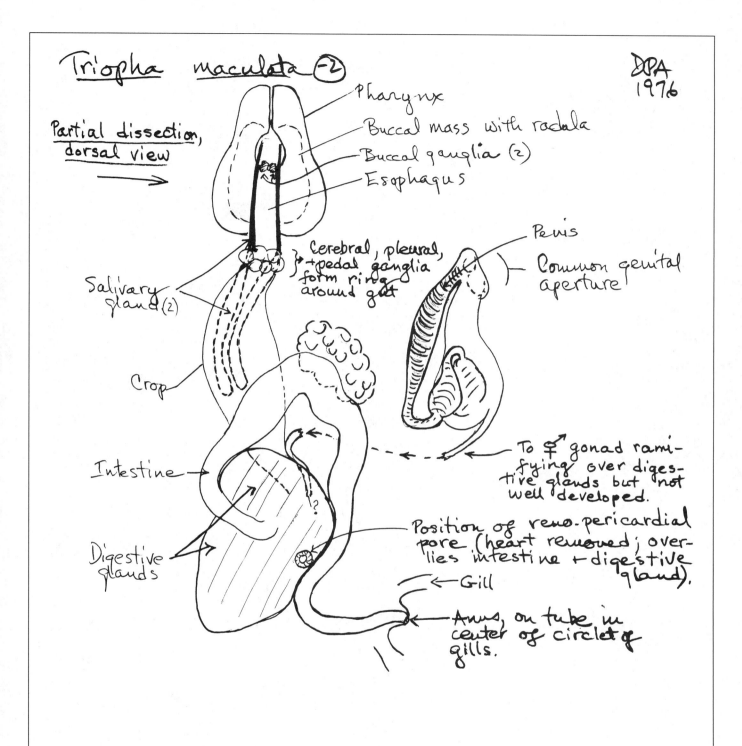

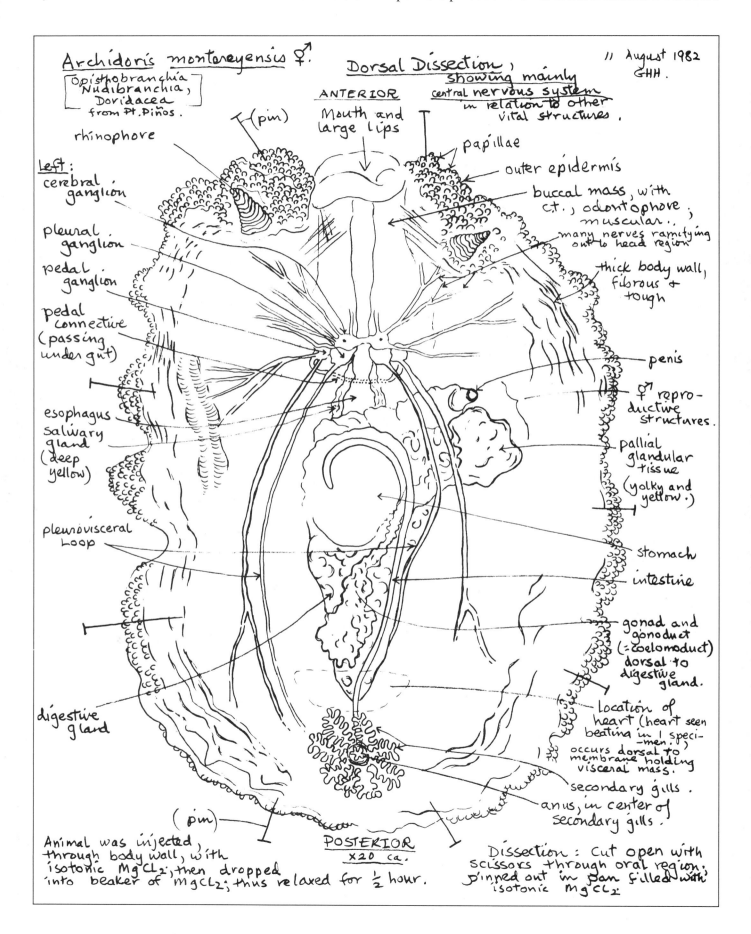

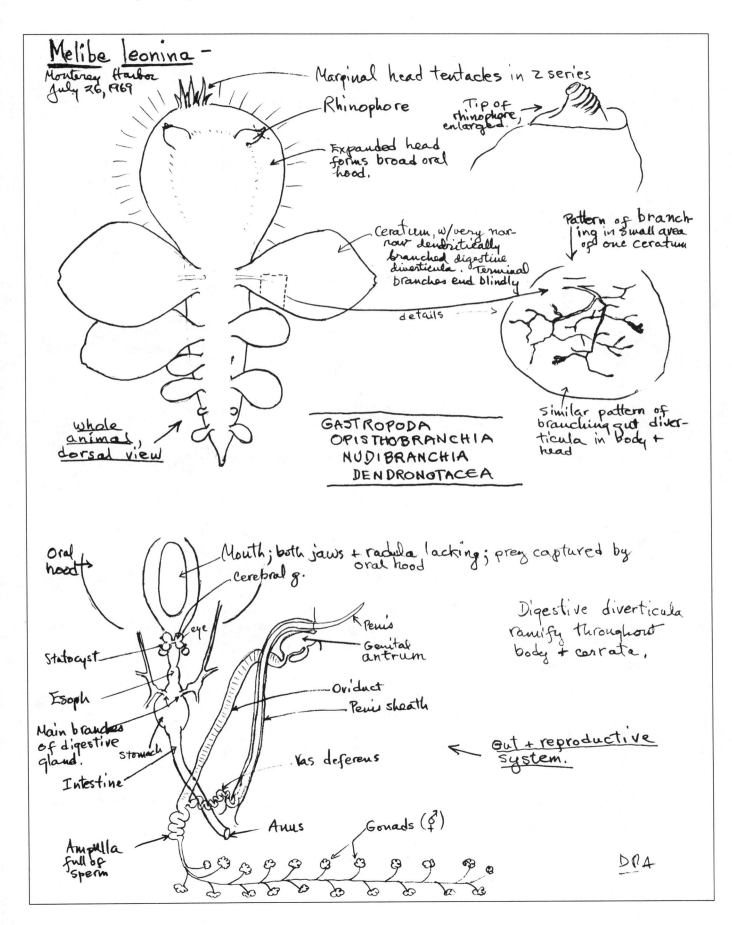

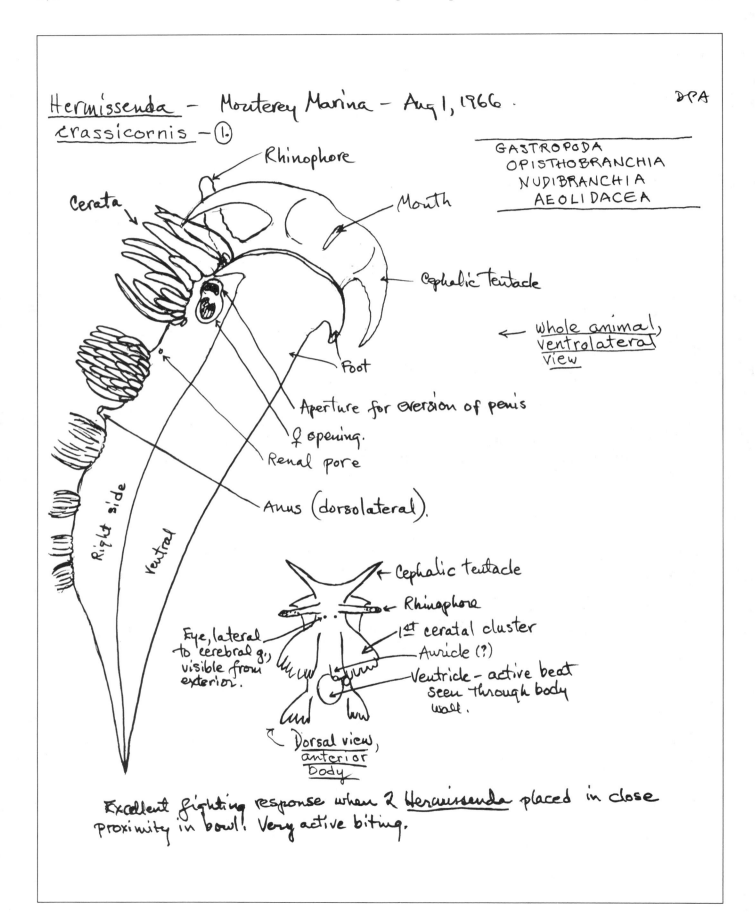

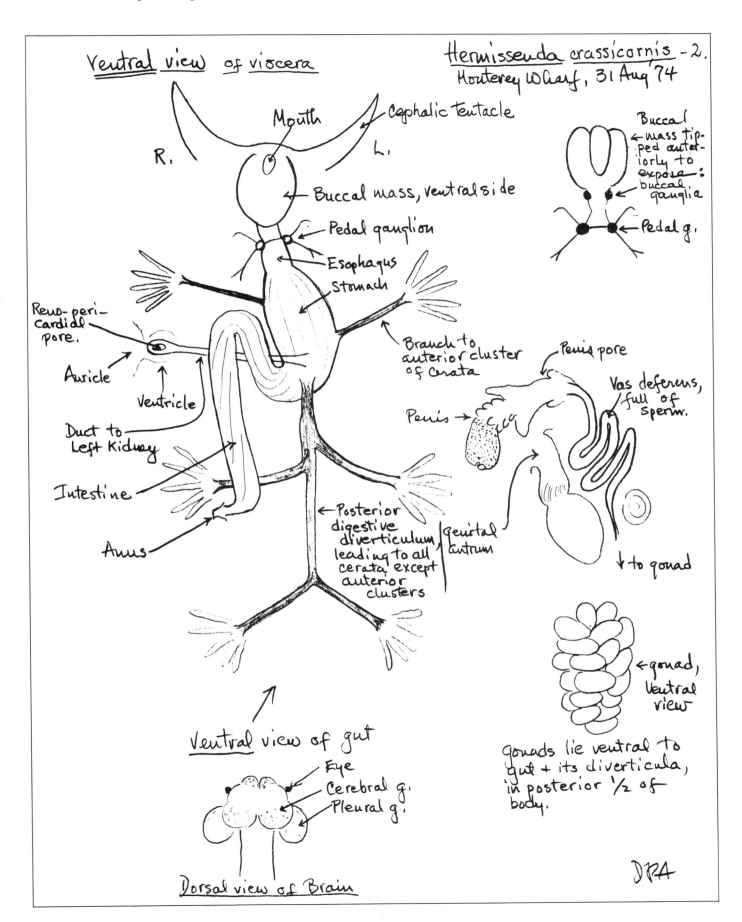

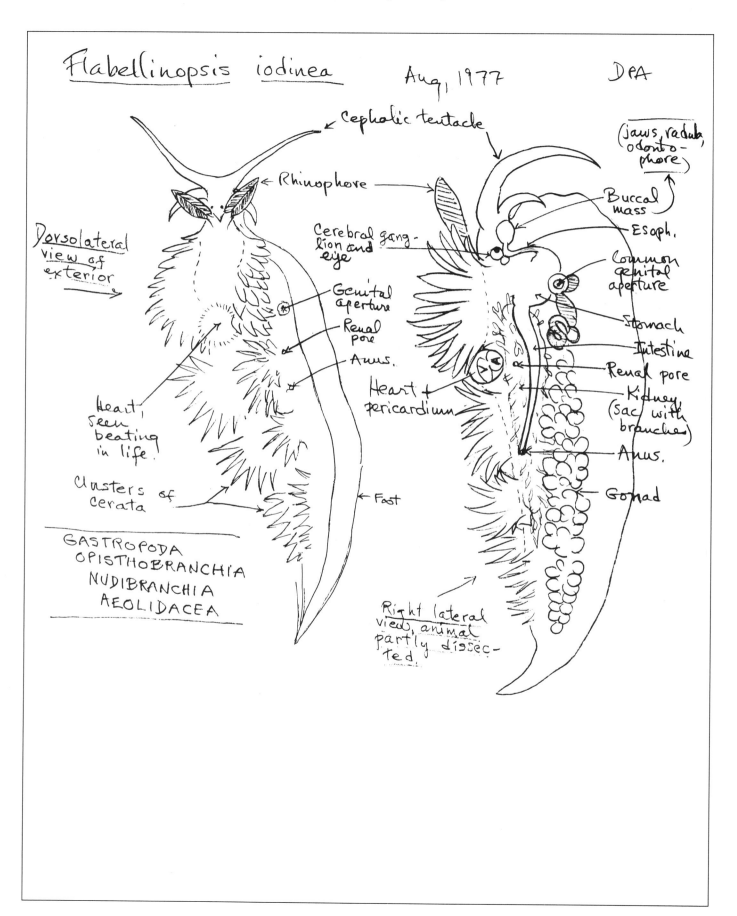

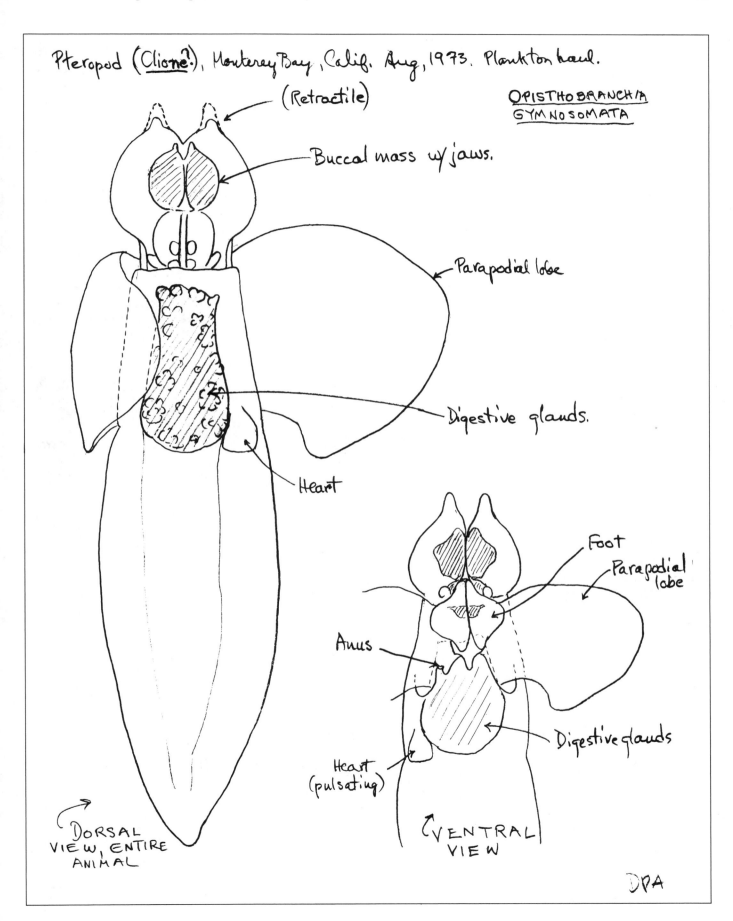

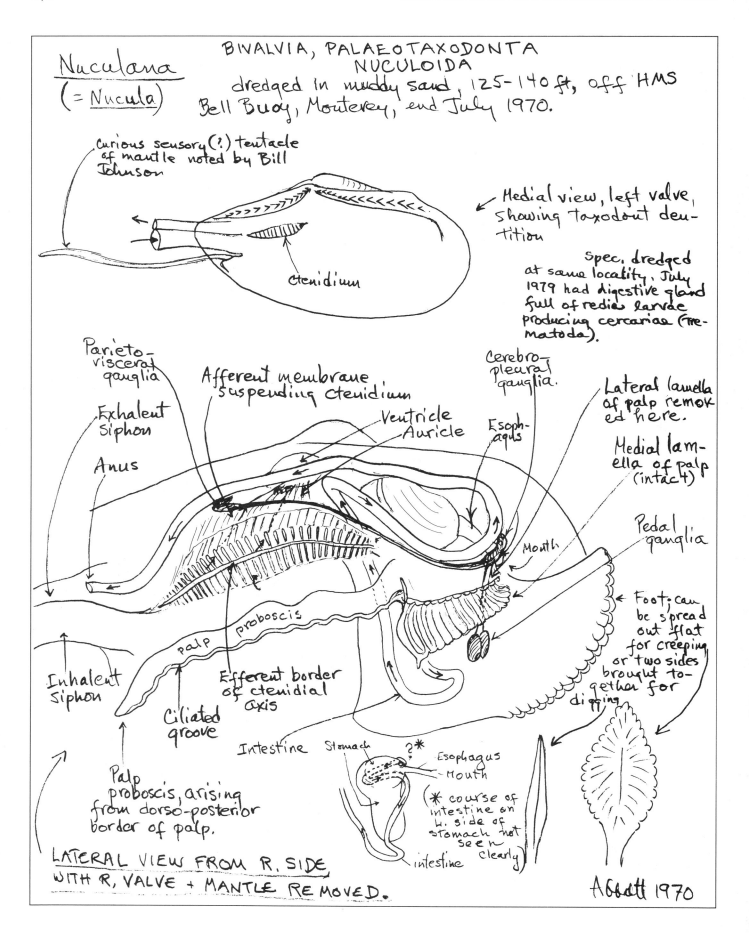

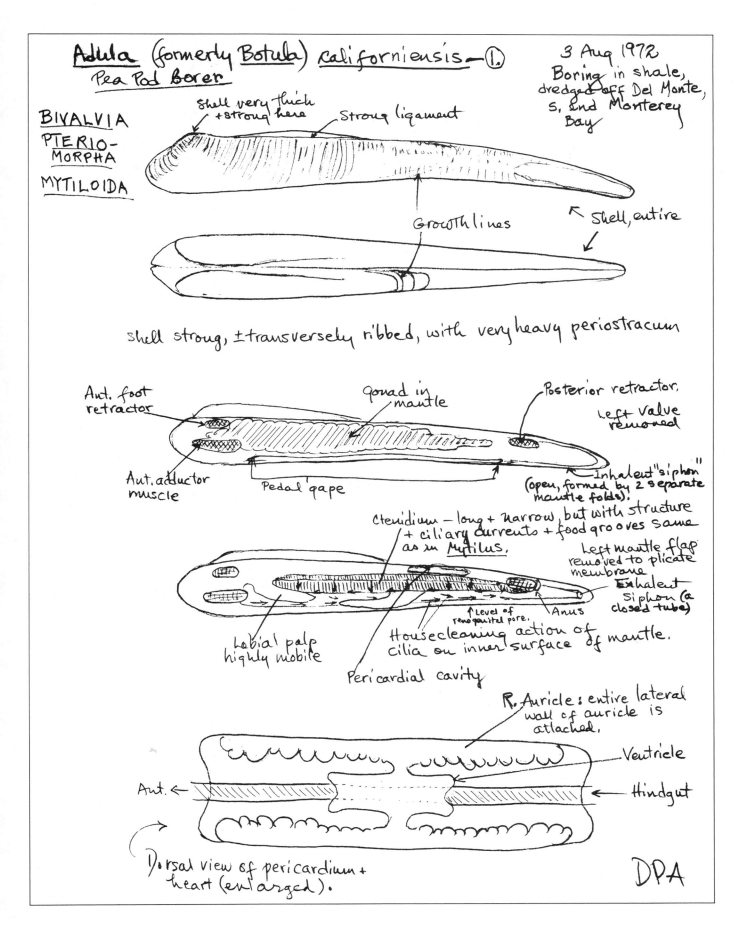

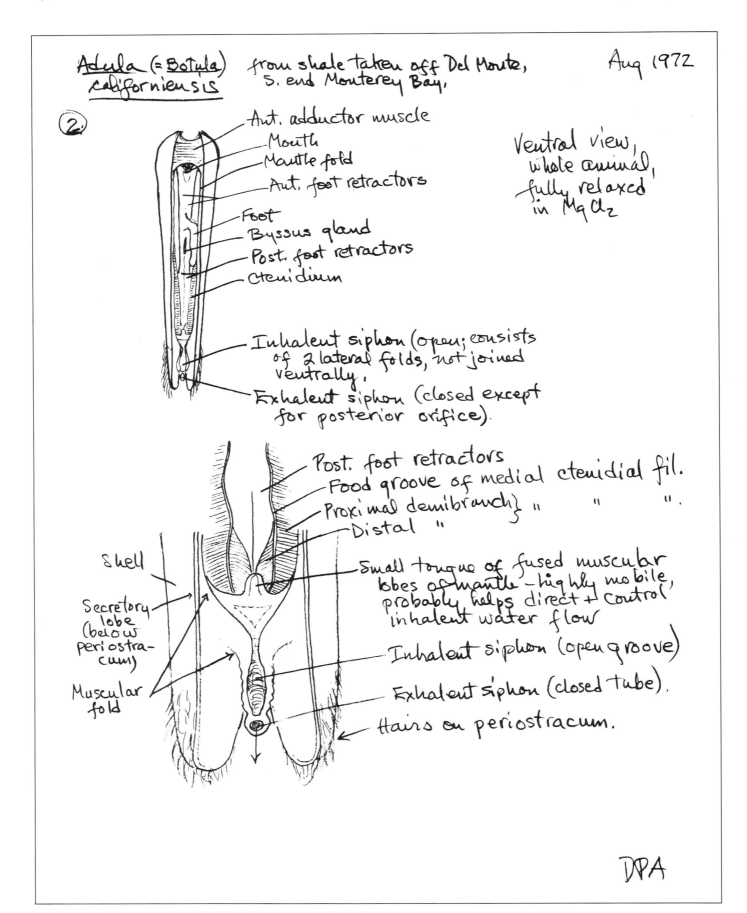

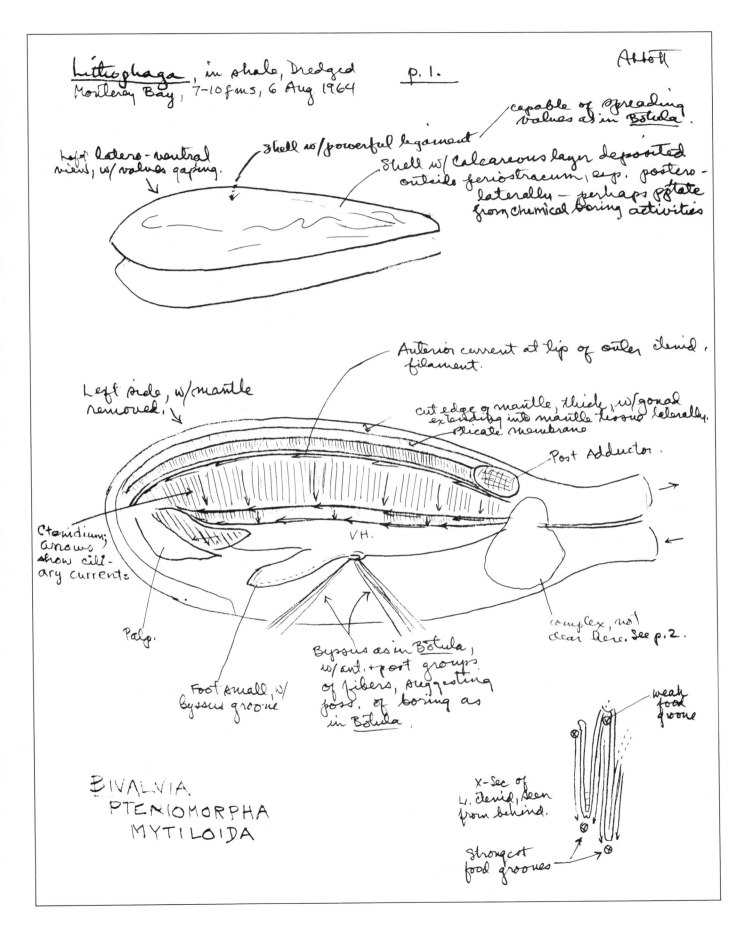

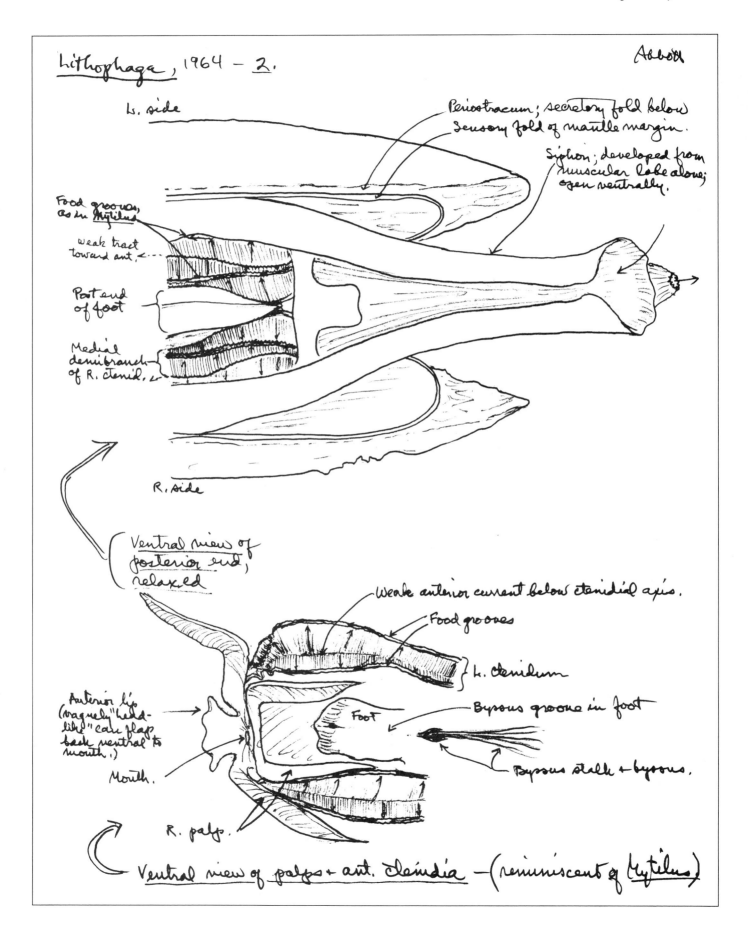

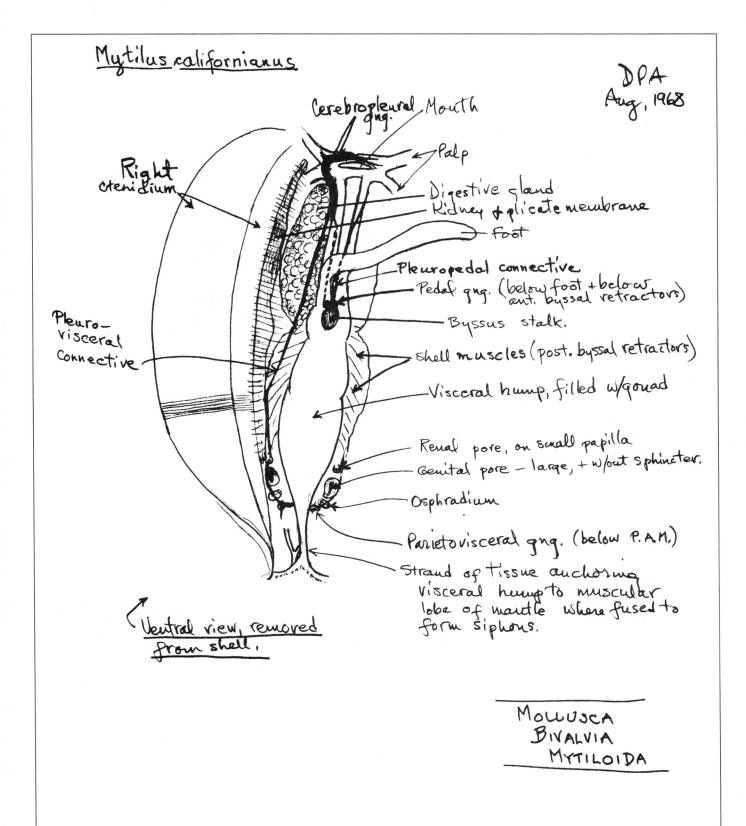

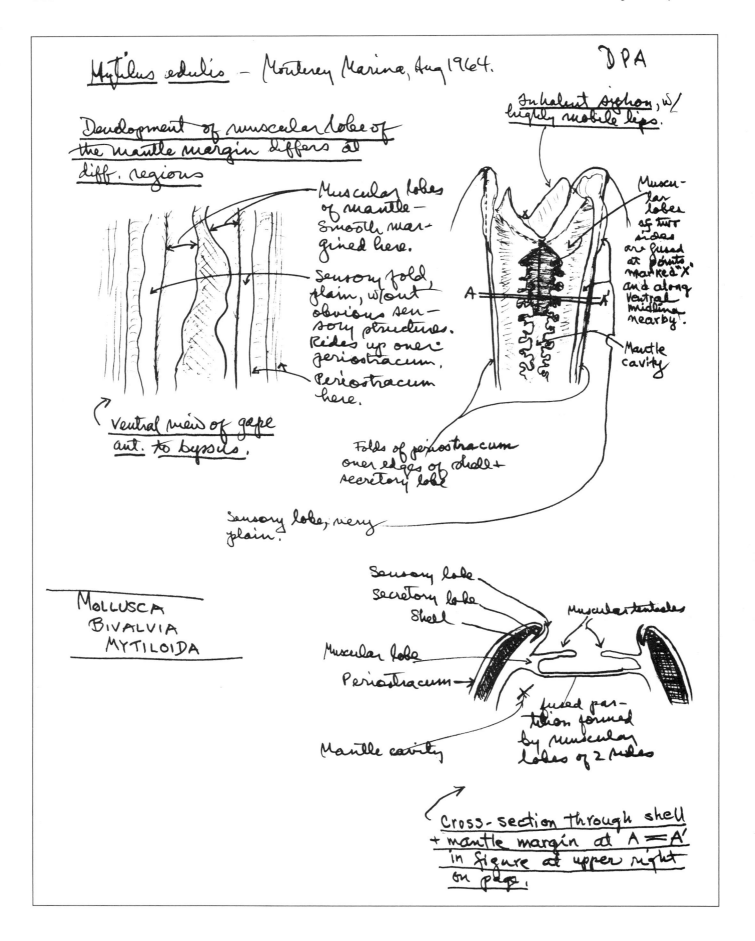

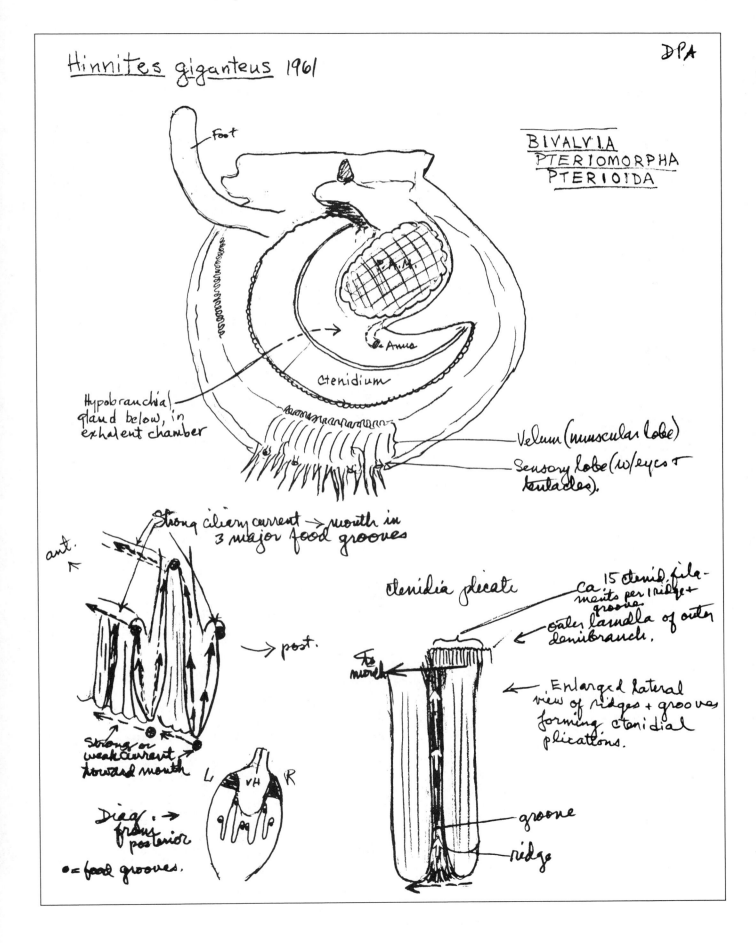

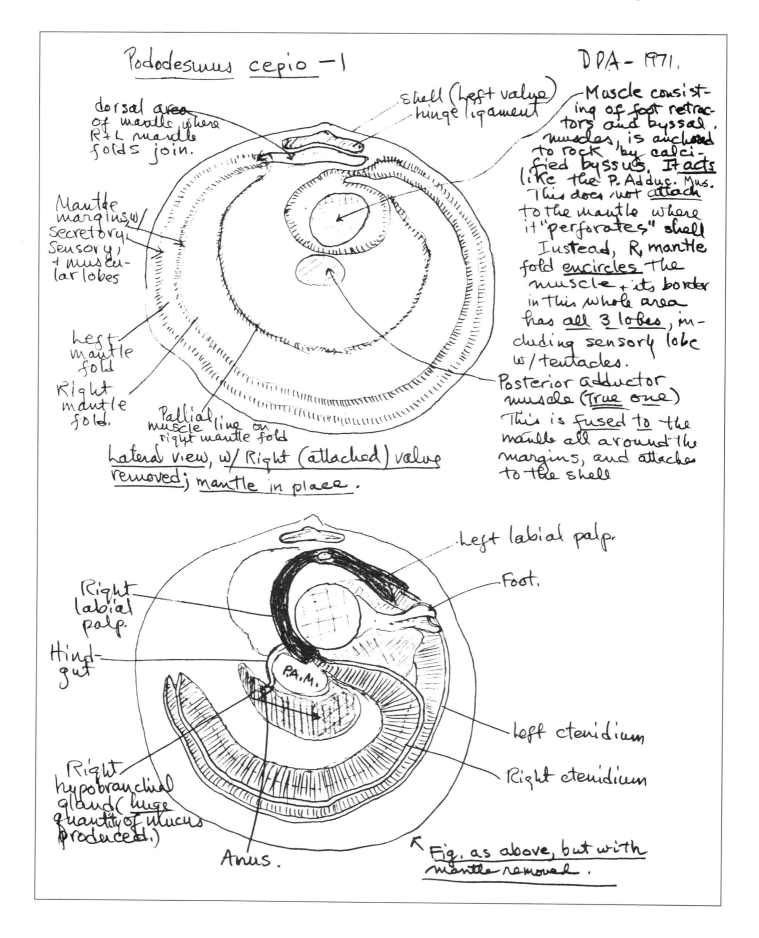

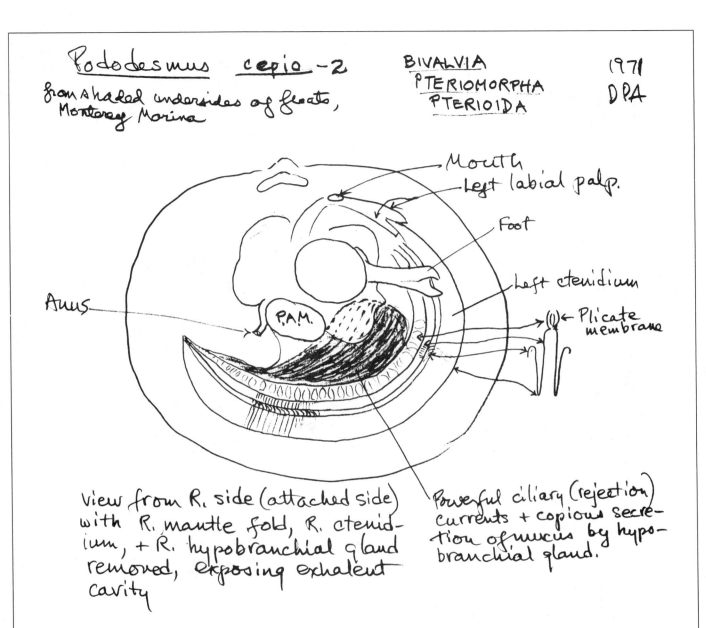

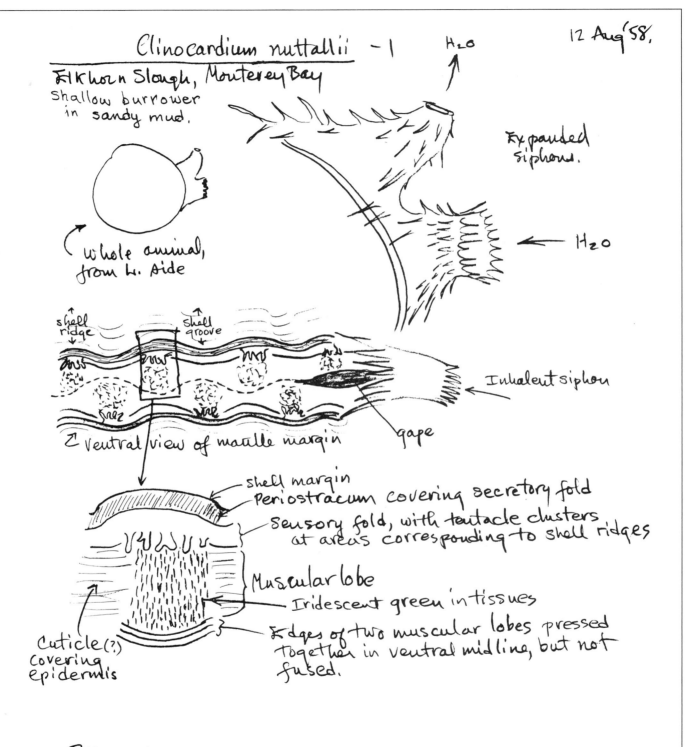

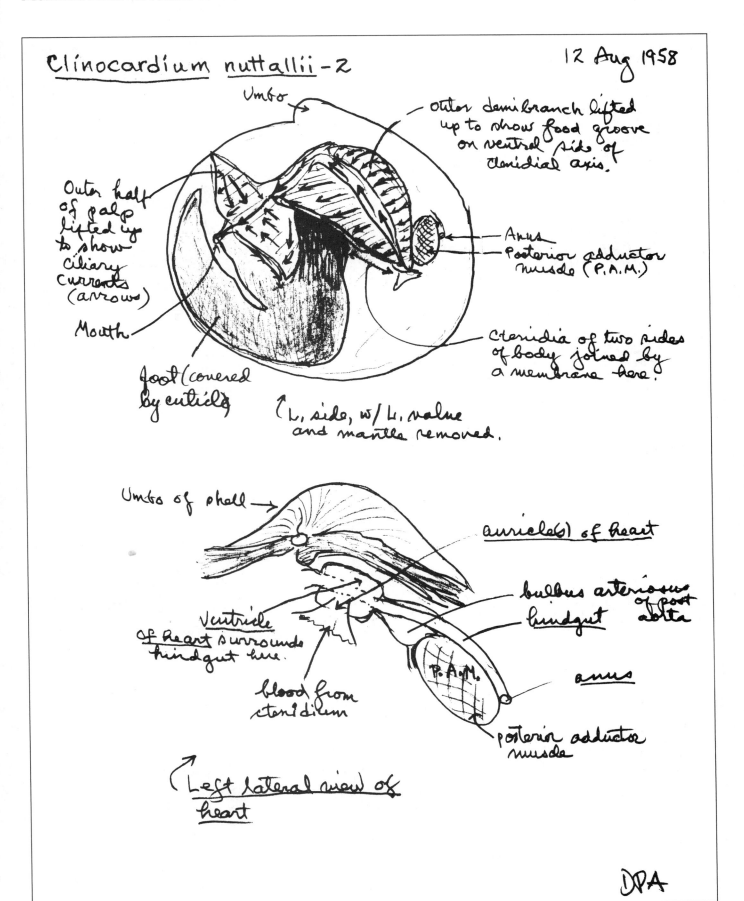

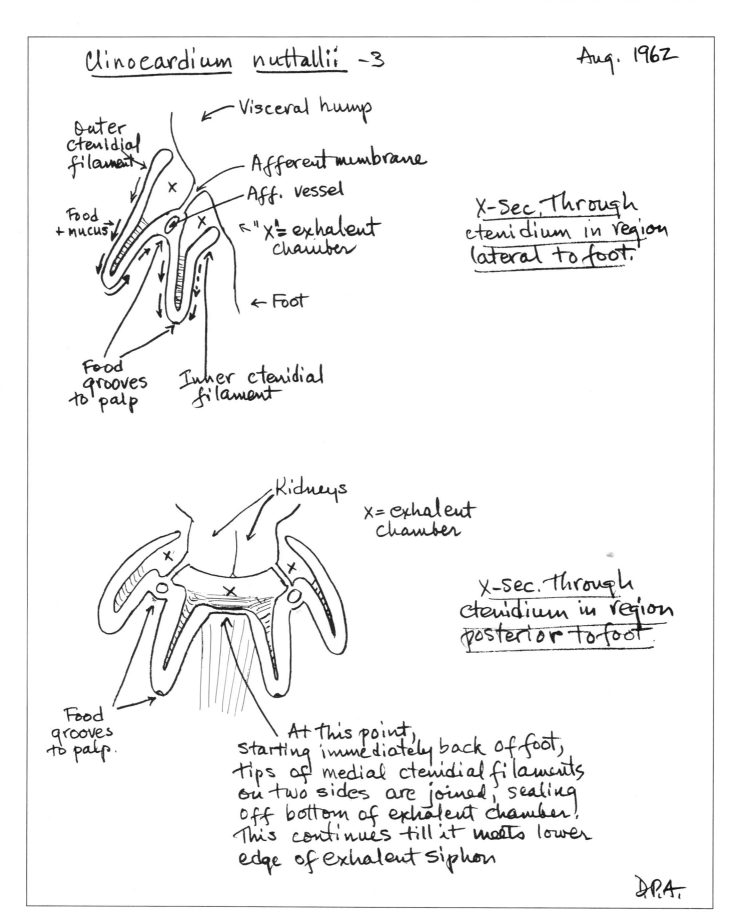

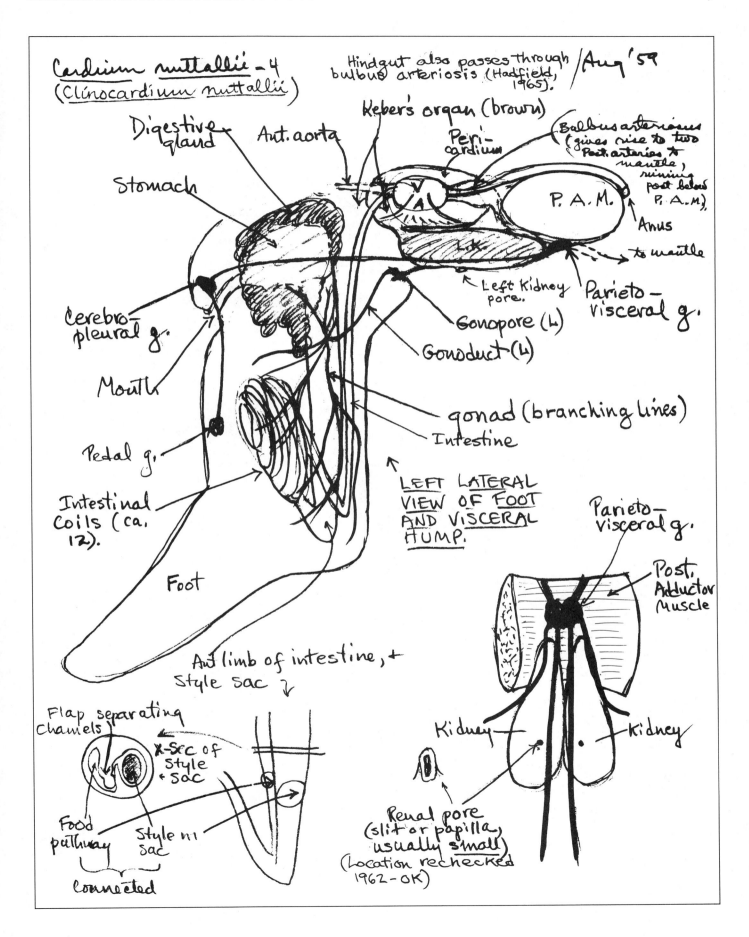

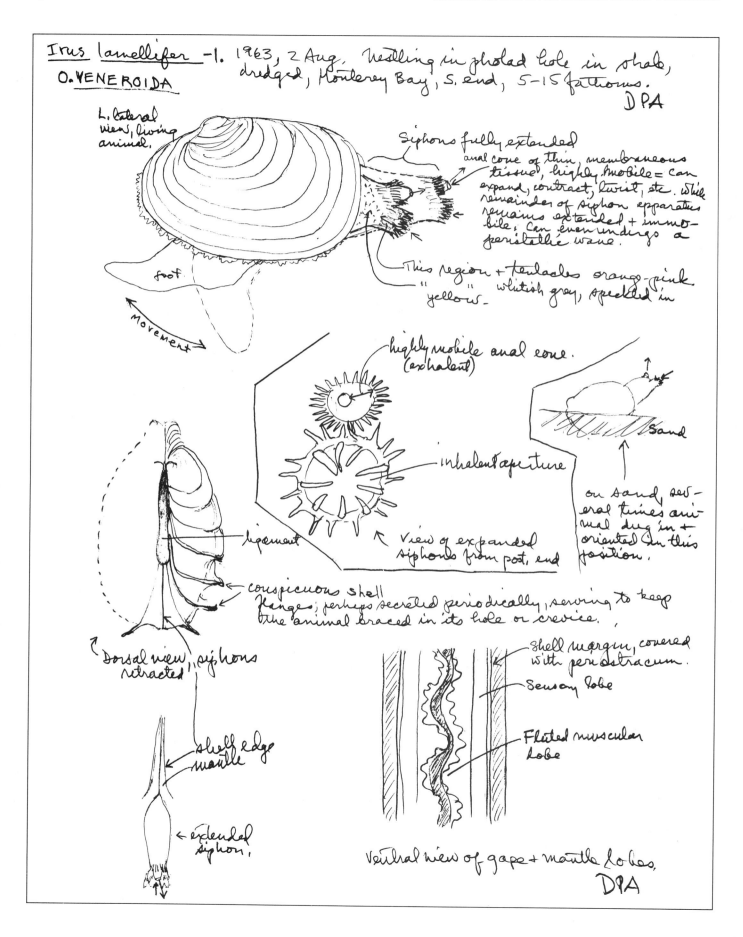

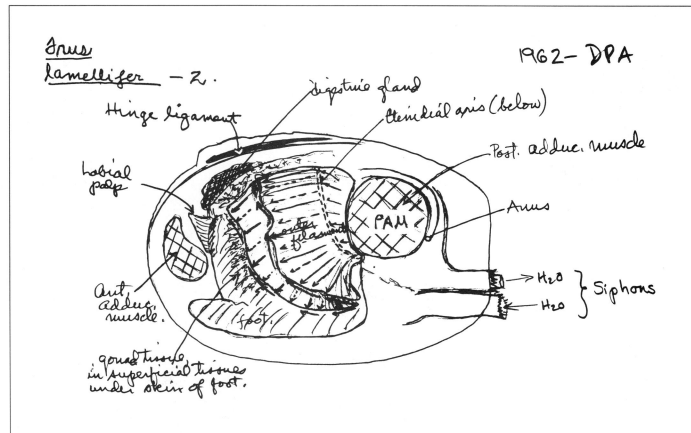

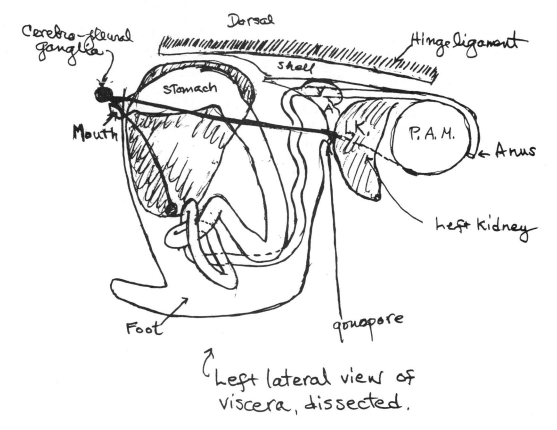

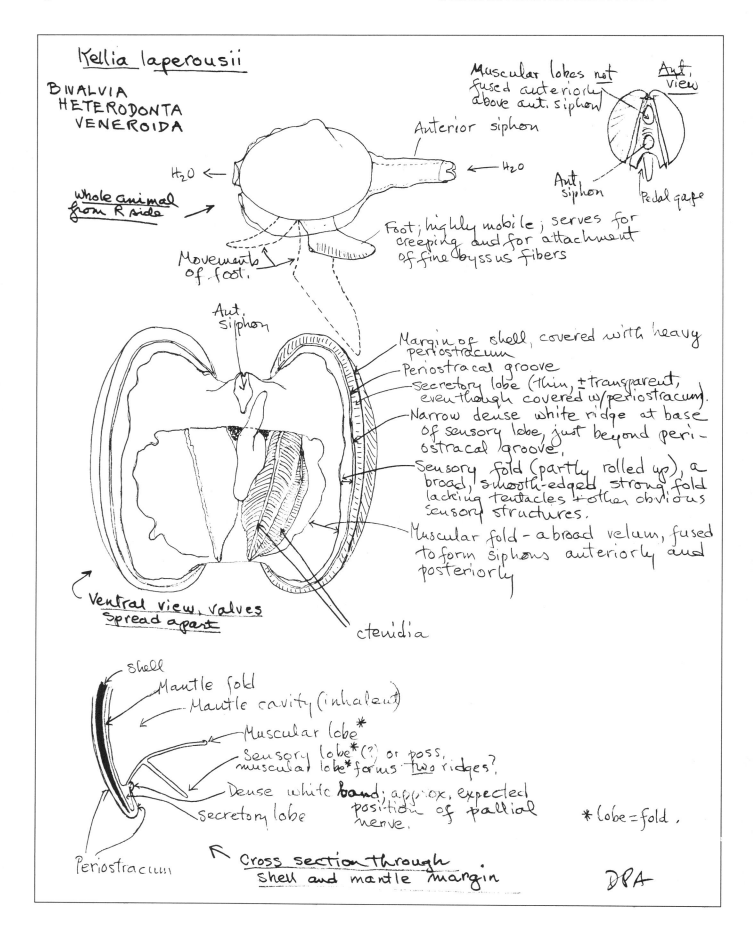

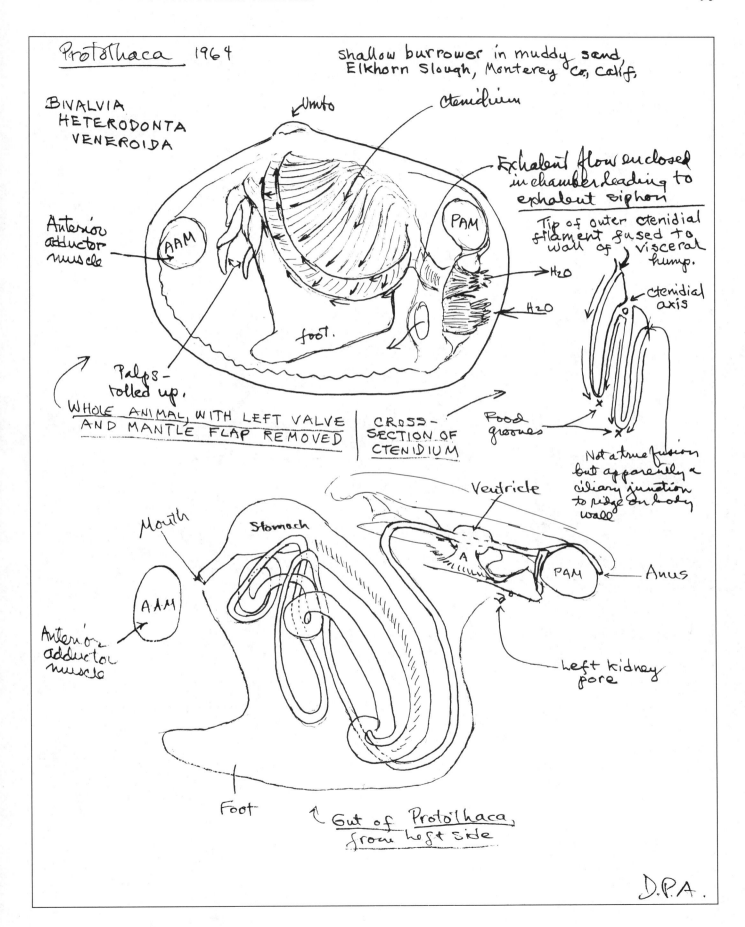

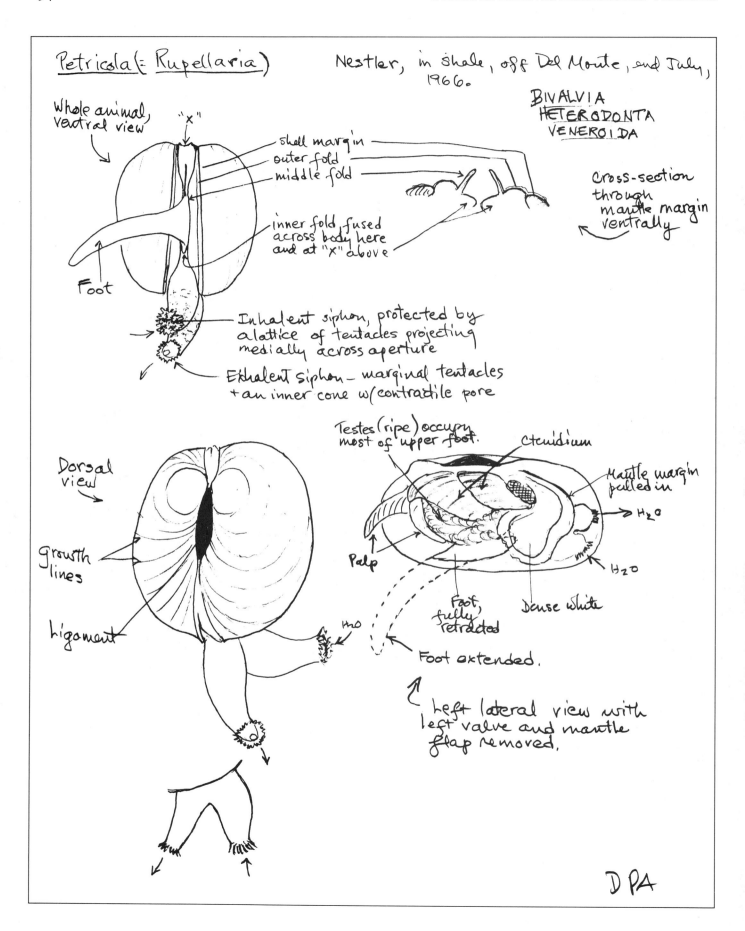

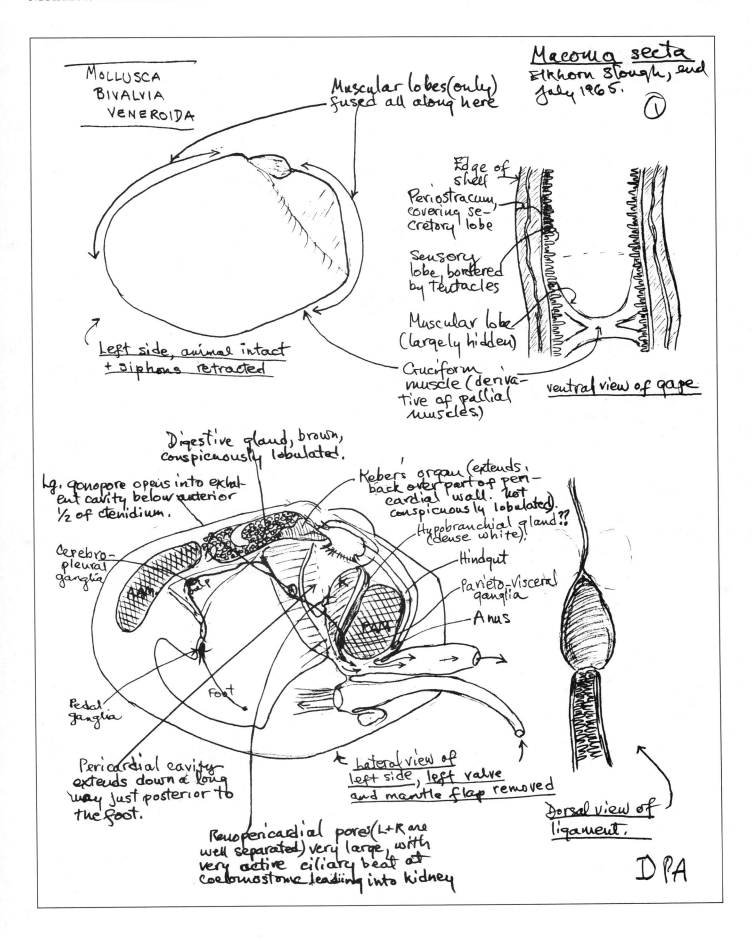

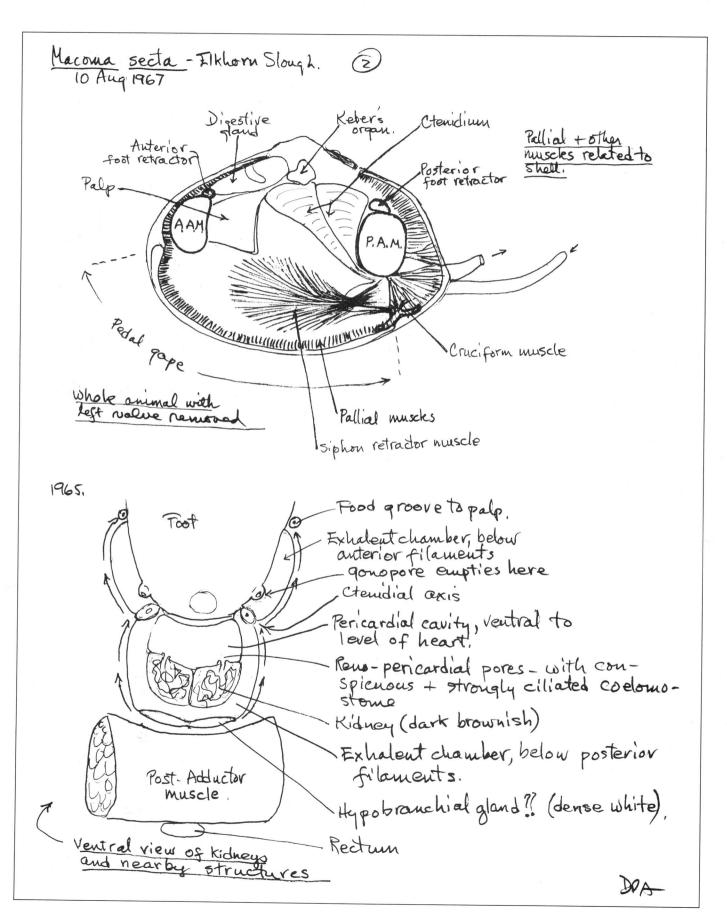

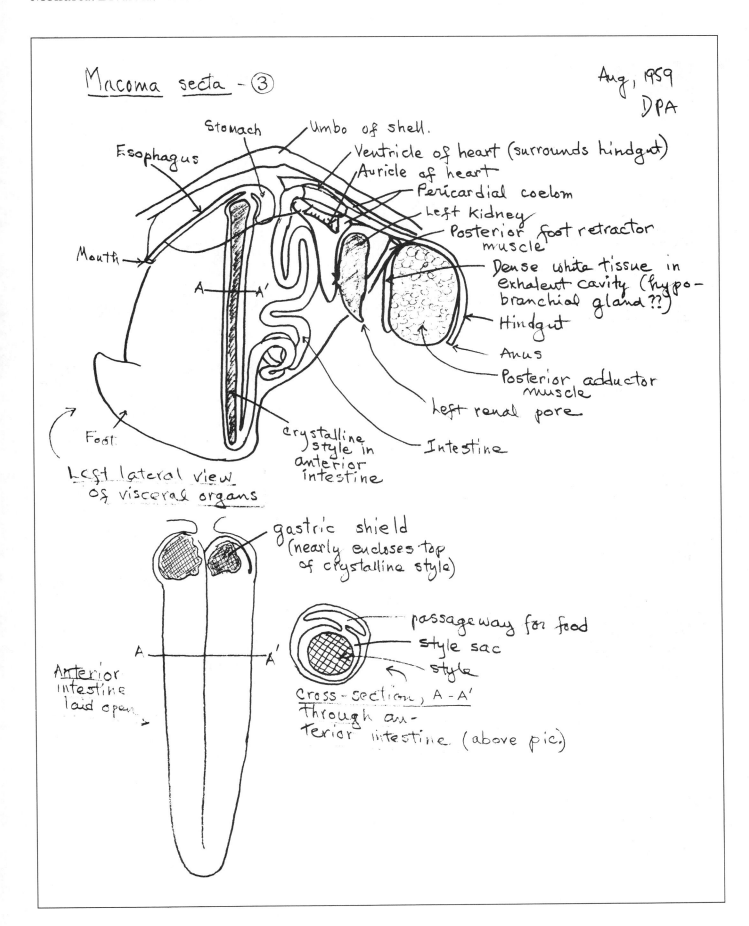

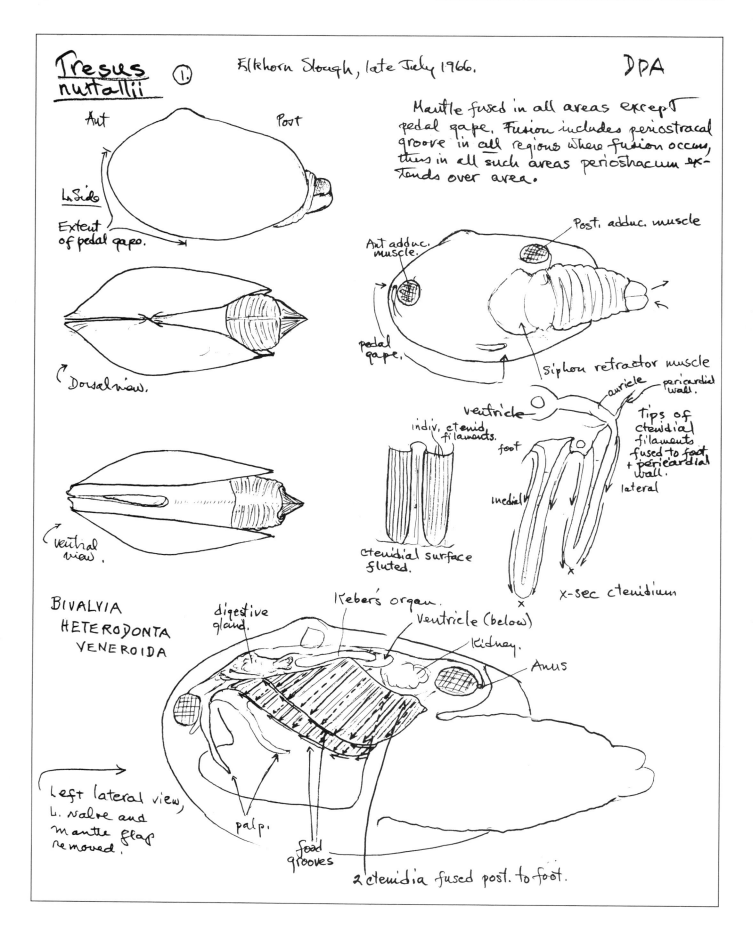

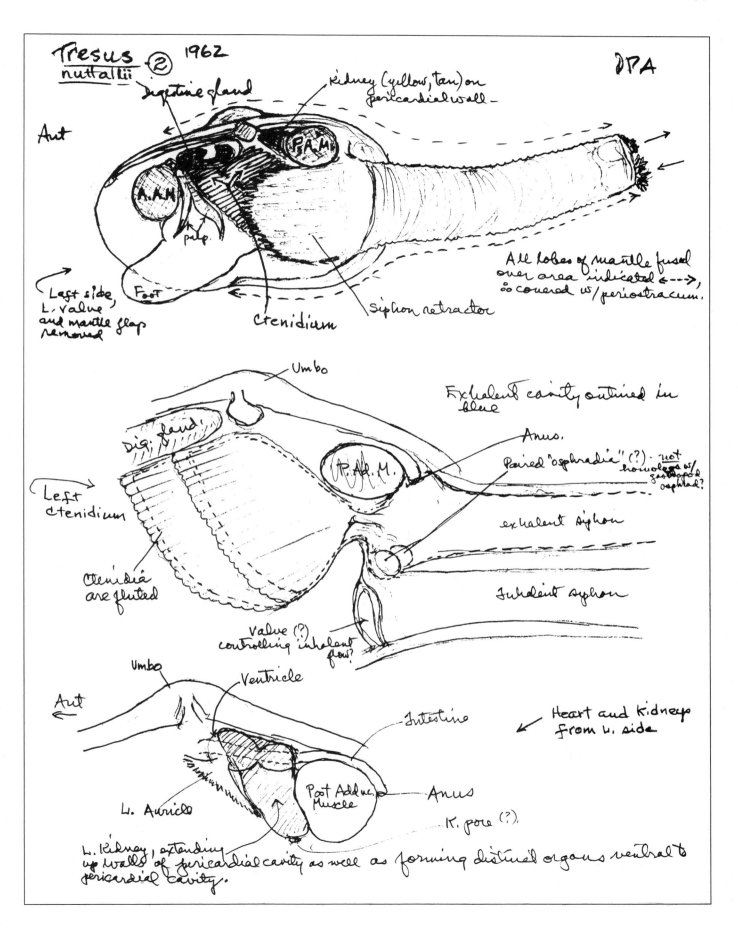

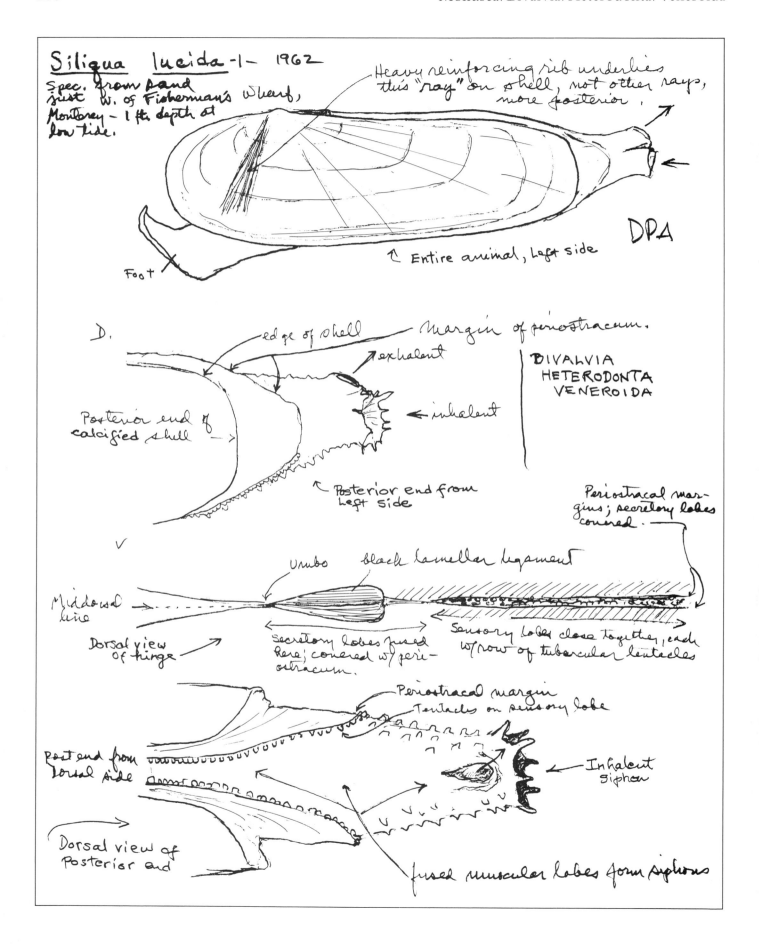

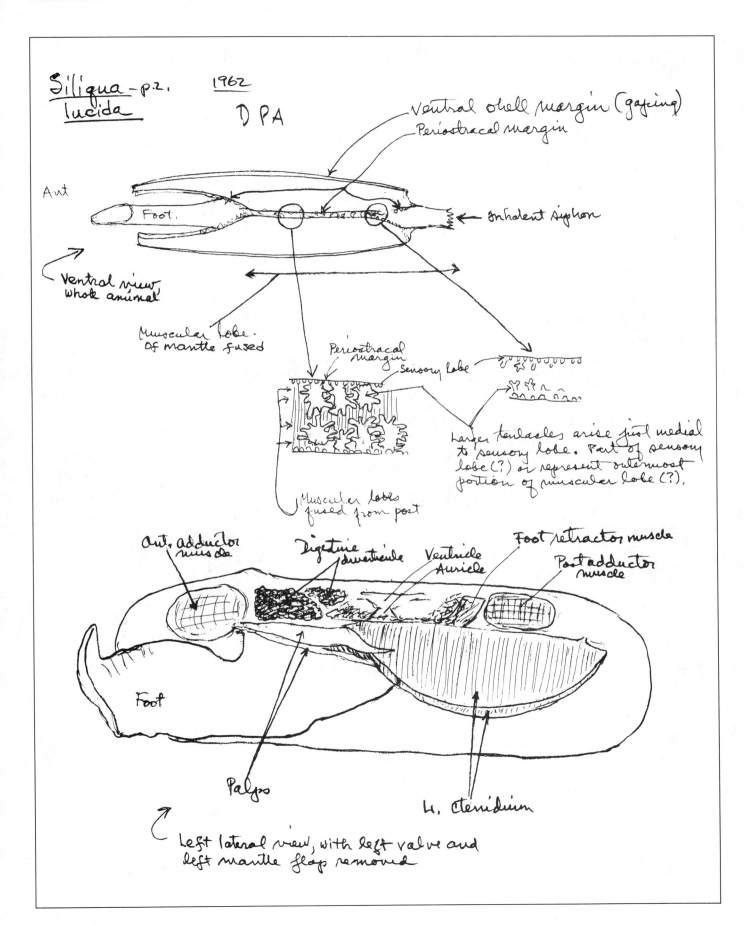

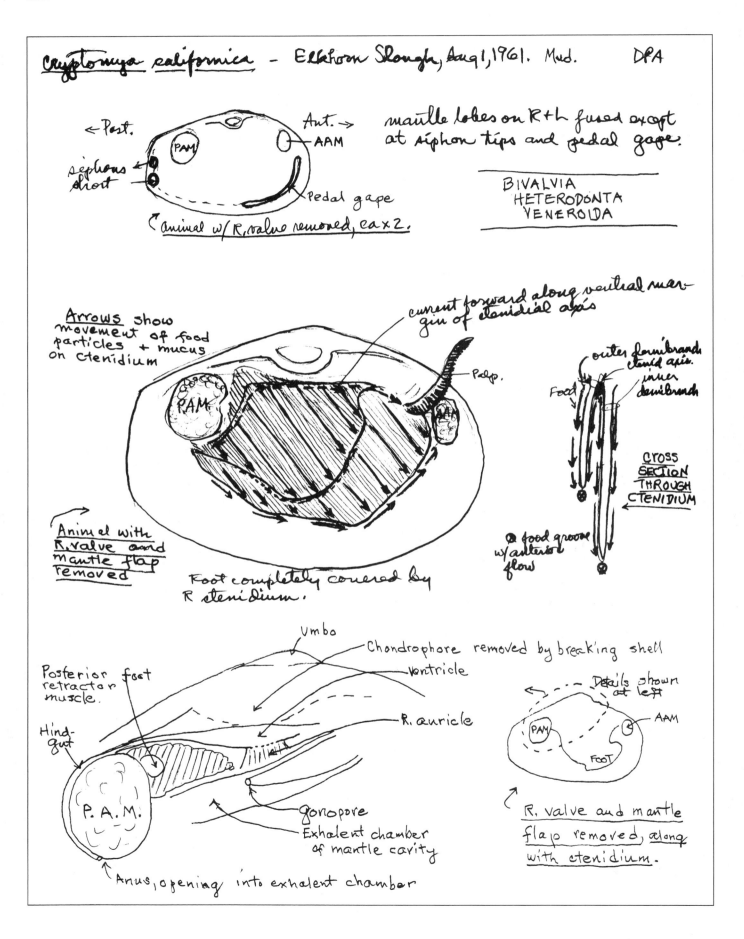

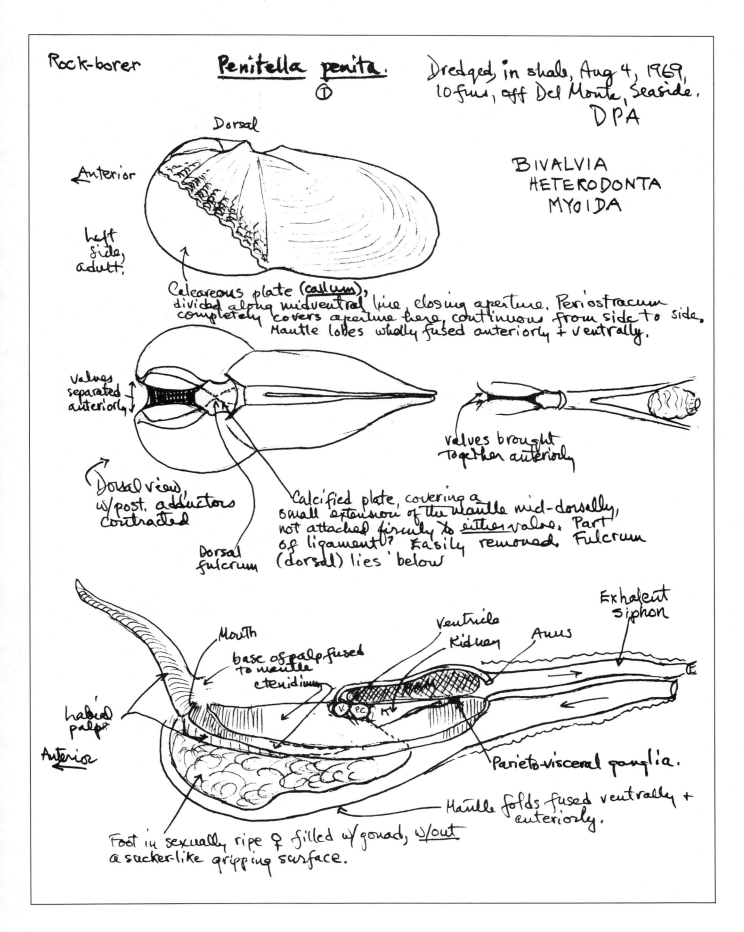

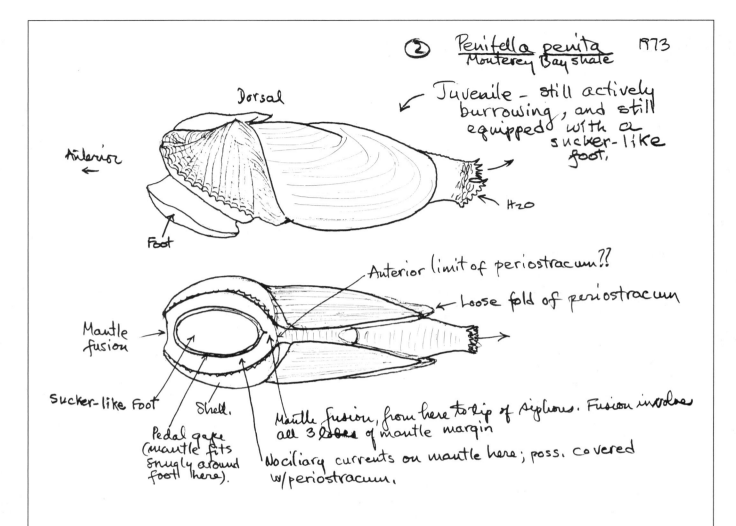

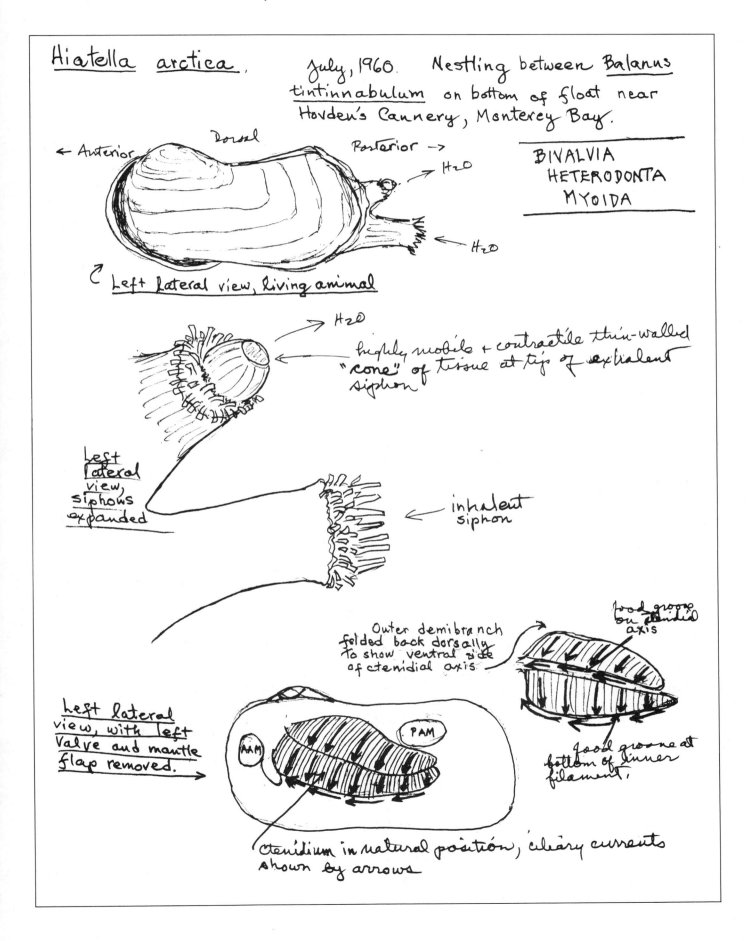

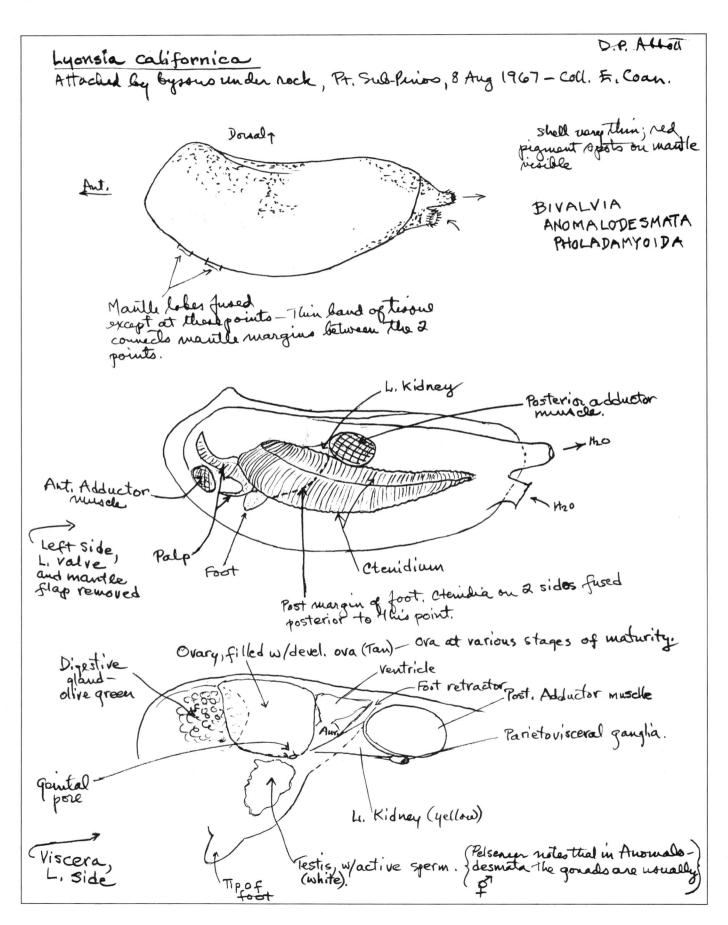

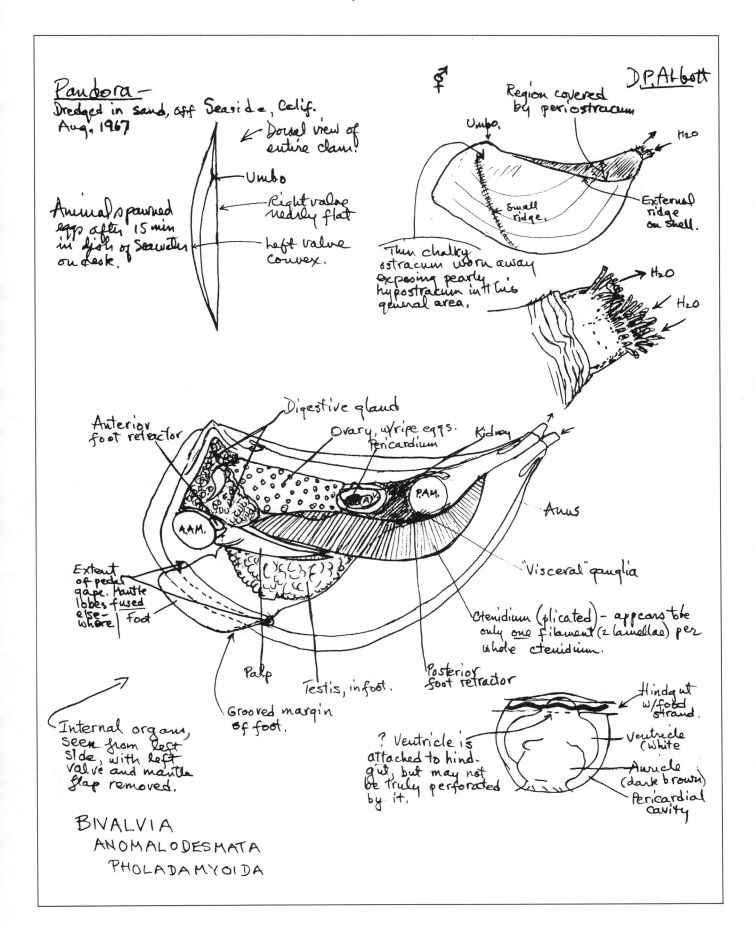

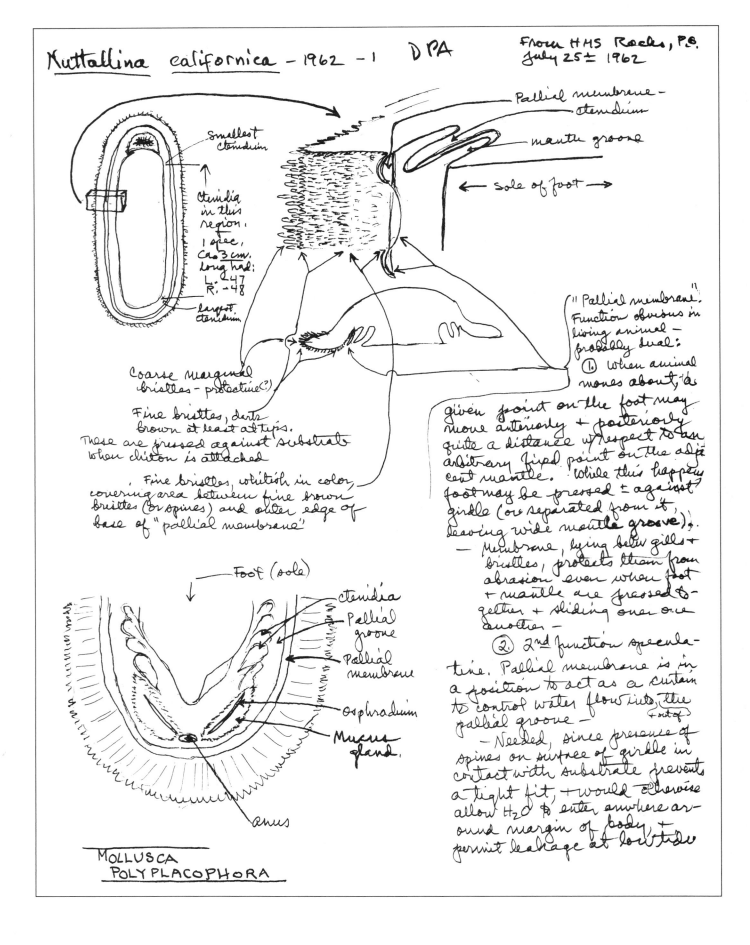

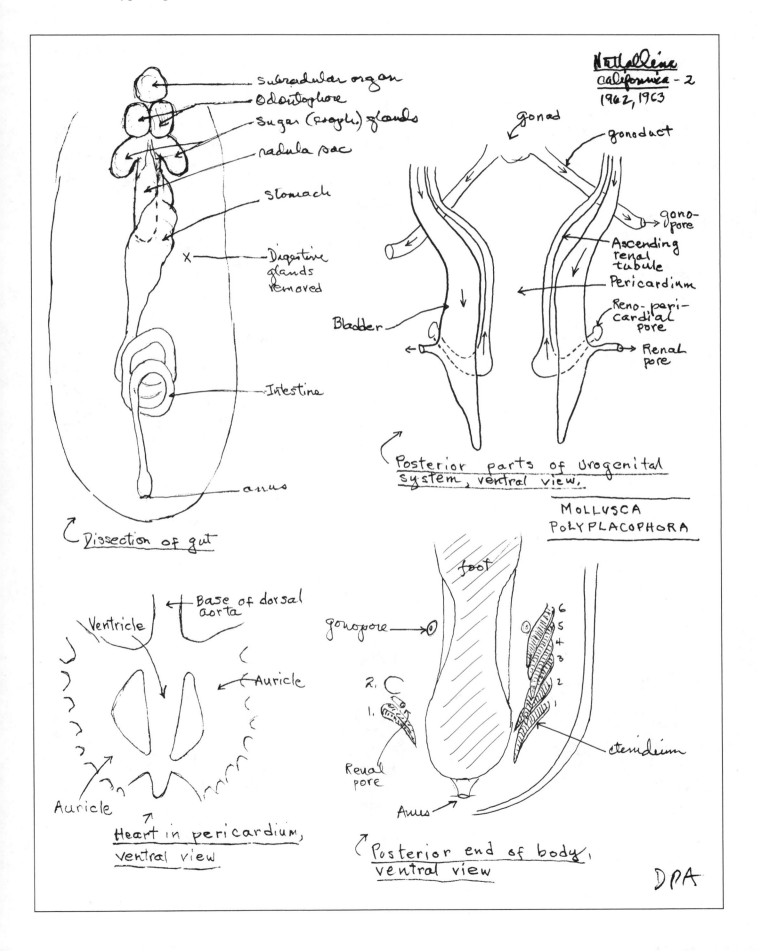

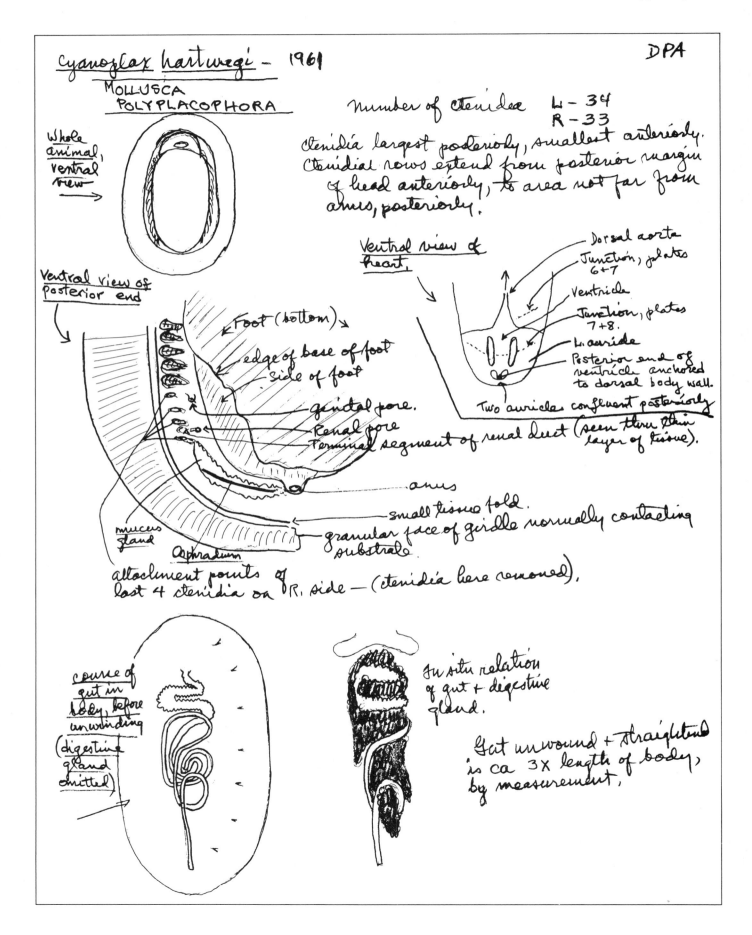

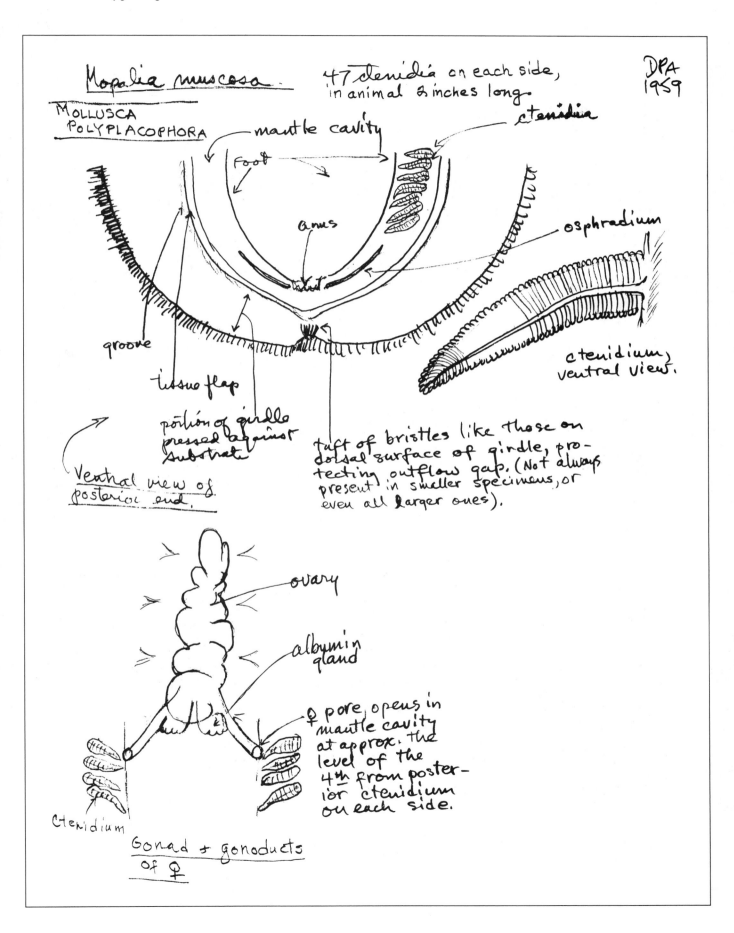

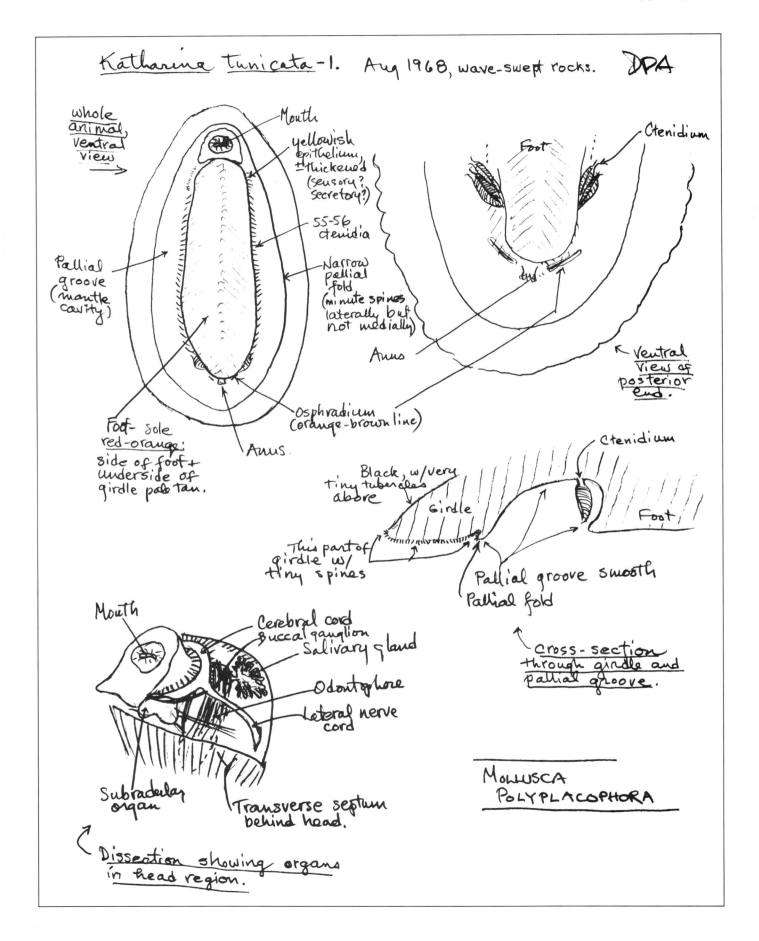

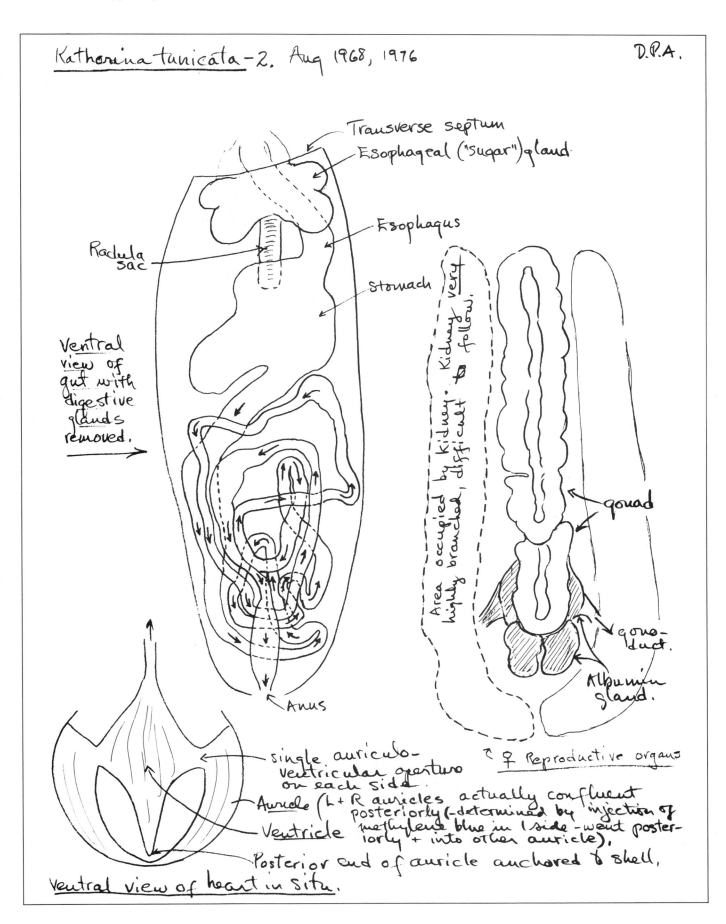

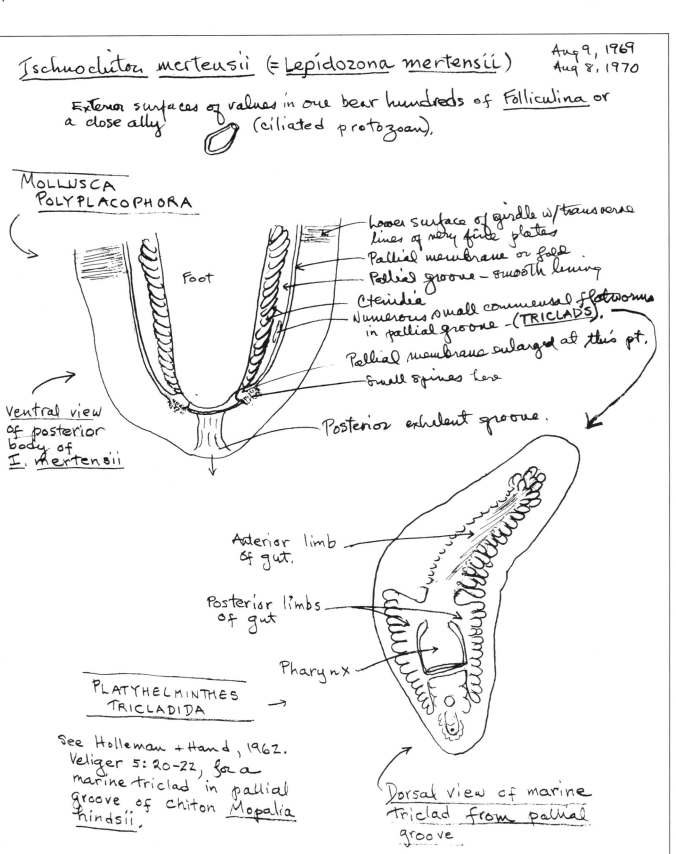

Chaetoderma erudita.
length ca. 3 cm, preserved.

"Albatross" Sta. 4524, DPA dredged on soft gray mud at 213-228 fms, 9.9 miles off Point Pinos, Monterey Bay, 26 May 1904.

MOLLUSCA
APLACOPHORA

- Slit-like opening to vestibule
- Body covered with spine-like spicules which are largest + shaggiest posteriorly
- Pallial cavity or cloaca with gills (2).

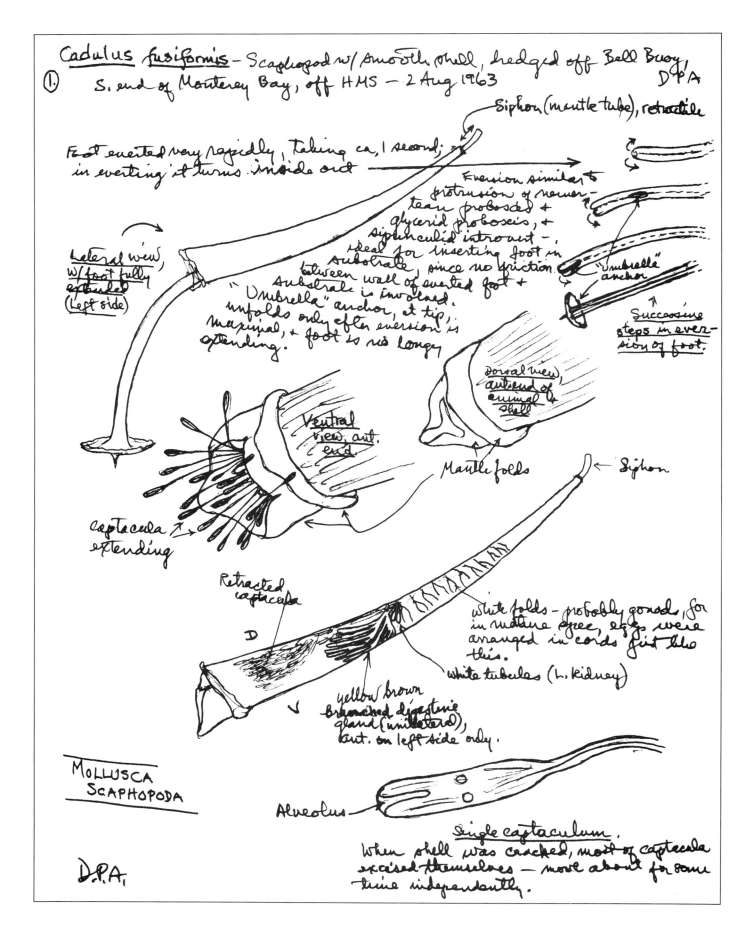

Mollusca/Scaphopoda

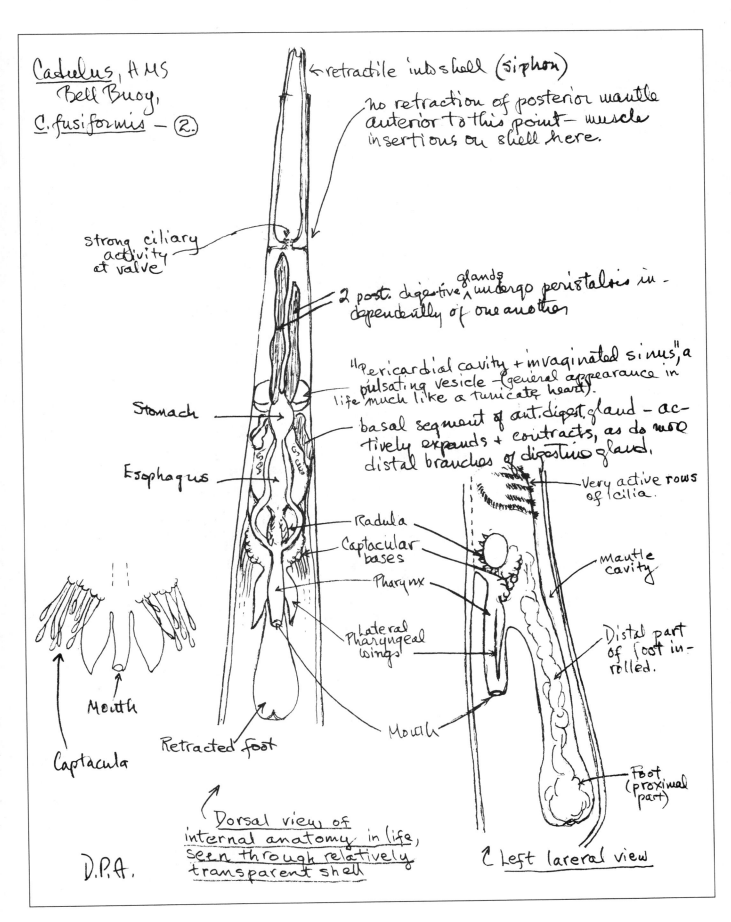

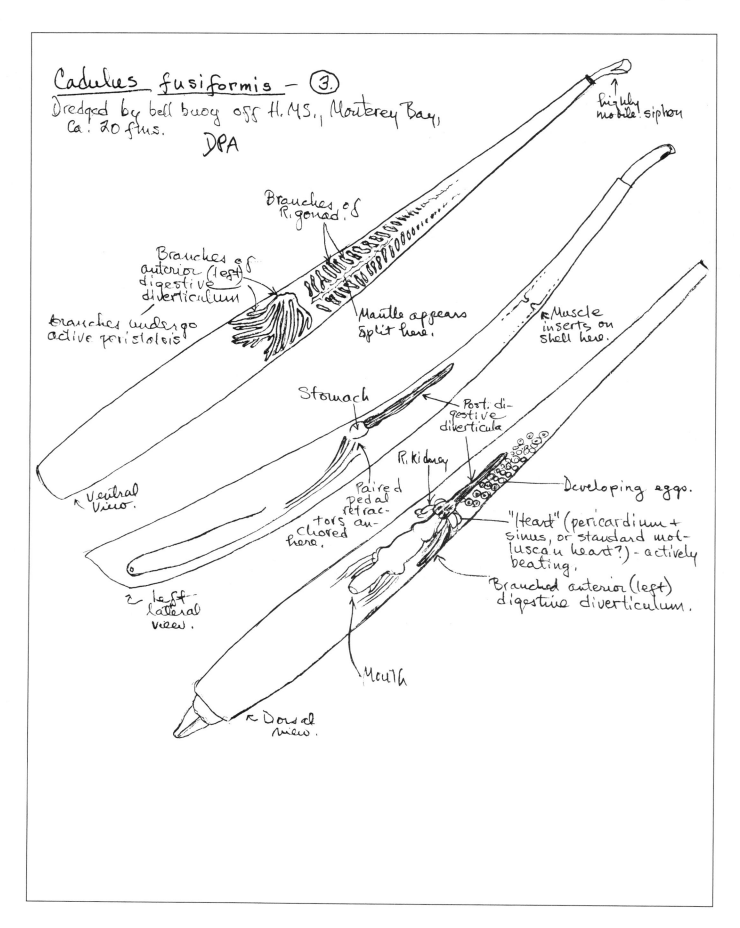

Mollusca/Scaphopoda

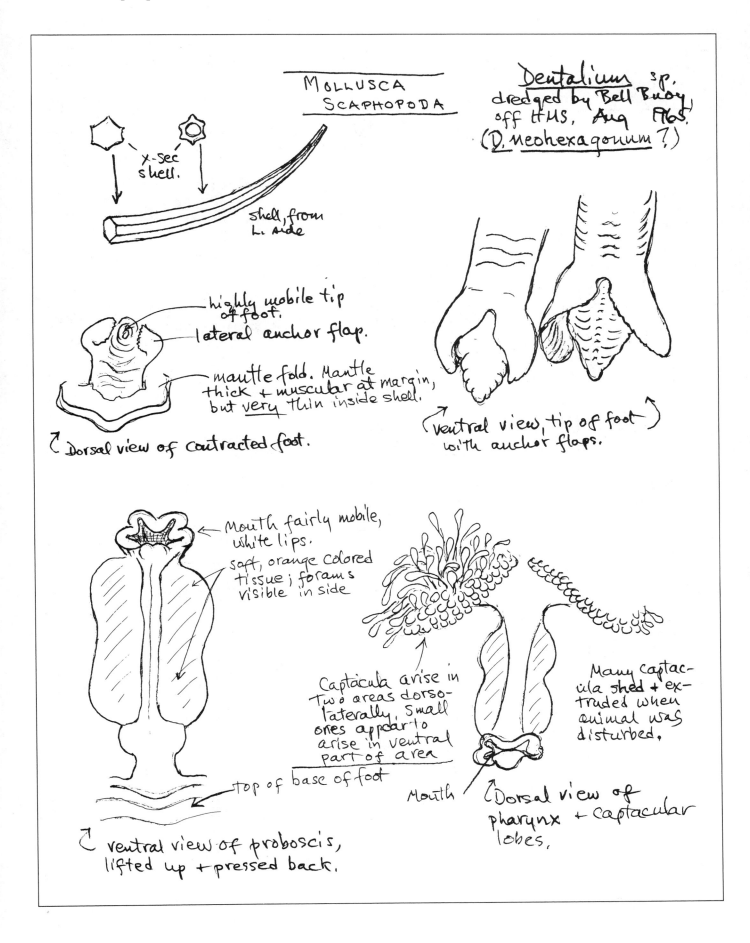

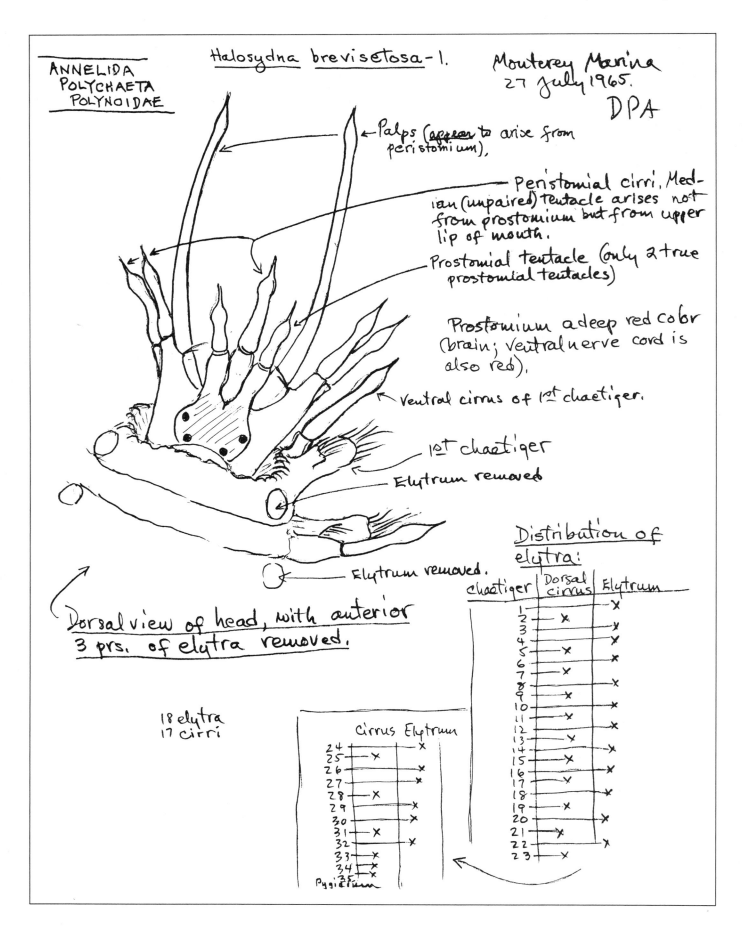

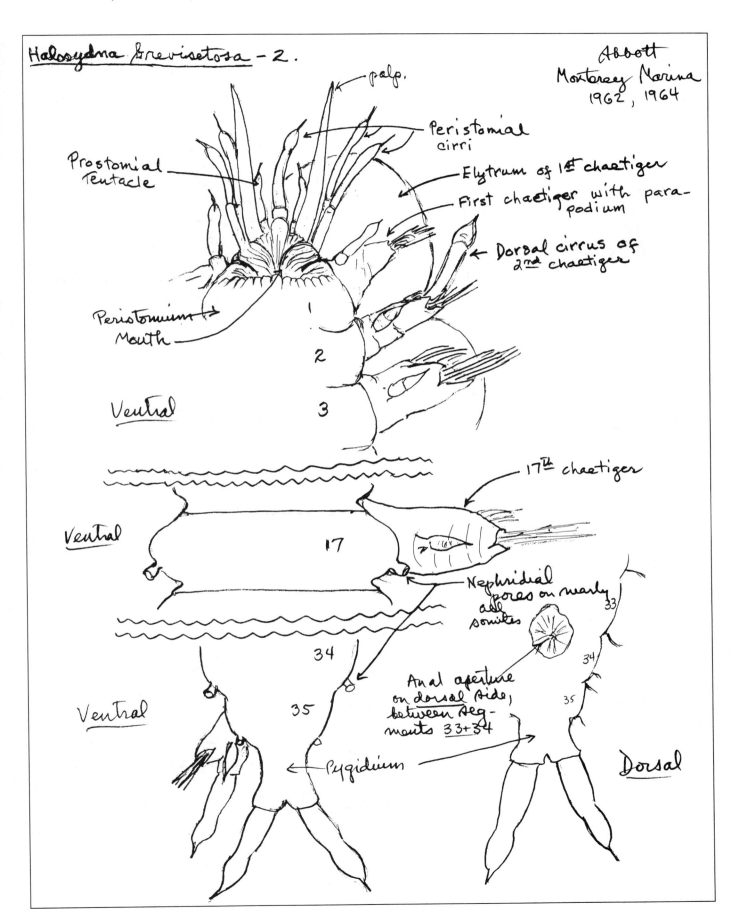

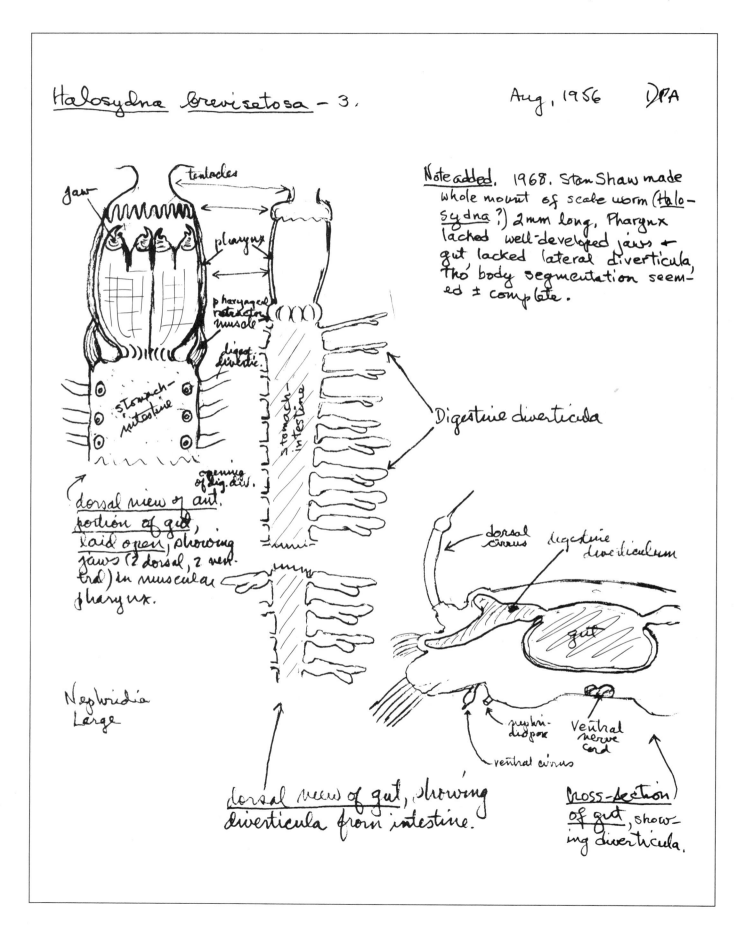

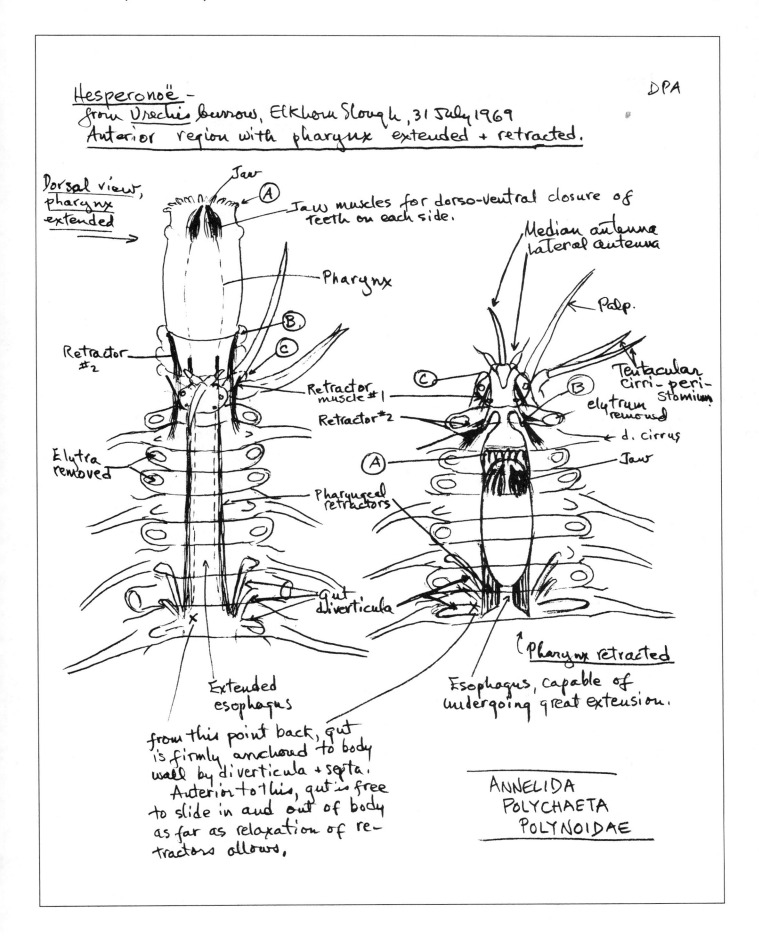

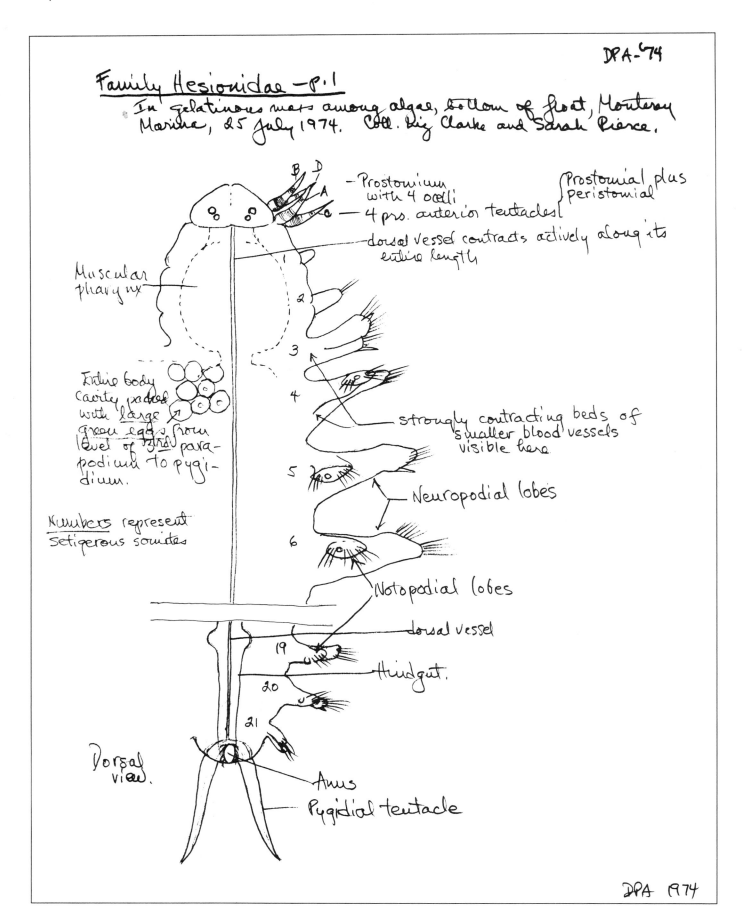

Hesionidae - p.2

DPA - 1974 p.2

A-D = 4 prs prostomial (?) plus peristomial tentacles.

Prostomial palps - capable of limited movement.

Mouth - can be opened <u>very</u> widely

First 2 parapodia are uniramous.

Ventral view of anterior end.

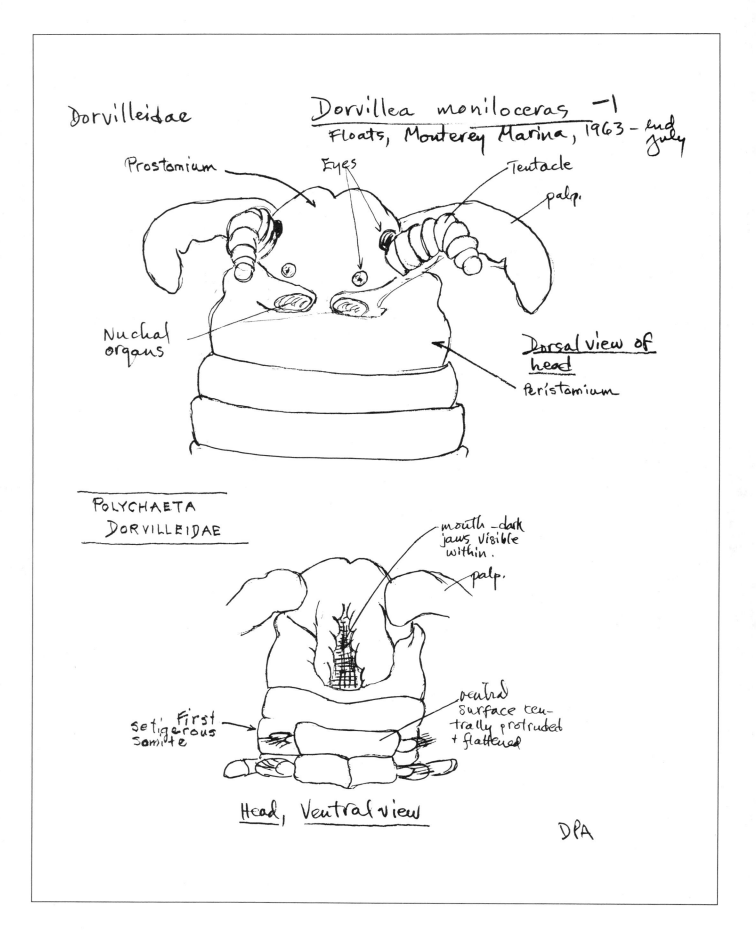

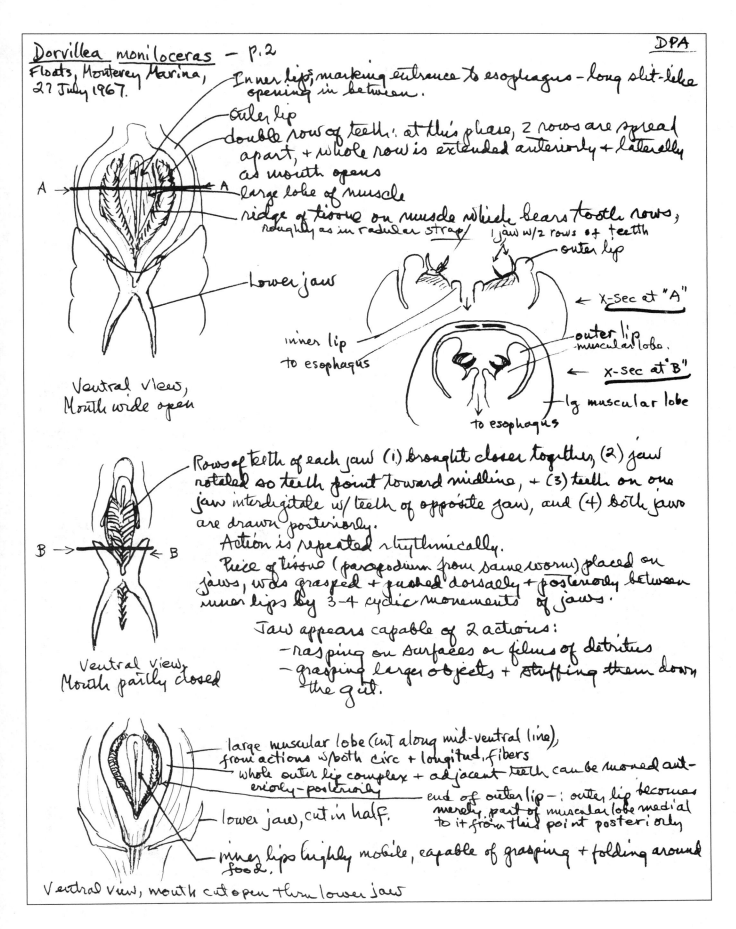

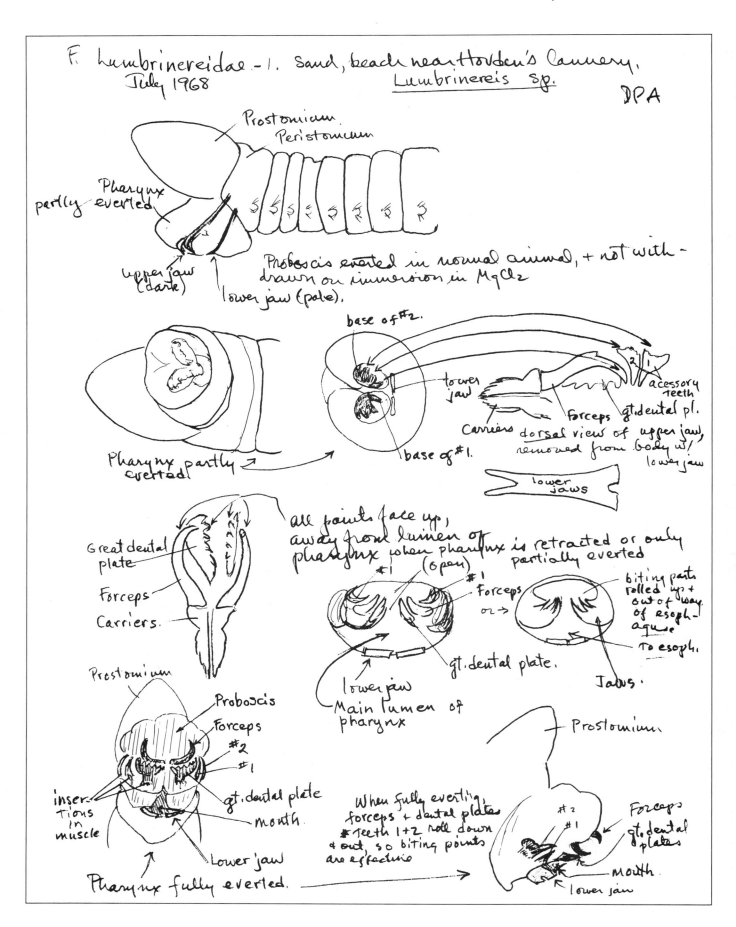

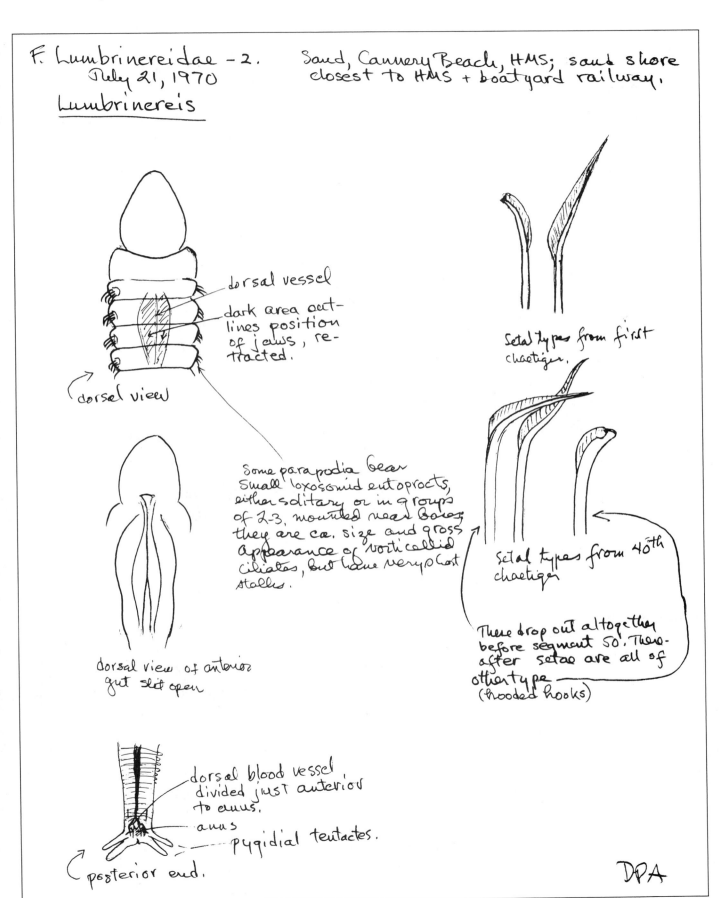

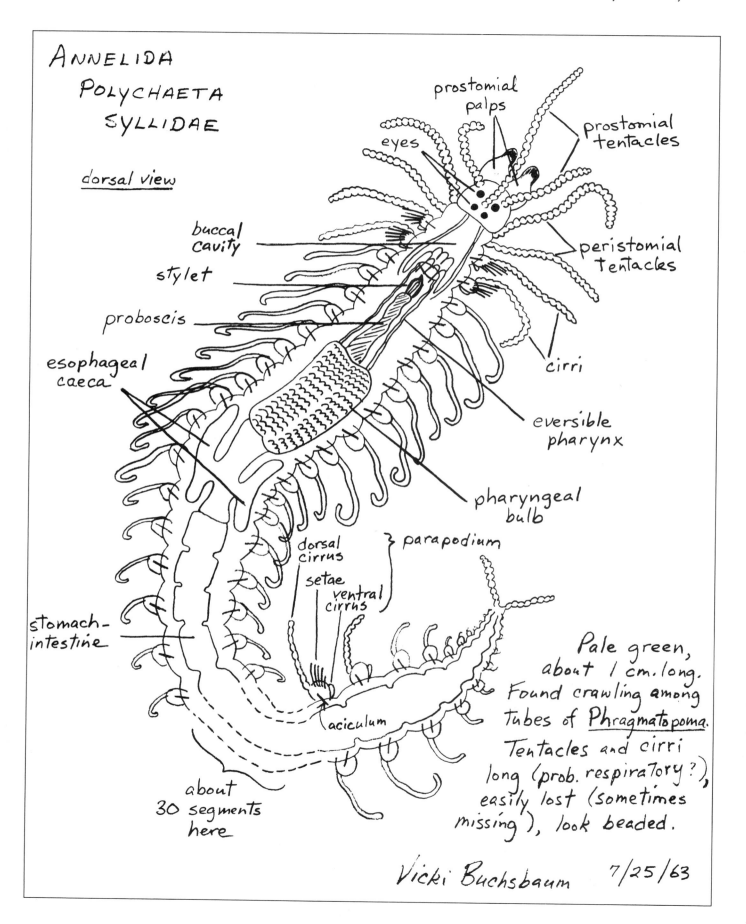

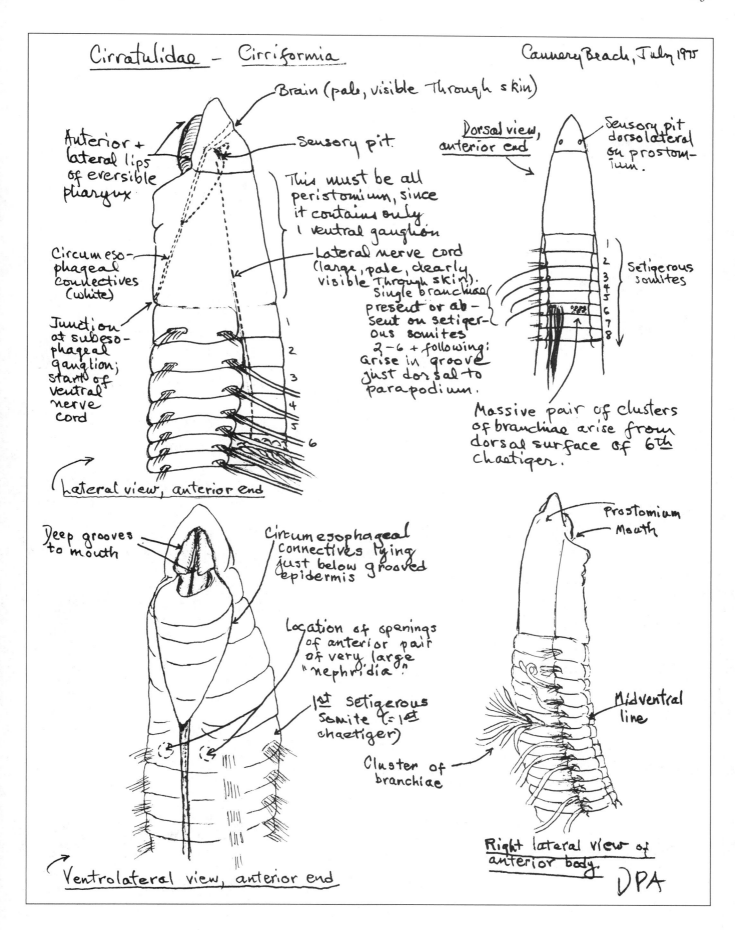

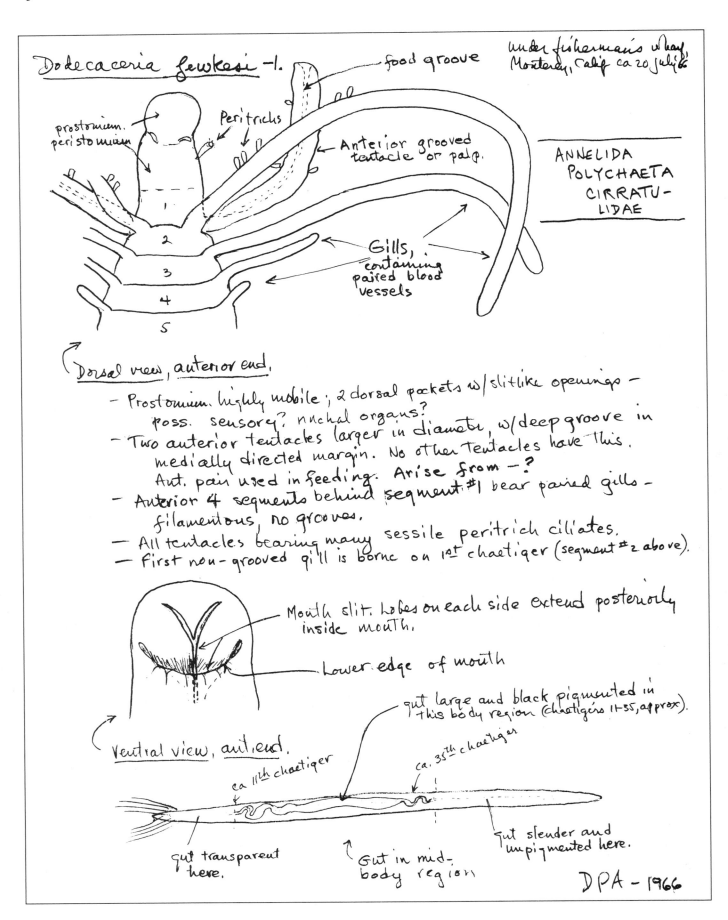

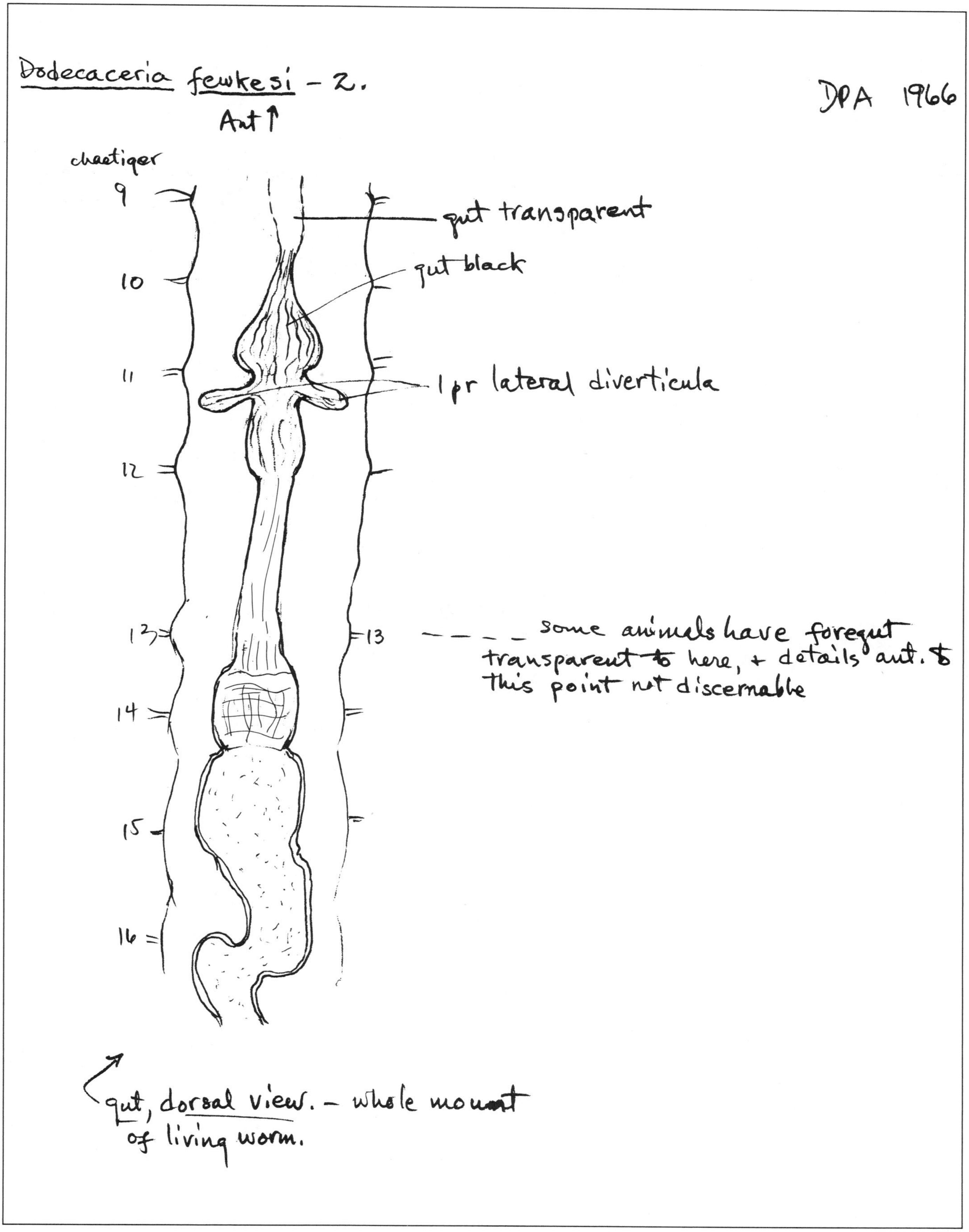

Annelida/Polychaeta/Cirratulidae

Dodecaceria fewkesi - 2.
Ant ↑

DPA 1966

chaetiger
9
10 — gut transparent
— gut black
11 — 1 pr lateral diverticula
12
13 — — — — — some animals have foregut transparent to here, + details ant. to this point not discernable
14
15
16

↰ gut, dorsal view. — whole mount of living worm.

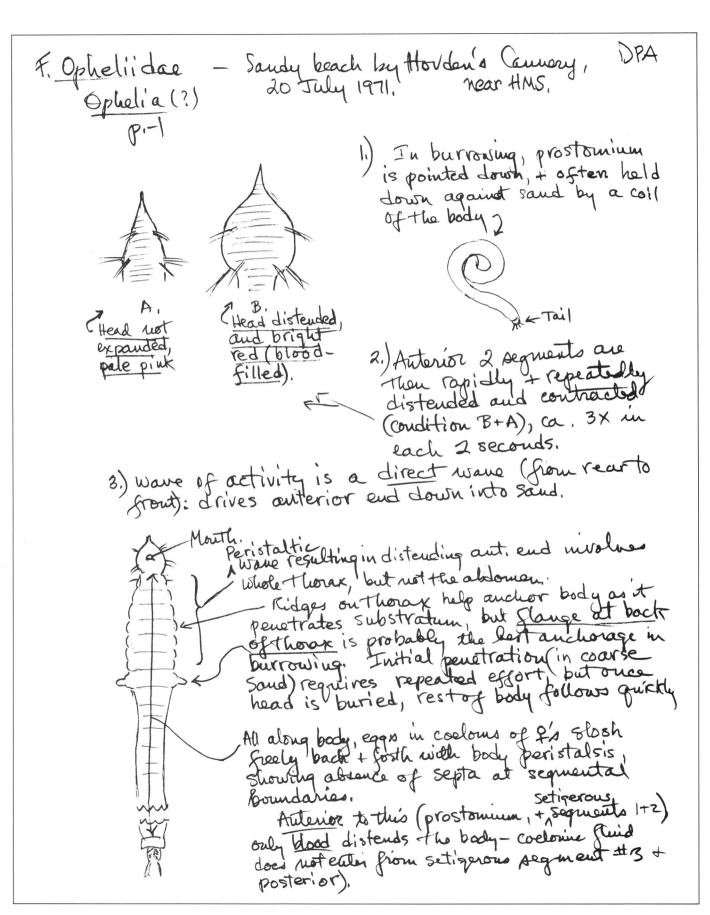

F. Opheliidae — Sandy beach by Hovden's Cannery, near HMS. DPA
Ophelia (?) 20 July 1971.
p. 1

1.) In burrowing, prostomium is pointed down, + often held down against sand by a coil of the body.

← Tail

A. Head not expanded, pale pink
B. Head distended, and bright red (blood-filled).

2.) Anterior 2 segments are then rapidly + repeatedly distended and contracted (condition B + A), ca. 3× in each 2 seconds.

3.) Wave of activity is a direct wave (from rear to front): drives anterior end down into sand.

Mouth — Peristaltic wave resulting in distending ant. end involves whole thorax, but not the abdomen.

Ridges on thorax help anchor body as it penetrates substratum, but flange at back of thorax is probably the best anchorage in burrowing. Initial penetration (in coarse sand) requires repeated effort, but once head is buried, rest of body follows quickly.

All along body, eggs in coeloms of ♀'s slosh freely back + forth with body peristalsis, showing absence of septa at segmental boundaries.
Anterior to this (prostomium, + setigerous segments 1+2) only blood distends the body — coelomic fluid does not enter from setigerous segment #3 + posterior).

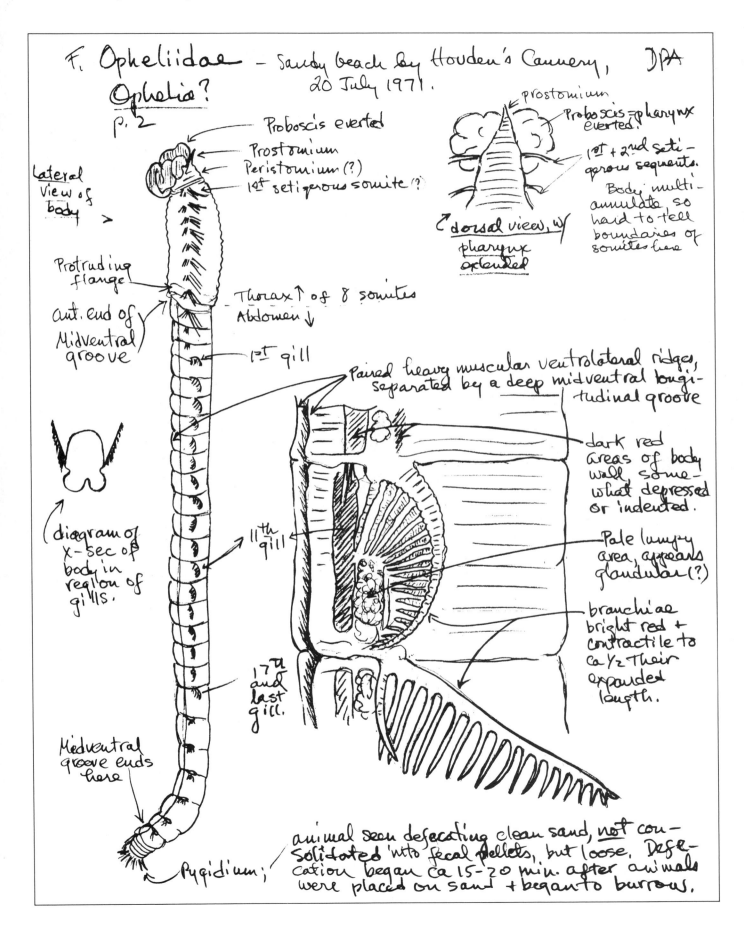

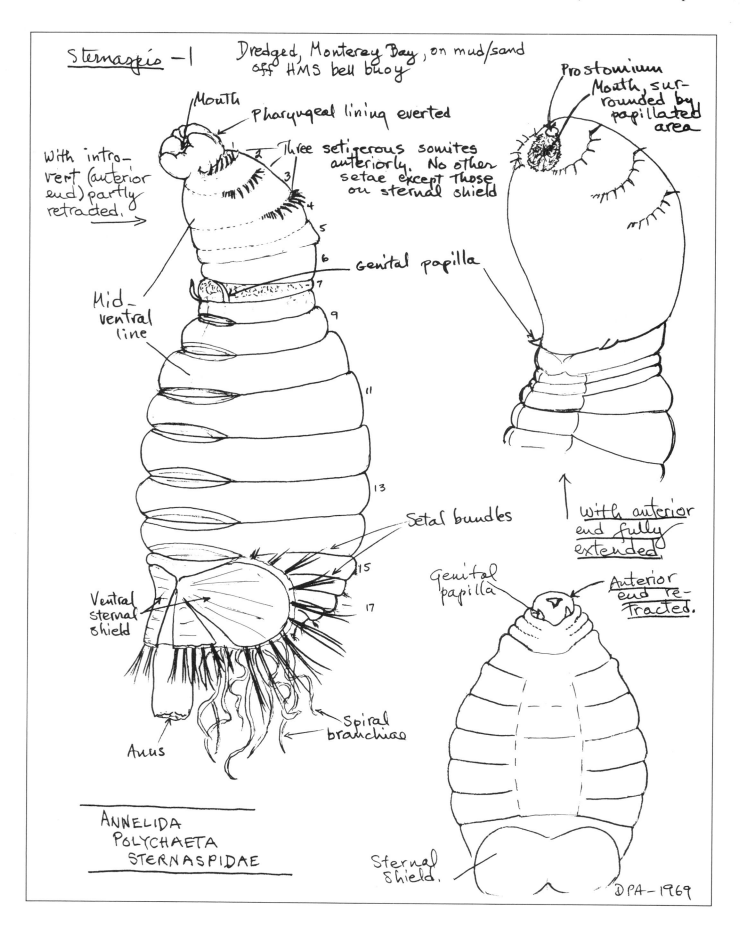

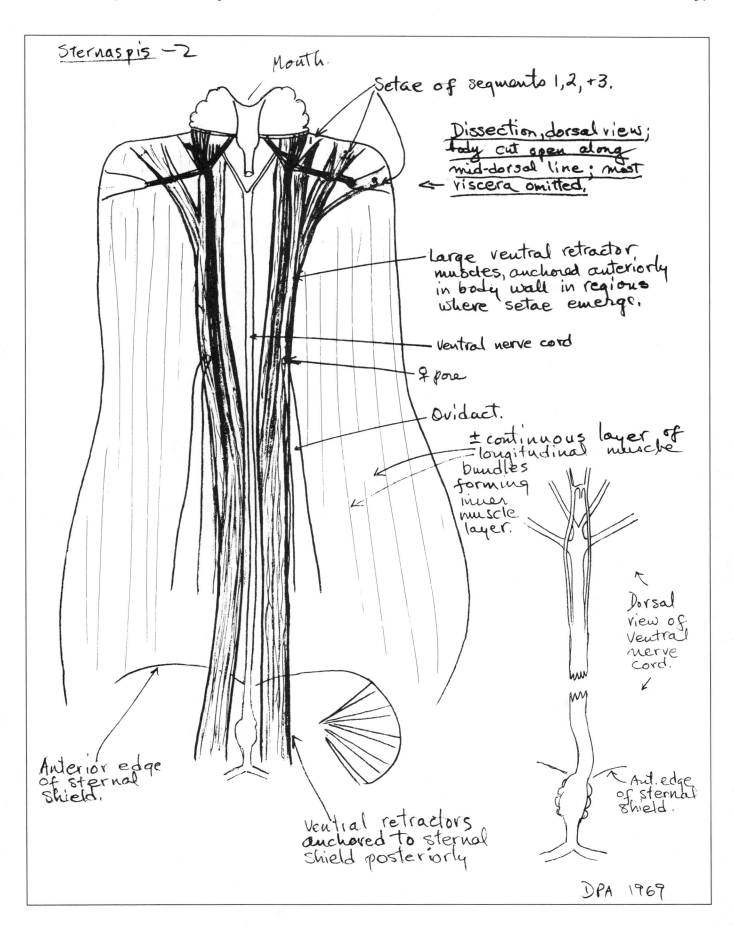

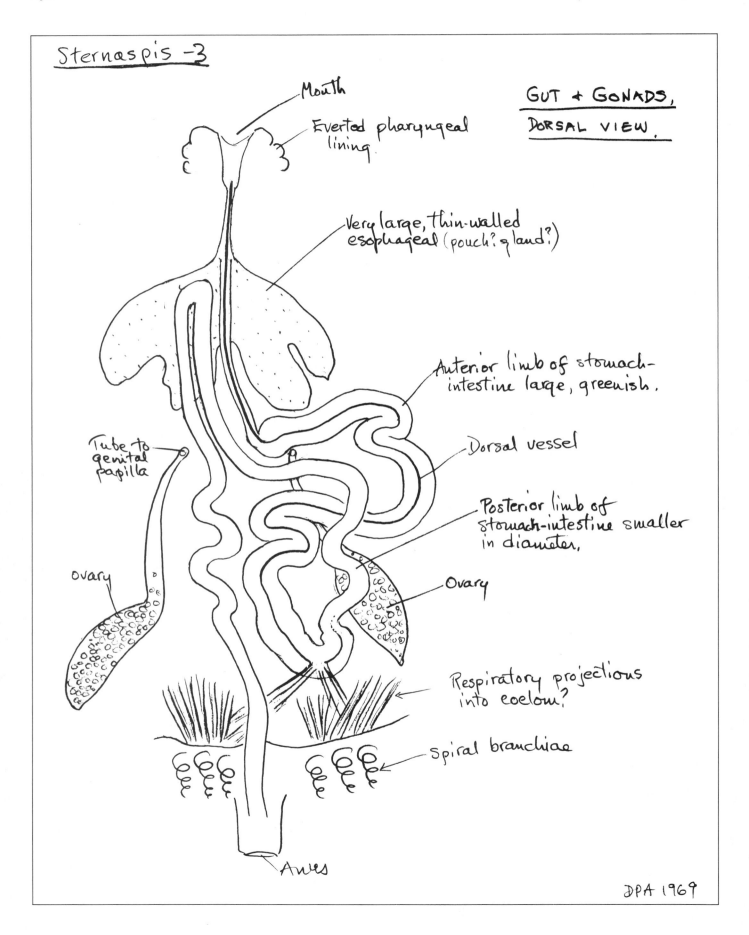

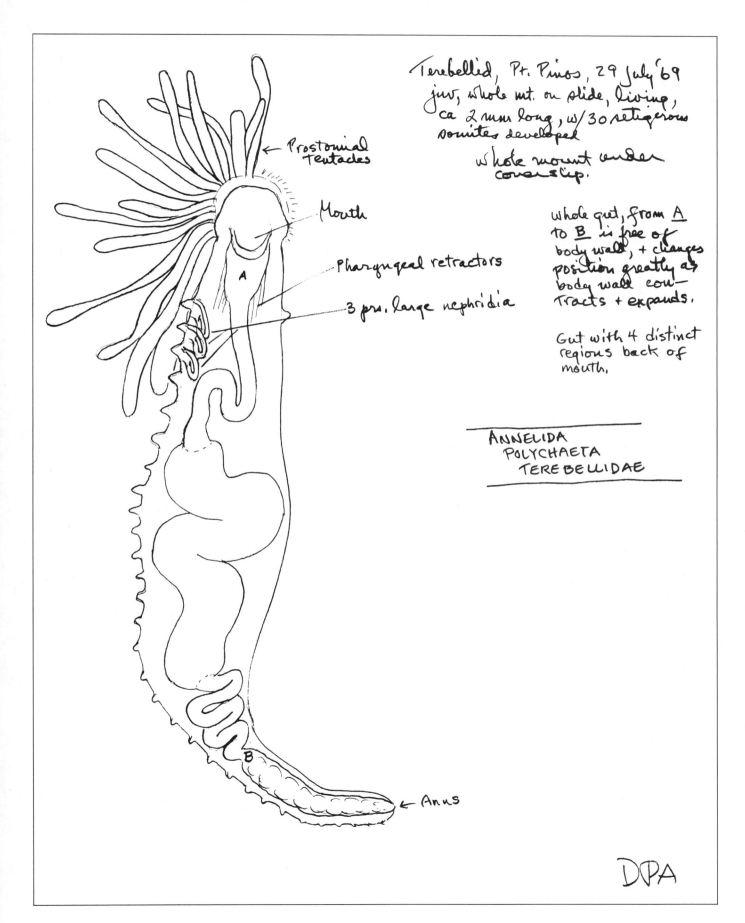

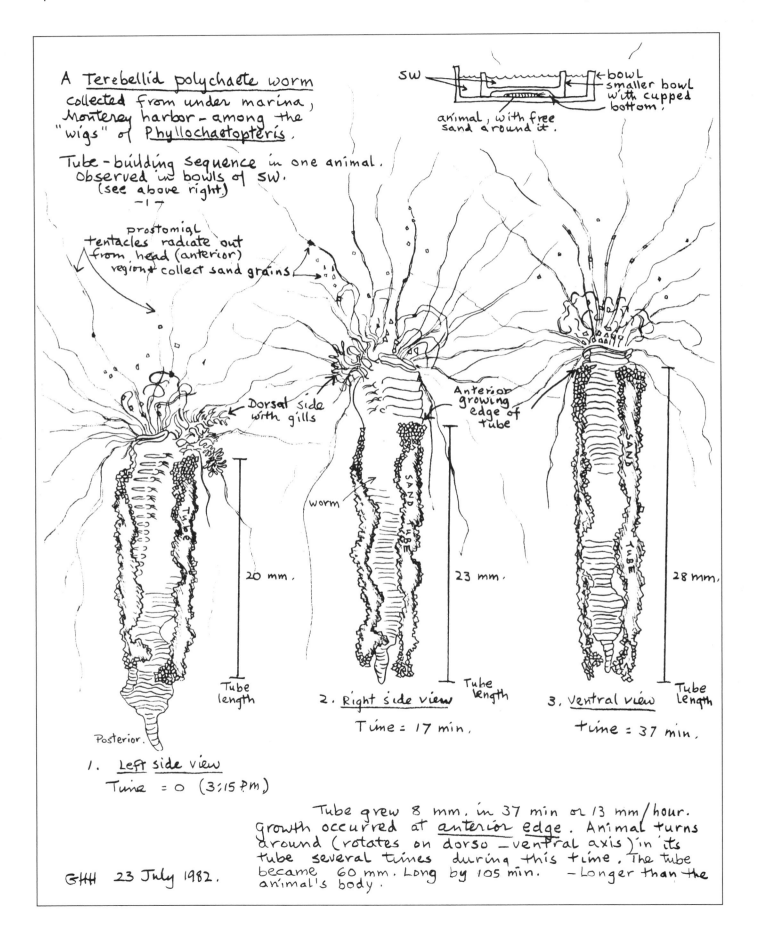

Annelida/Polychaeta/Terebellidae

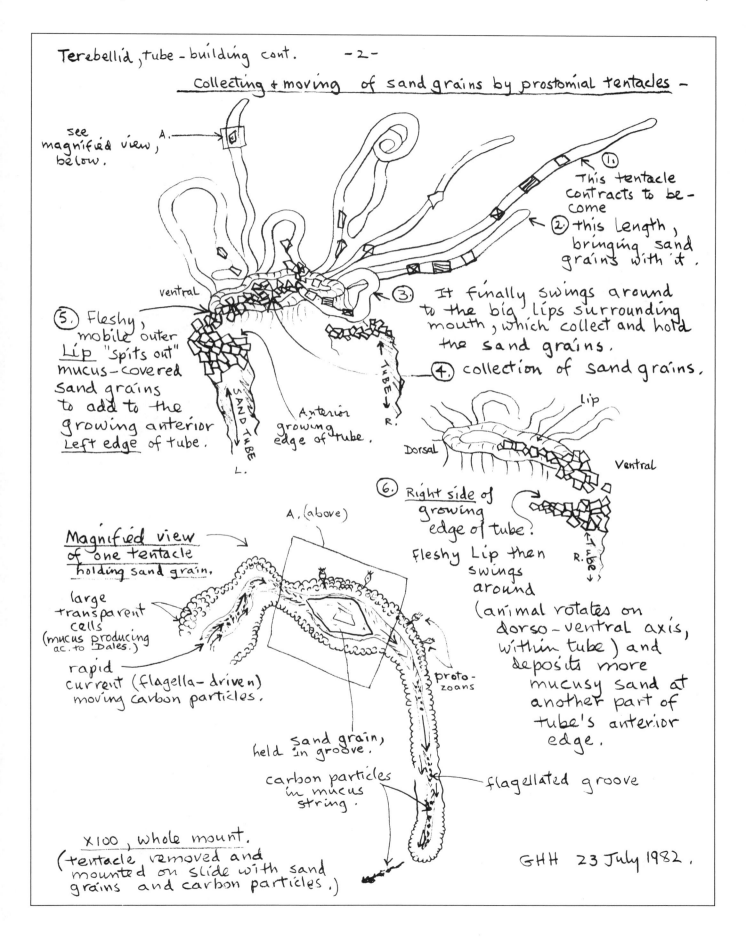

Phragmatopoma californica — HMS rocks by Cannery 26 July '73
DPA

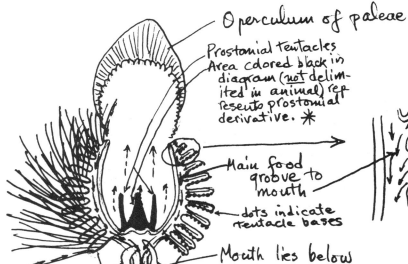

- Operculum of paleae
- Prostomial tentacles. Area colored black in diagram (not delimited in animal) represents prostomial derivative. *
- Main food groove to mouth
- dots indicate tentacle bases
- Mouth lies below lip, here

Ciliary-mucoid food tract down antero-medial face of each tentacle.

Slender posterior extension of body recurves forward on the <u>ventral</u> side of the body.

Ventral view, anterior end of *Phragmatopoma californica*

* ■ = parts derived from the prostomium in earlier ontogeny (see R.P. Dales, 1952. Quart. J. Microscop. Sci. <u>93</u>: 435-52).

Fertile ♂'s + ♀'s spawn when removed from tubes. High % of spawning.

For details of behavior see P.A. Roy, 1974. Bull. So. Calif. Acad. Sci. <u>73</u>: 117-25.

POLYCHAETA
SABELLARIIDAE

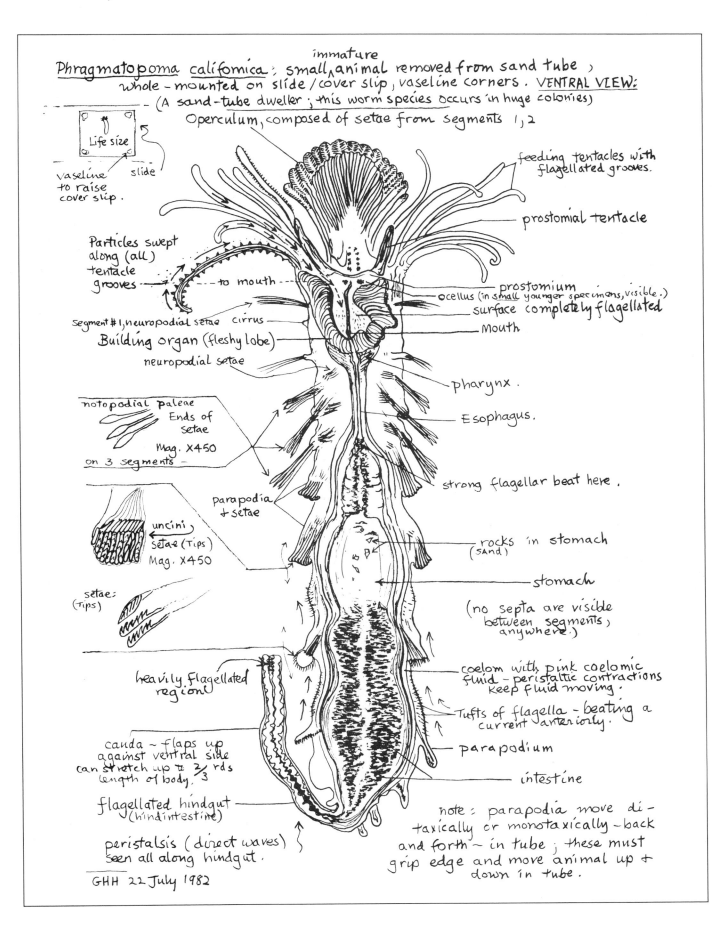

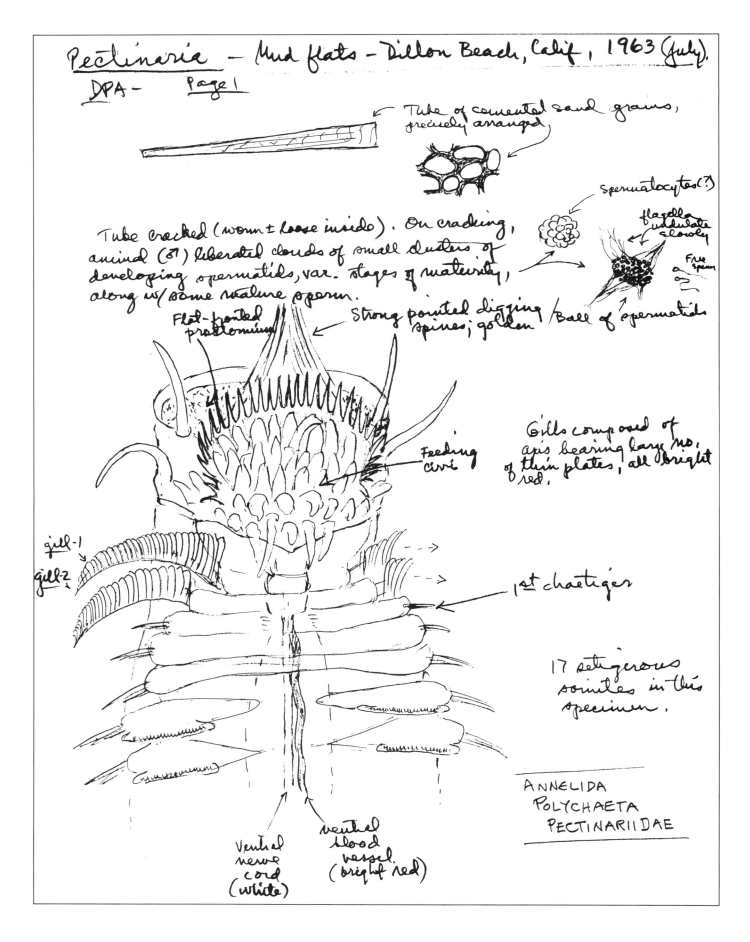

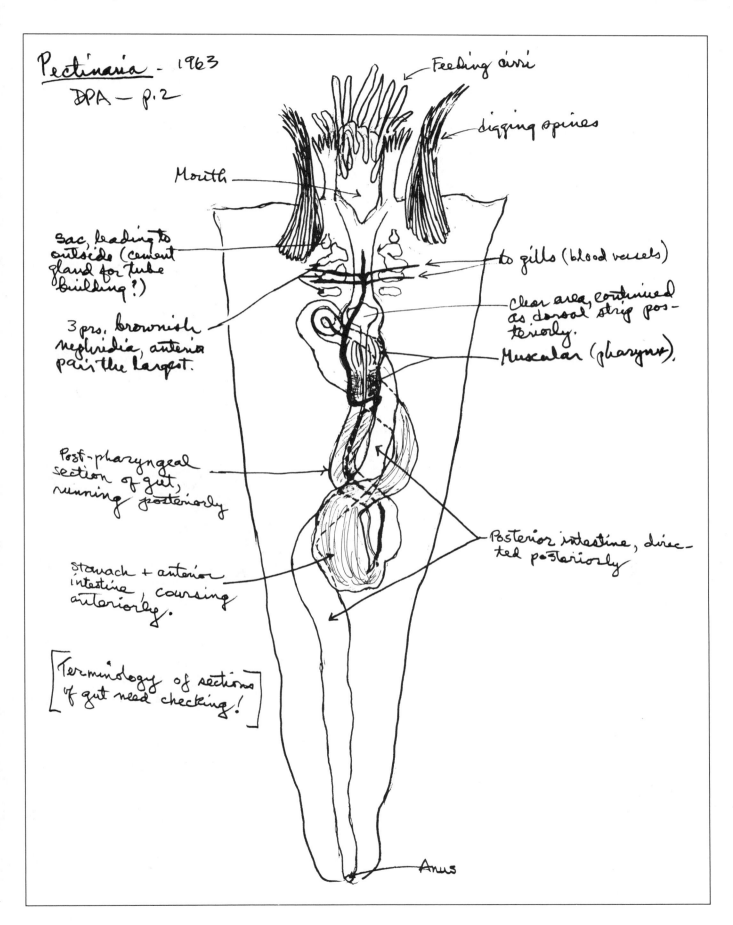

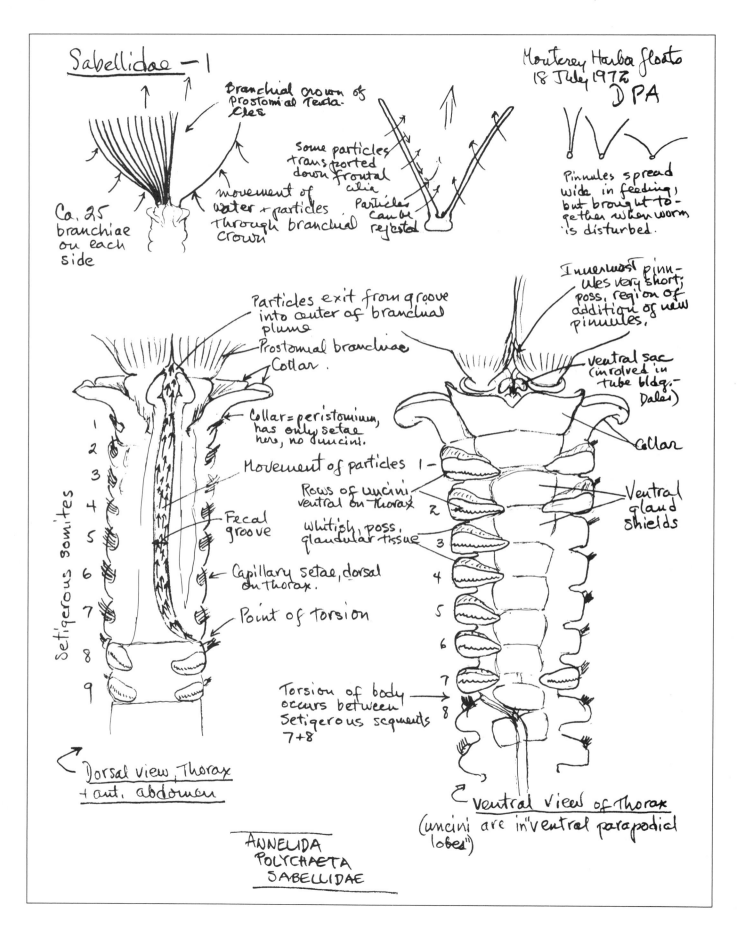

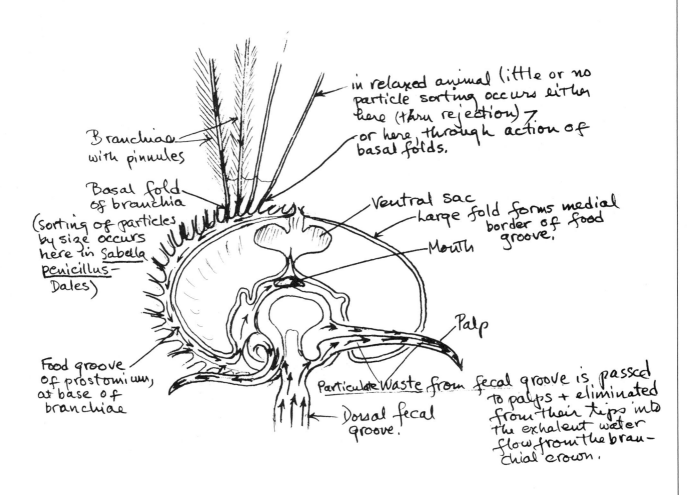

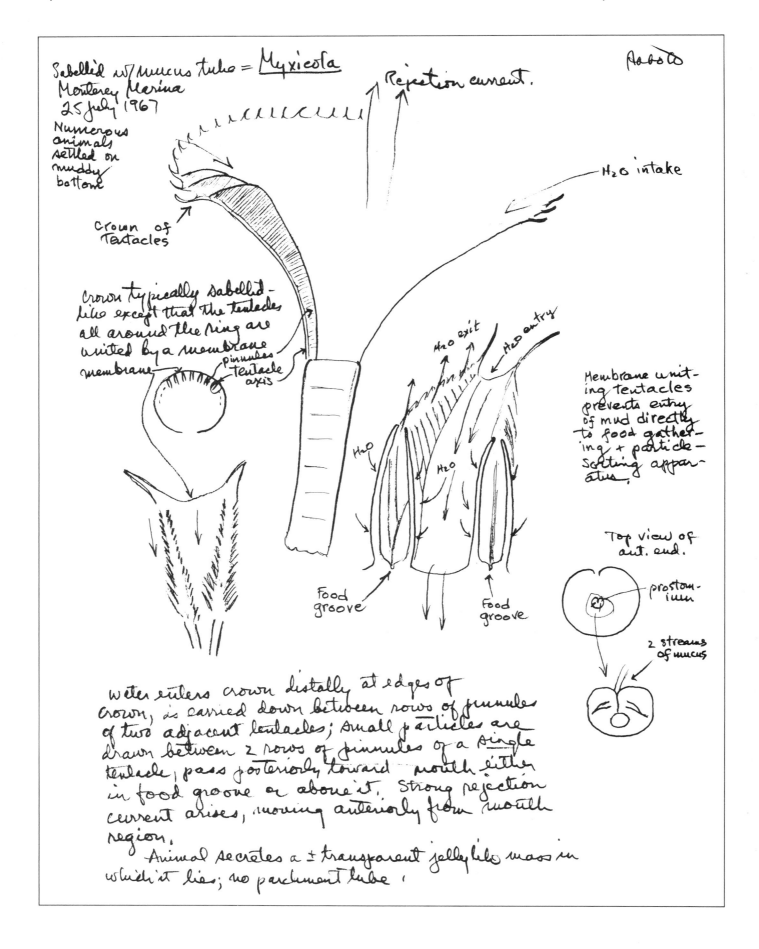

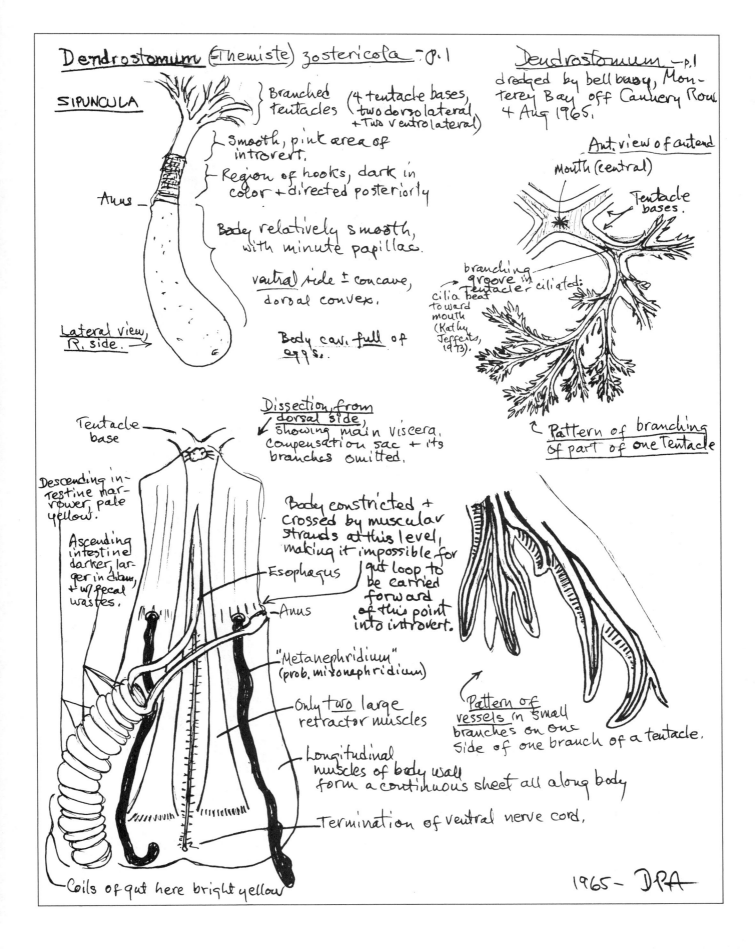

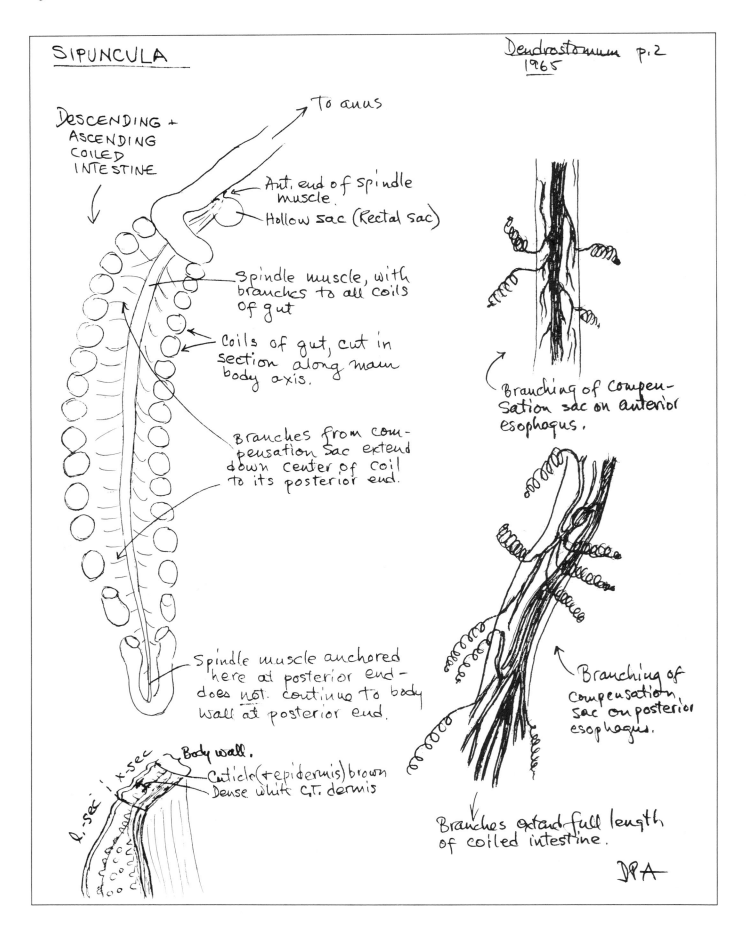

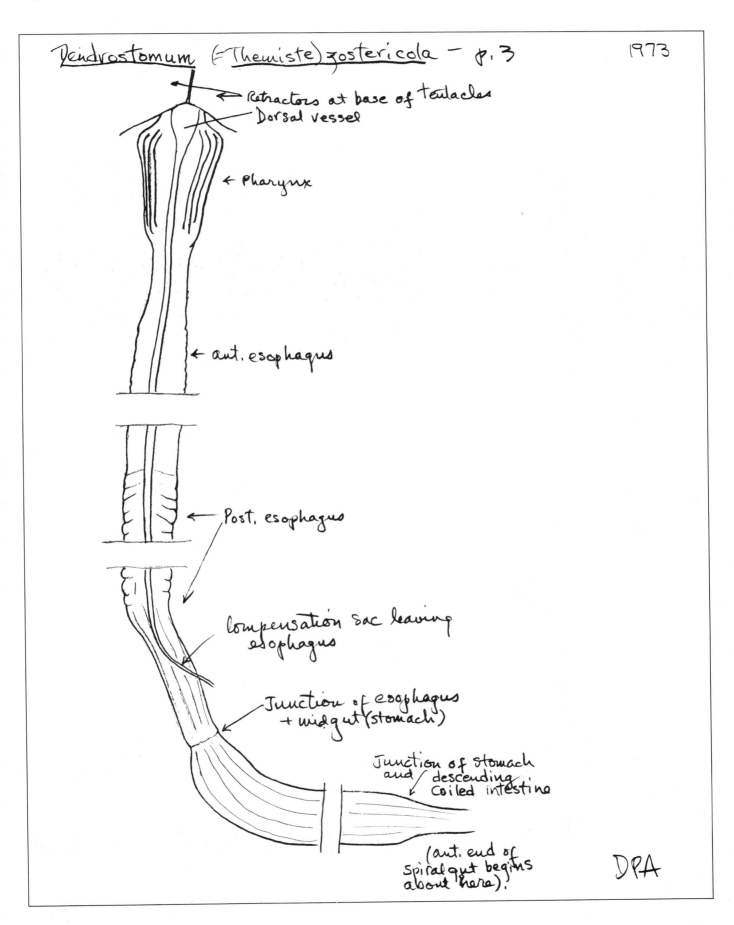

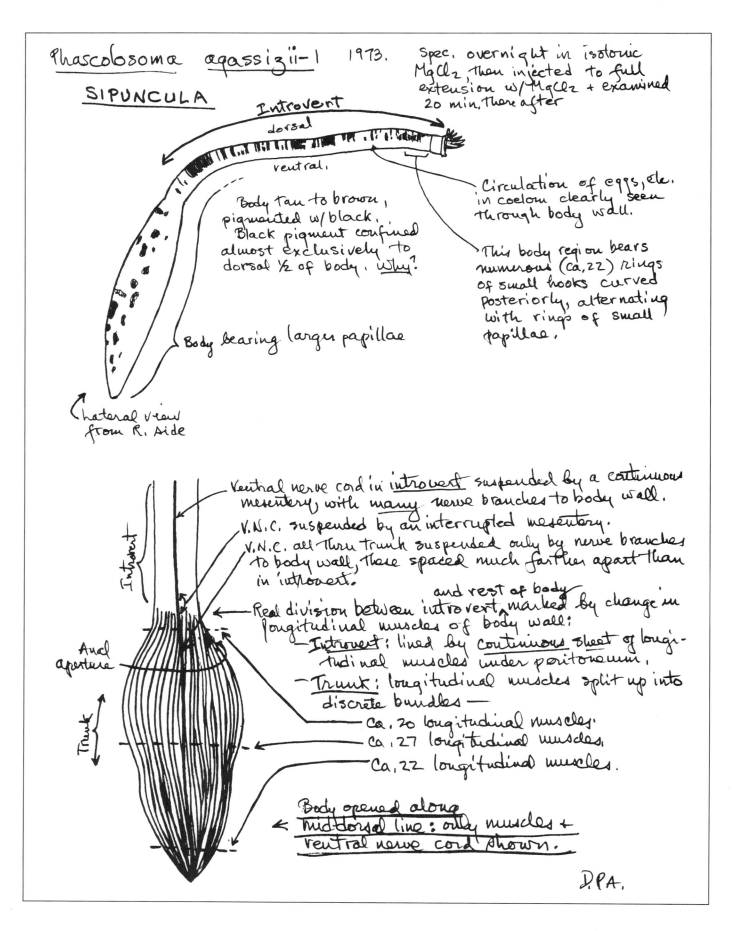

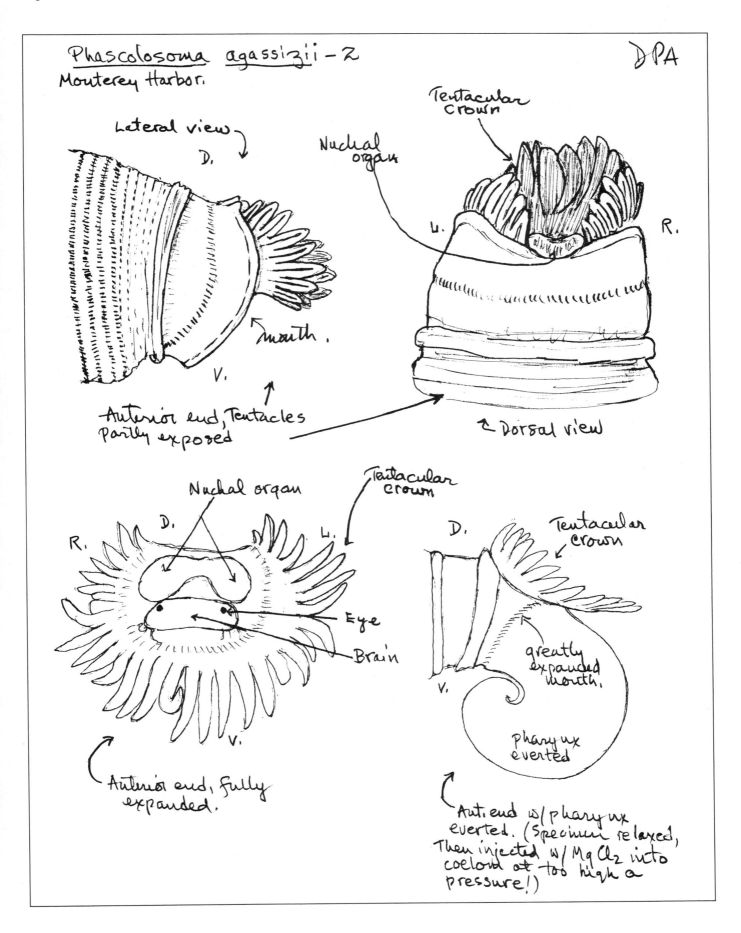

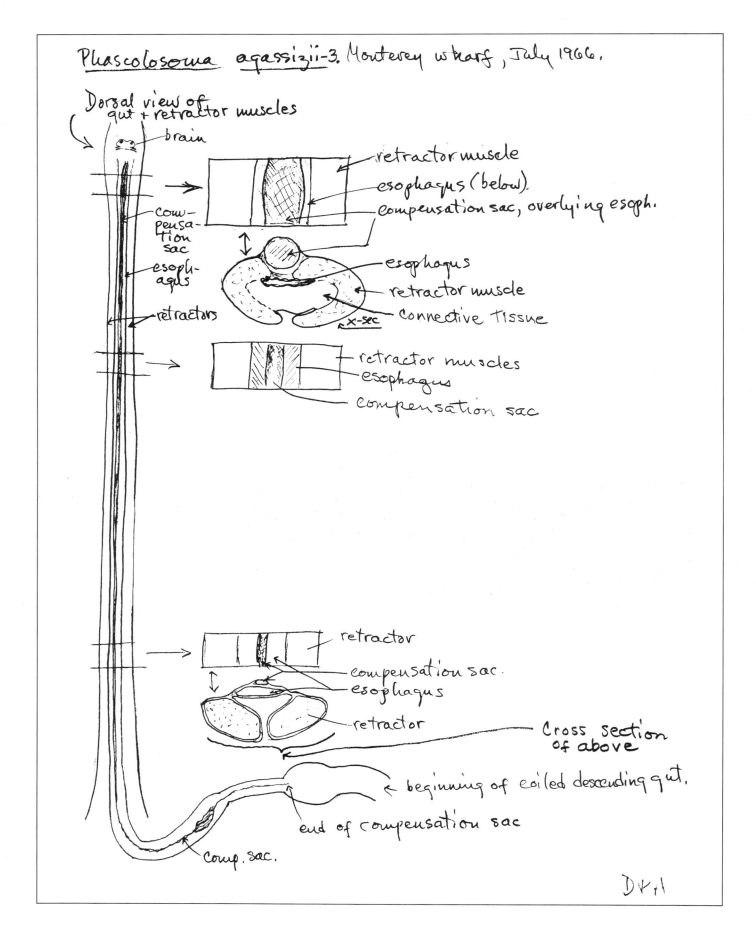

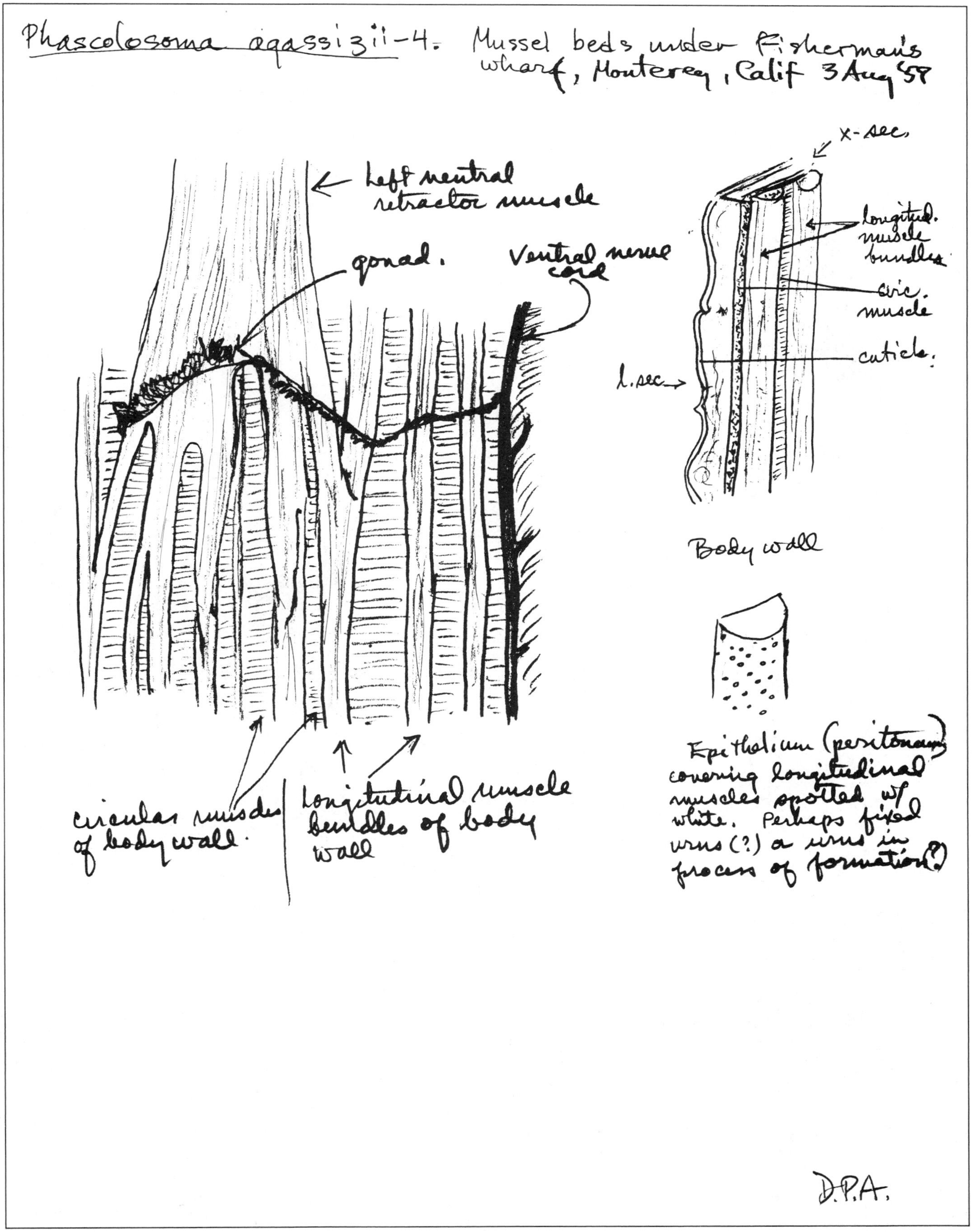

Sipuncula

Phascolosoma agassizii - 4. Mussel beds under Fisherman's wharf, Monterey, Calif. 3 Aug 58

← left ventral retractor muscle

gonad.

ventral nerve cord

circular muscles of body wall.

longitudinal muscle bundles of body wall

x-sec.

longitud. muscle bundles

circ. muscle

cuticle.

l. sec →

Body wall

Epithelium (peritoneum) covering longitudinal muscles spotted w/ white. Perhaps fixed urns (?) or urns in process of formation(?)

D.P.A.

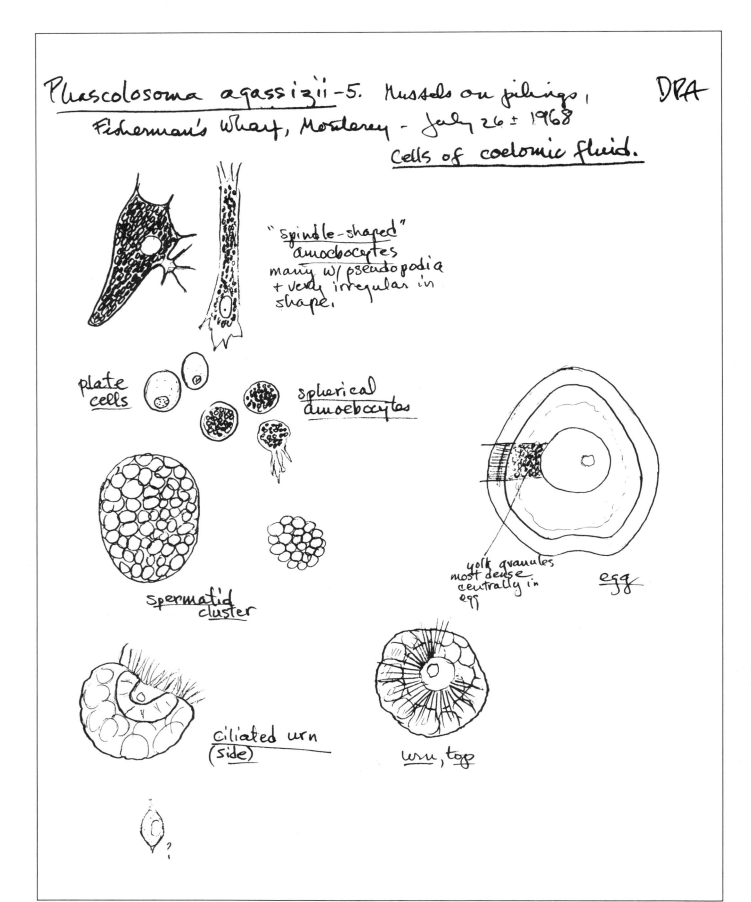

Sipuncula

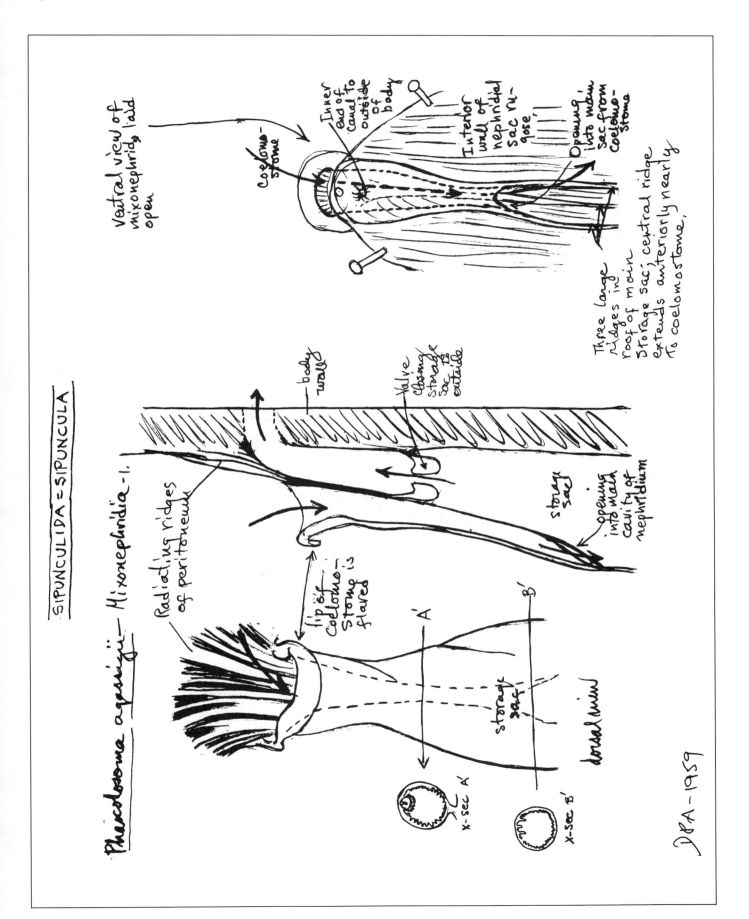

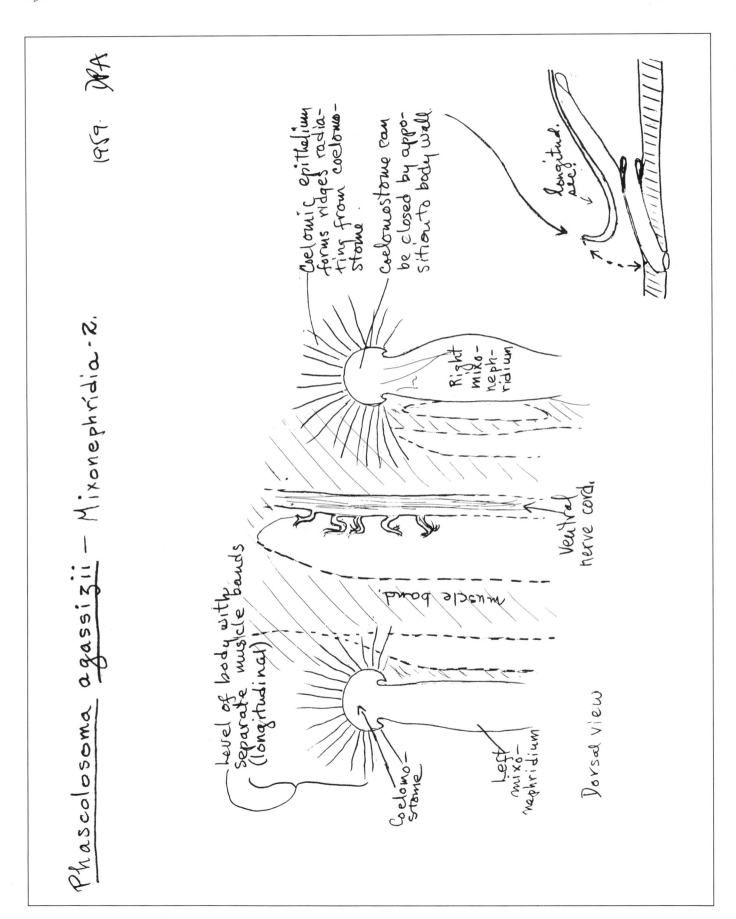

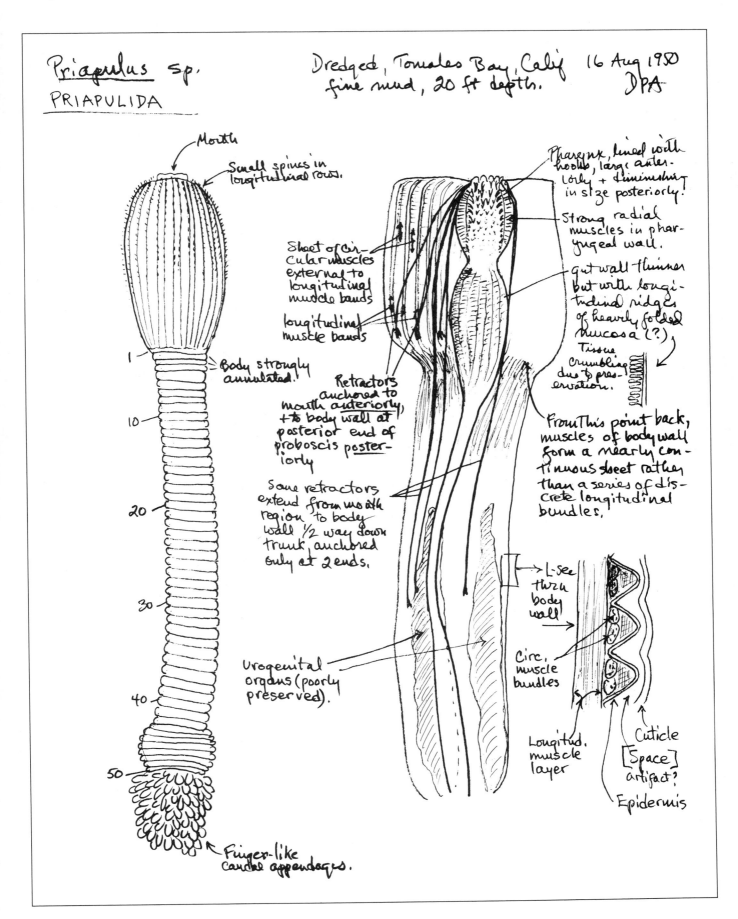

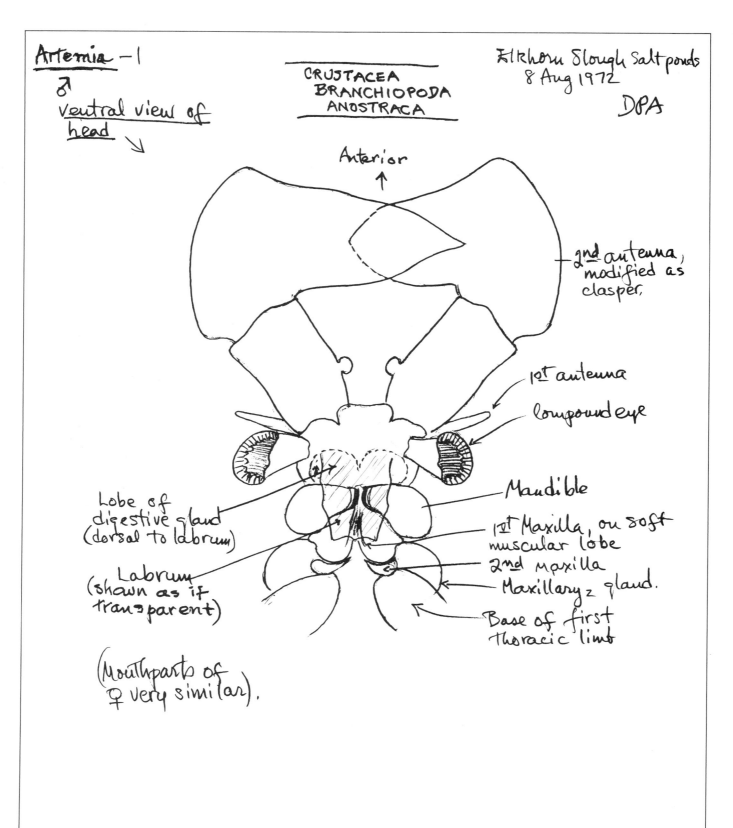

Arthropoda/Crustacea/Branchiopoda/Anostraca

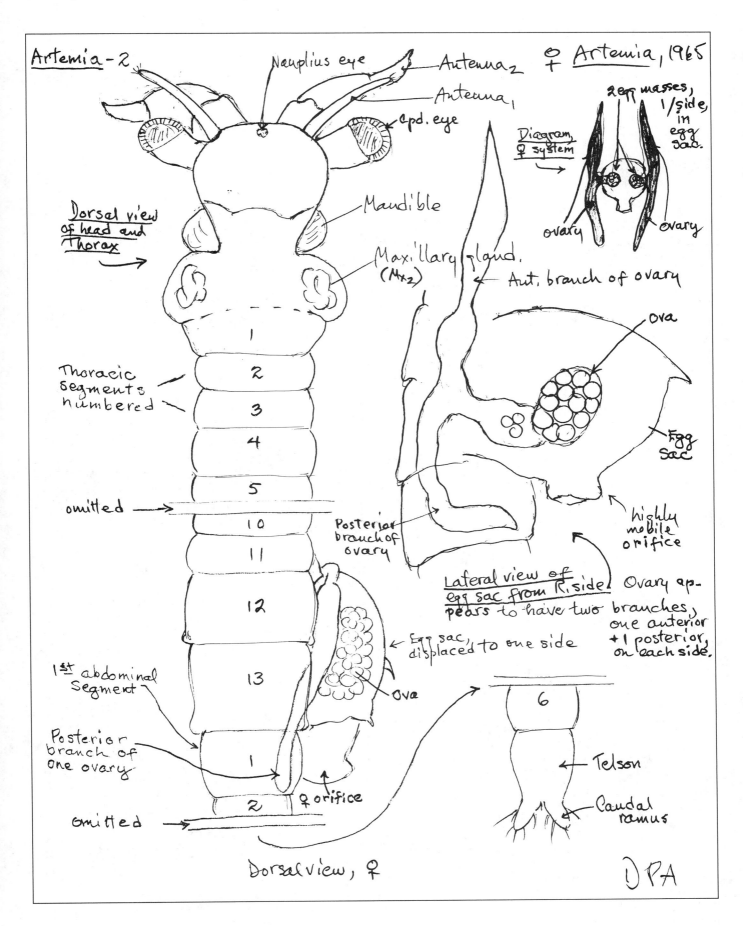

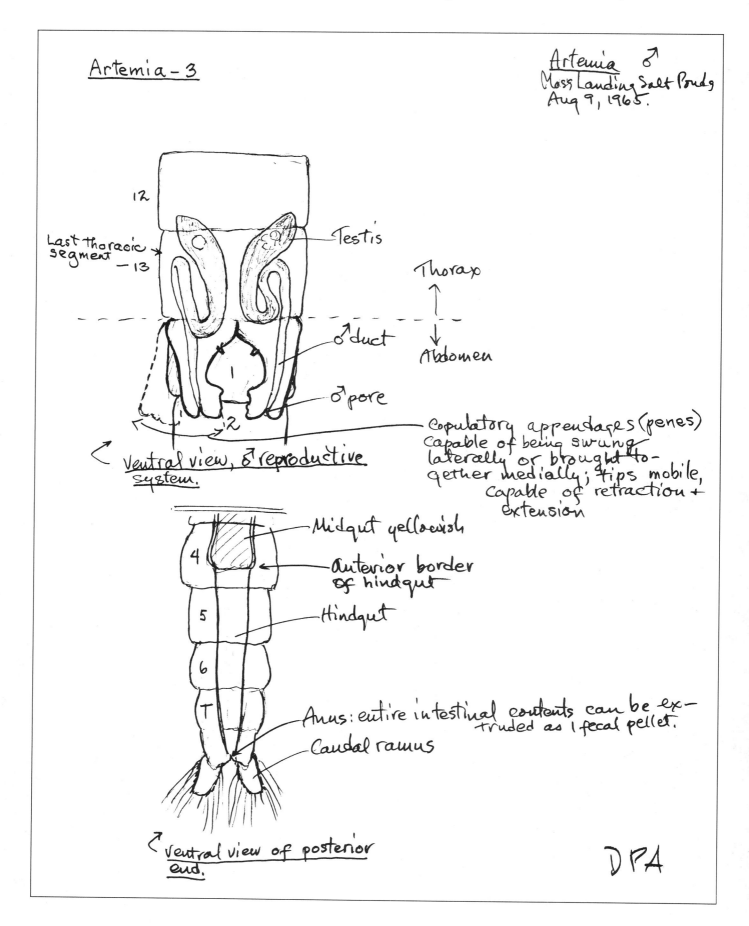

Arthropoda/Crustacea/Branchiopoda/Anostraca

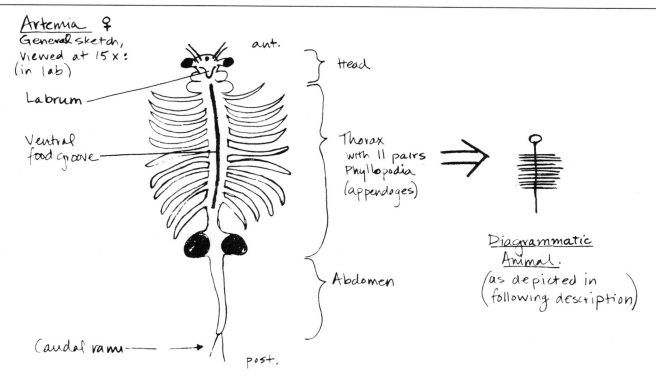

Artemia ♀
General sketch, viewed at 15x: (in lab)

- Labrum
- ant.
- Head
- Ventral food groove
- Thorax with 11 pairs Phyllopodia (appendages)
- Abdomen
- Caudal rami
- post.

Diagrammatic Animal. (as depicted in following description)

Swimming (7x)

Artemia swims in the direction with its anterior end going first. It can swim with either its lateral, ventral, or dorsal side upward. Swimming is executed by the beating of the leaf-like thoracic appendages, phyllopodia. These beat in metachronal rhythms. A large number of beating variations were observed. Every couple of minutes the phyllopodia may (or may not) stop beating for a period ranging from a split second to a full 1-2 seconds. Freely swimming in a large dish of sea water, Artemia moves quickly & with agility. The animals swim almost constantly, often changing direction & depth. They can make sharp turns or swim straight ahead.

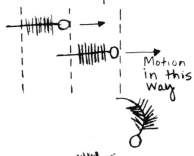

Motion in this way

- Streamlined swimming.
- Very agile, body can bend along its entire length.

Sharp turn!

- Overall swimming movement through water may be jerky or even (smooth).
- Power stroke is backward.

Eva Aladjem 11 Aug 82

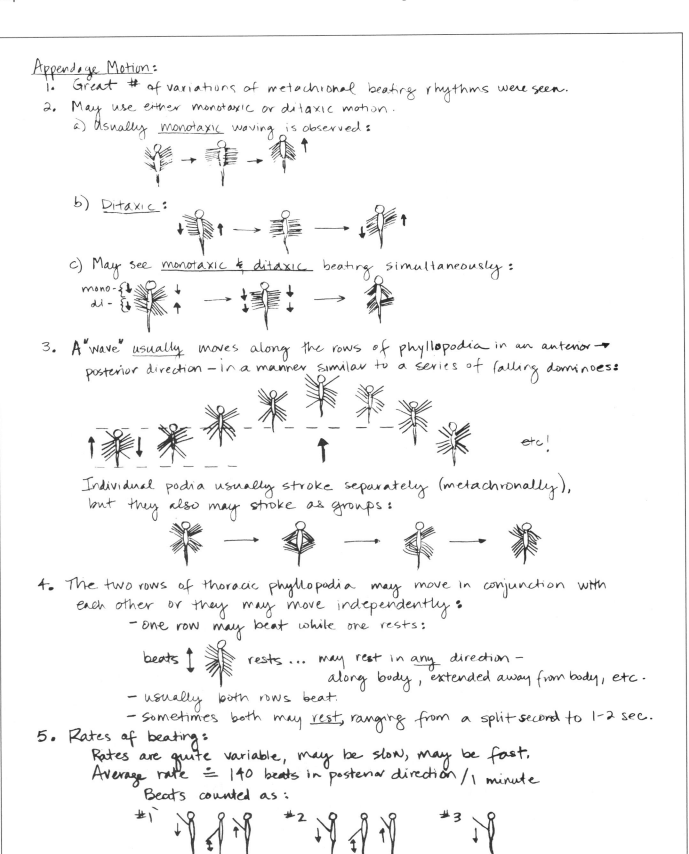

Appendage Motion:
1. Great # of variations of metachronal beating rhythms were seen.
2. May use either monotaxic or ditaxic motion.
   a) Usually _monotaxic_ waving is observed:
   b) _Ditaxic_:
   c) May see _monotaxic & ditaxic_ beating simultaneously:
3. A "wave" _usually_ moves along the rows of phyllopodia in an anterior → posterior direction — in a manner similar to a series of falling dominoes:

   Individual podia usually stroke separately (metachronally), but they also may stroke as groups:

4. The two rows of thoracic phyllopodia may move in conjunction with each other or they may move independently:
   — one row may beat while one rests:

   beats ↕ rests... may rest in _any_ direction — along body, extended away from body, etc.

   — usually both rows beat.
   — sometimes both may _rest_, ranging from a split second to 1-2 sec.

5. Rates of beating:
   Rates are quite variable, may be slow, may be fast.
   Average rate ≅ 140 beats in posterior direction / 1 minute
   Beats counted as:

E.V.A. 8/11/82

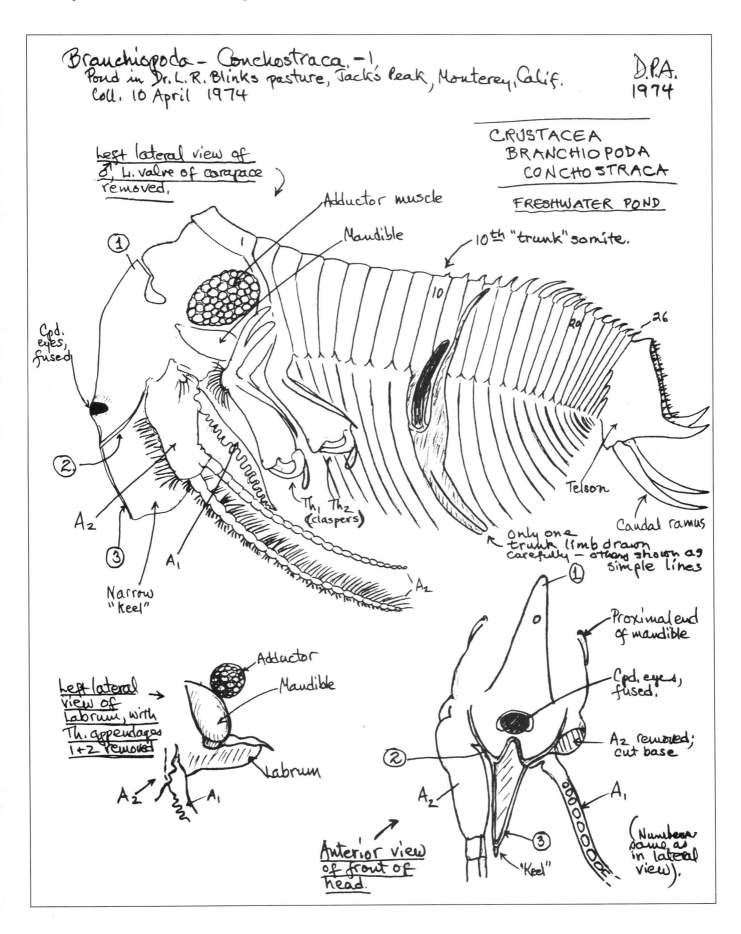

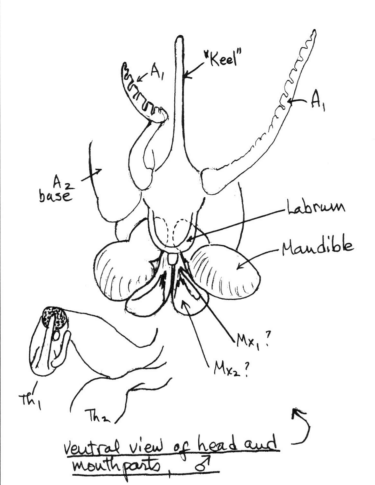

Conchostraca - 2.

DPA
10 April, 1974
Monterey, Calif.
Freshwater pond

Ventral view of head and mouthparts, ♂

Arthropoda/Crustacea/Branchiopoda/Cladocera

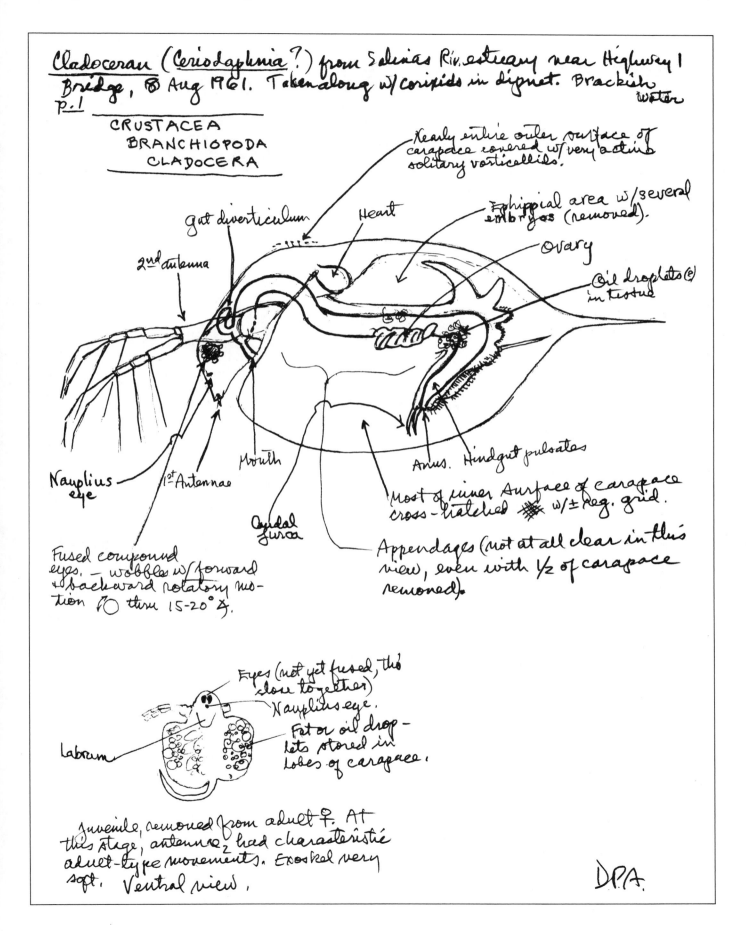

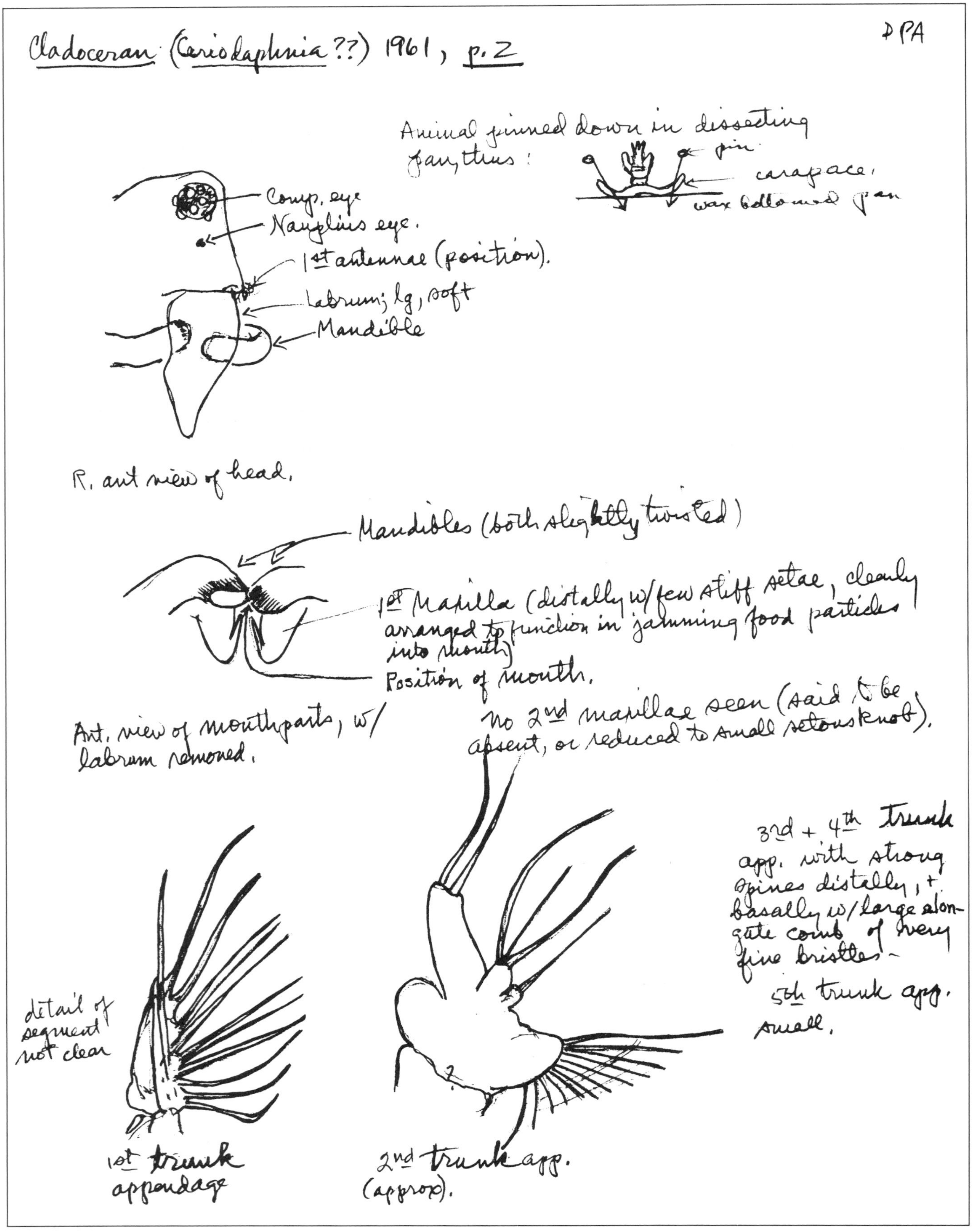

Cladoceran (Ceriodaphnia??) 1961, p. 2

PPA

Animal pinned down in dissecting pan, thus: — pin, carapace, wax bottom of pan

- Comp. eye
- Nauplius eye
- 1st antennae (position)
- Labrum; lg, soft
- Mandible

R. ant view of head.

— Mandibles (both slightly twisted)

1st Maxilla (distally w/ few stiff setae, clearly arranged to function in jamming food particles into mouth)

Position of mouth.

Ant. view of mouthparts, w/ labrum removed.

No 2nd maxillae seen (said to be absent, or reduced to small setous knob).

3rd + 4th trunk app. with strong spines distally, + basally w/ large elongate comb of very fine bristles.

5th trunk app. small.

detail of segment not clear

1st trunk appendage

2nd trunk app. (approx).

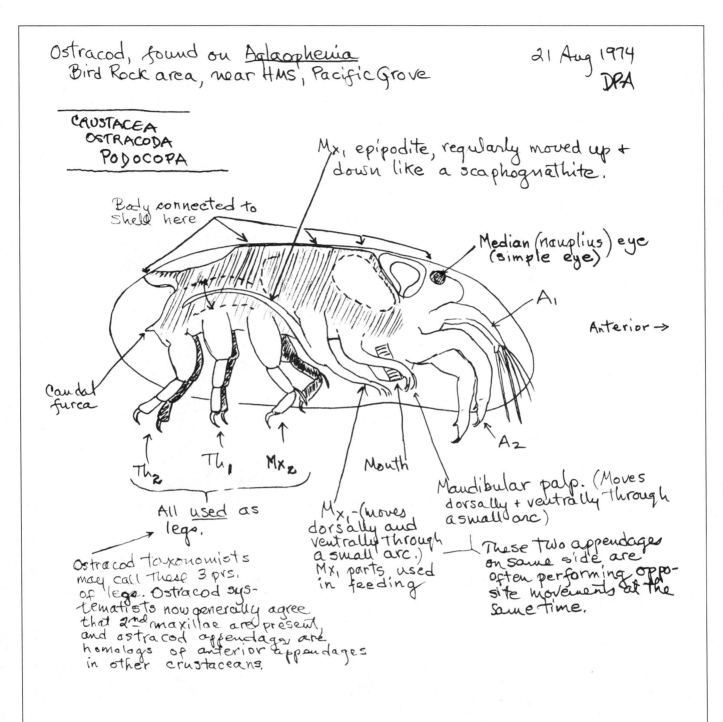

Ostracod —

microfauna, Agassiz beach + land edge of channel by Pete's Island, HMS, 17 Aug 1969.

A_1 — extended anteriorly from shell as animal moves, + alternately tapped on substratum. Appear used as sensory appendages rather than as walking appendages, tho' potentially usable in locomotion.

Eye

Action of bailer visible here

CRUSTACEA
OSTRACODA

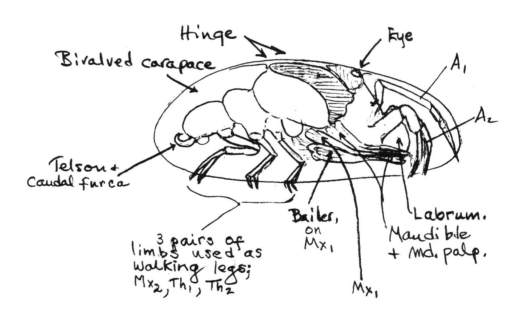

Hinge
Bivalved carapace
Eye
A_1
A_2
Telson + Caudal furca
3 pairs of limbs used as walking legs; Mx_2, Th_1, Th_2
Bailer, on Mx_1
Mx_1
Labrum.
Mandible + Md. palp.

Ostracods have 5–7 pr. limbs; homologies of posterior ones best left to specialists. For appendage nomenclature see — Cohen, A.C., 1982. Ostracoda, pp. 181–202, in: S. P. Parker, ed., Synopsis + classification of living organisms, vol. 2. McGraw Hill, N.Y. see esp. p. 181.

With thanks for help from Anne Cohen

DPA.

Arthropoda/Crustacea/Copepoda

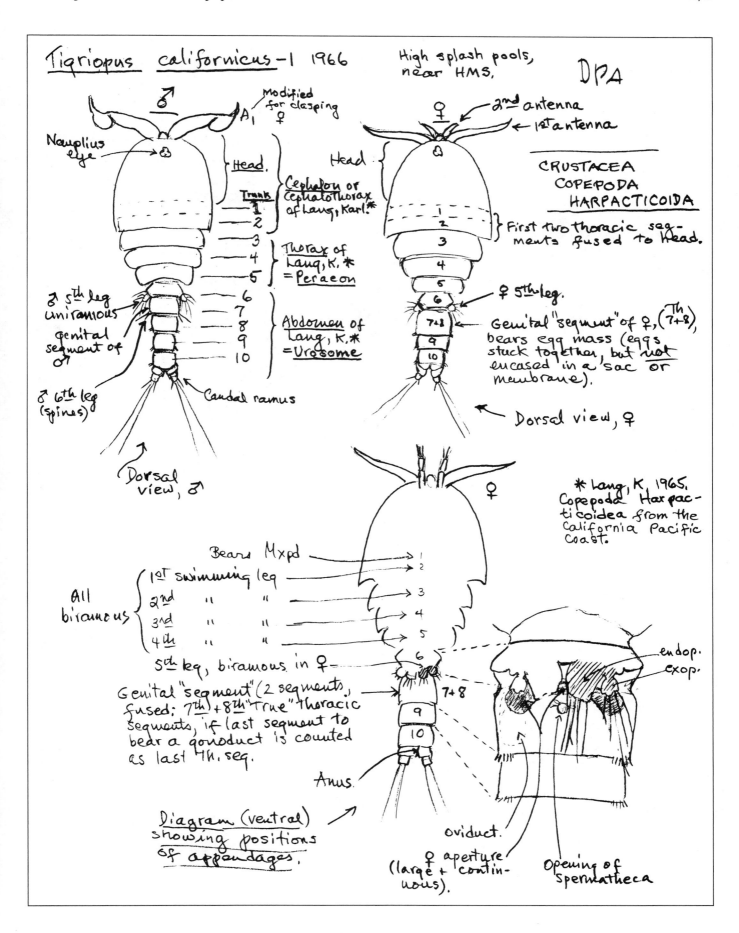

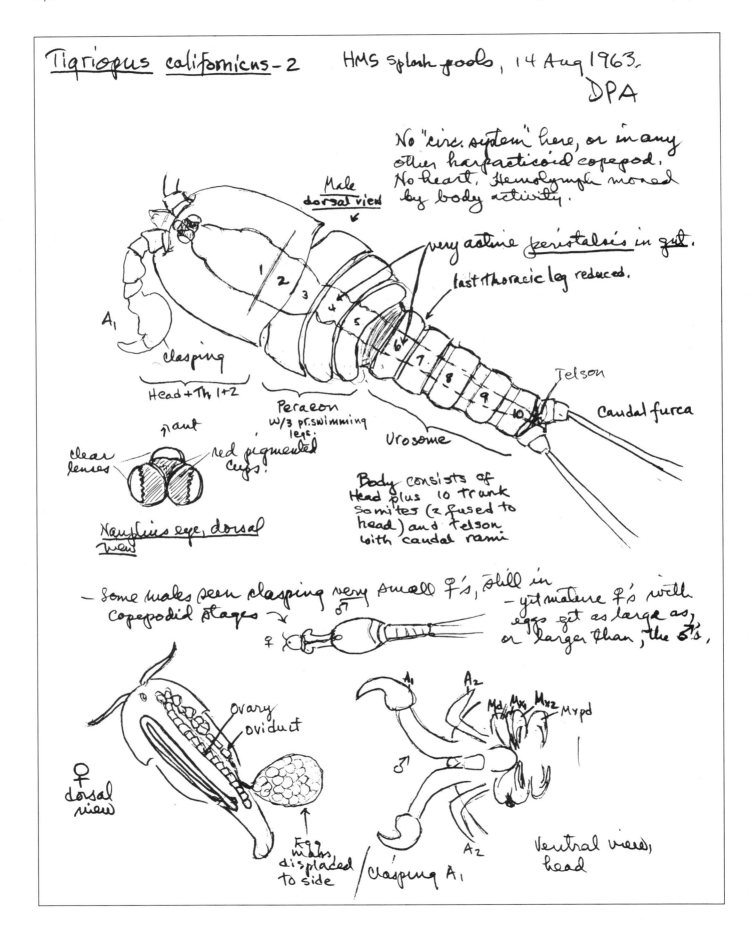

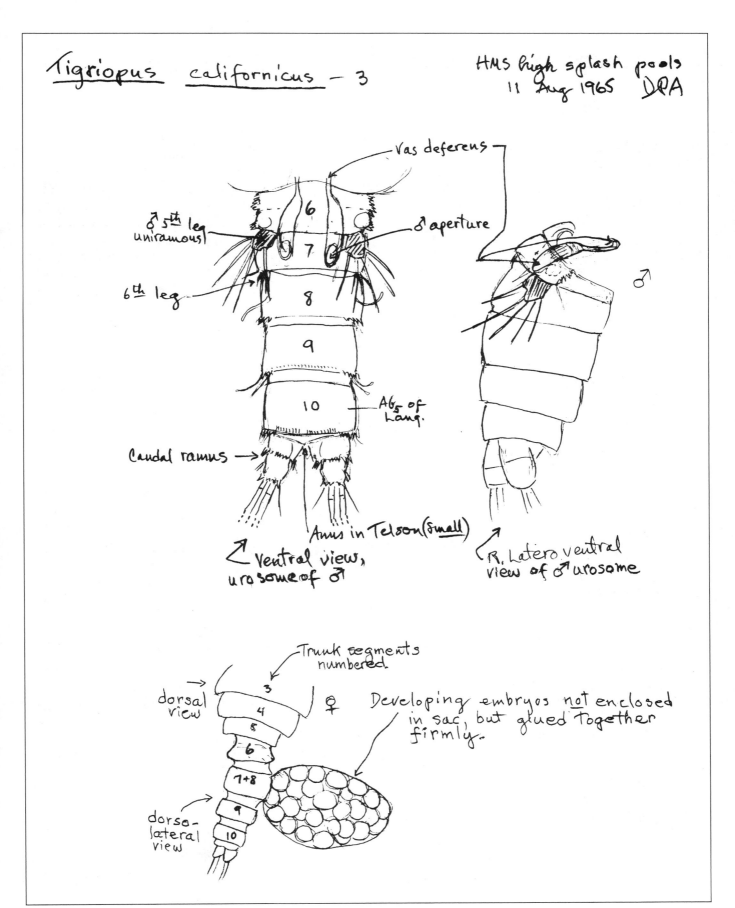

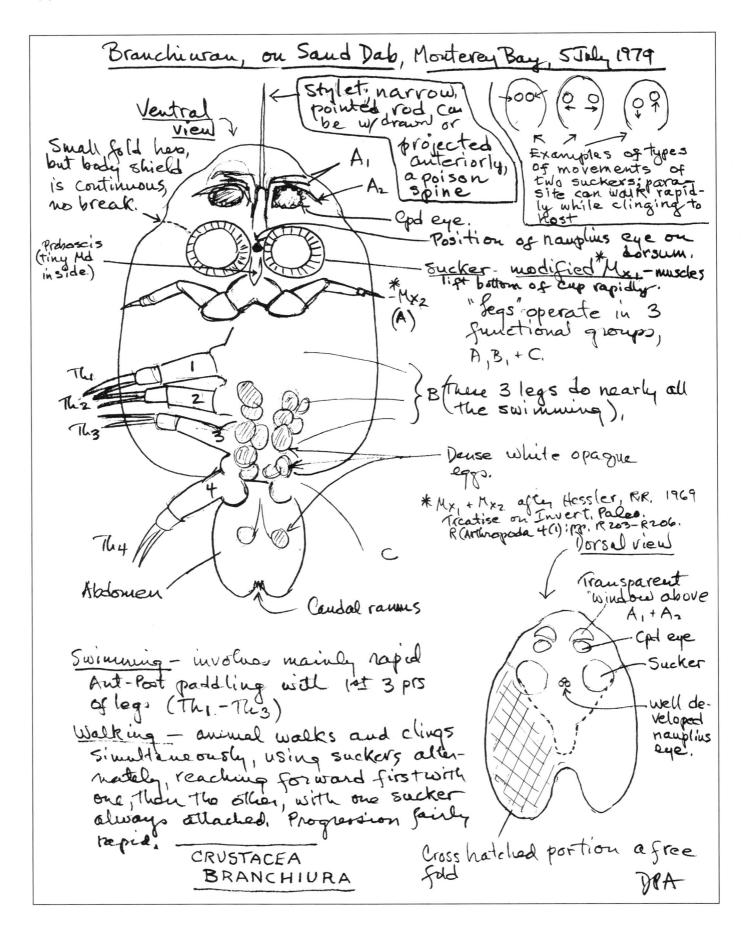

Arthropoda/Crustacea/Cirripedia/Lepadomorpha

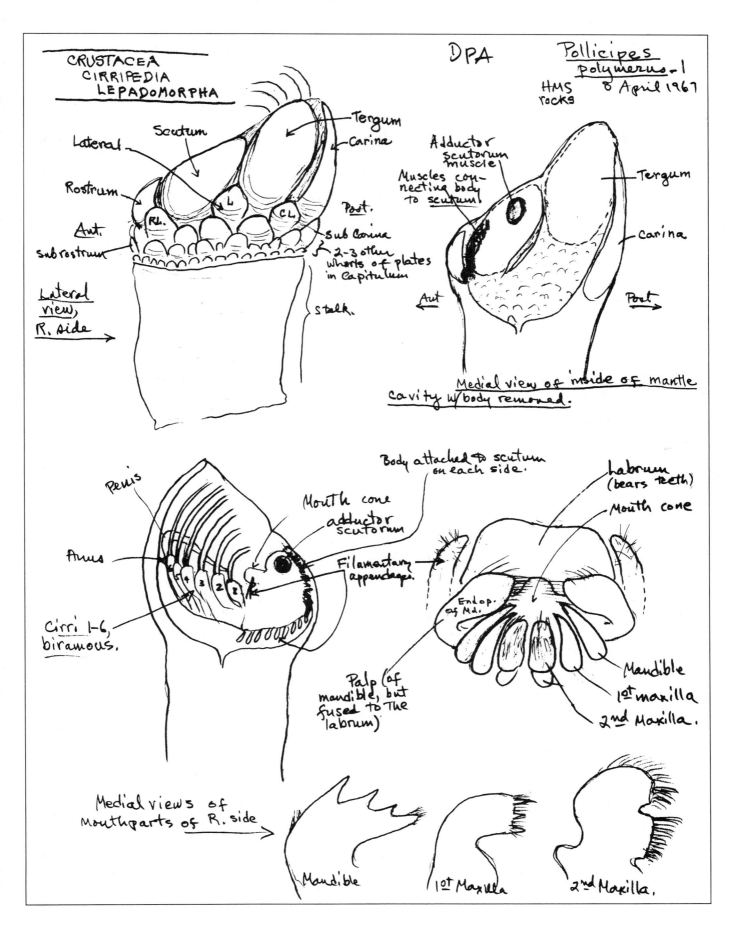

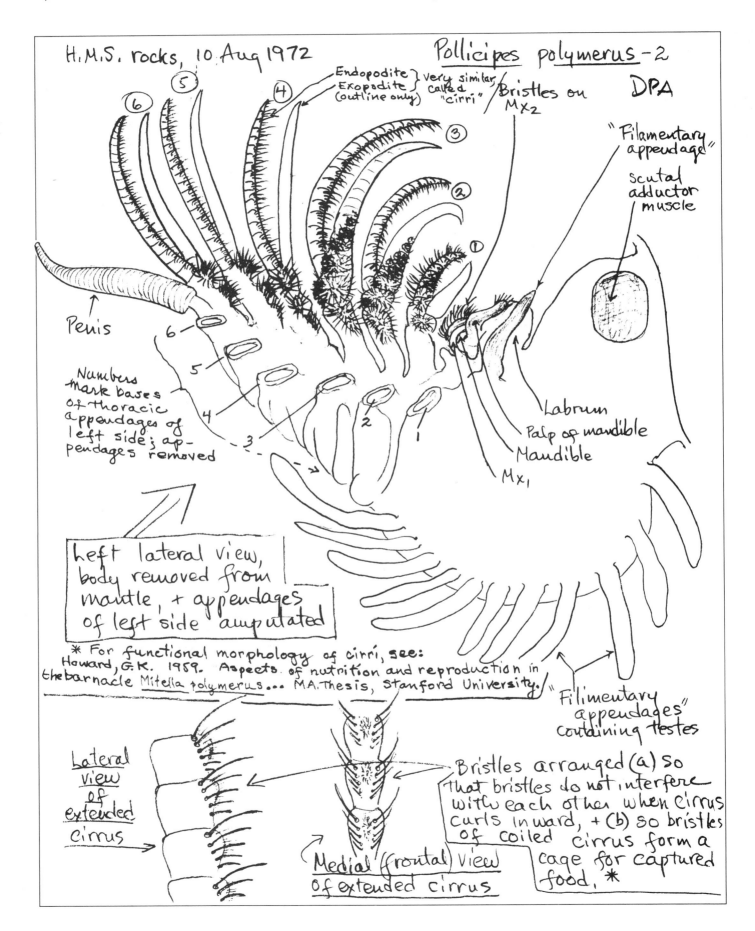

Arthropoda/Crustacea/Cirripedia/Lepadomorpha

Pollicipes polymerus — 3   DPA

Food. Gut contains larger crustaceans (amphipods, shrimps) as well as smaller organisms. Under lab conditions *Pollicipes* captures + eats *Artemia* as does *Lepas*, though latter is a more active predator.

Lateral view, gut + gonads: body removed from mantle cavity

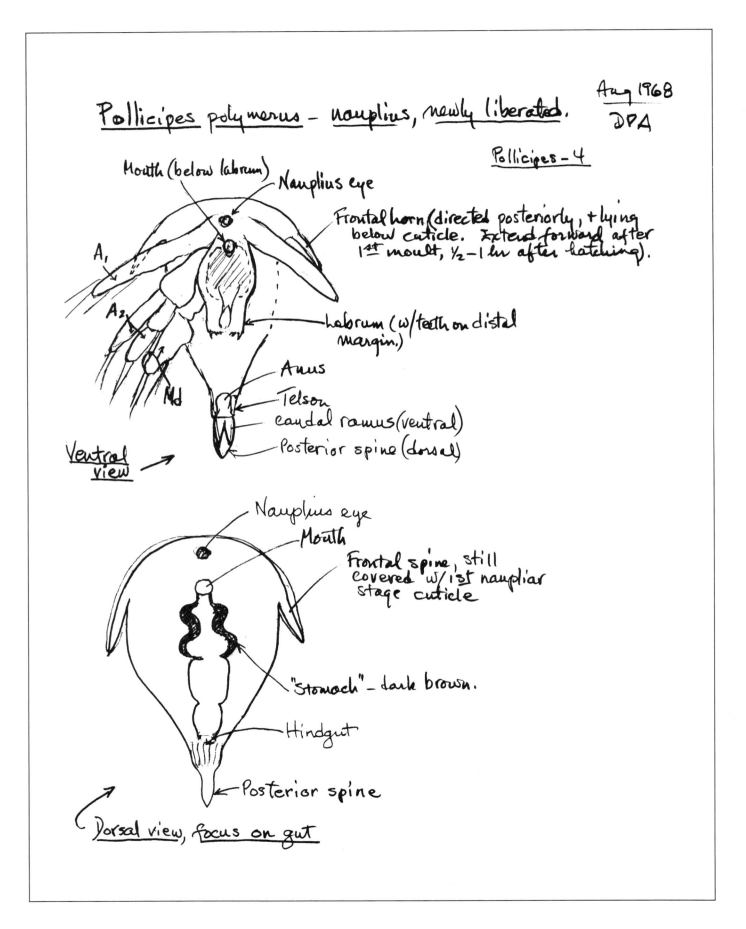

Arthropoda/Crustacea/Cirripedia/Balanomorpha

Catophragmus

CRUSTACEA
CIRRIPEDIA
BALANOMORPHA
CHTHAMALOIDEA

Taken 13 May 1968, rocky intertidal, small islet in Brasilito Bay, west coast of Costa Rica, Te Vega Cruise 18, Sta. 18. JPA

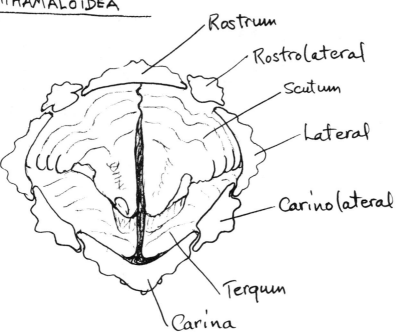

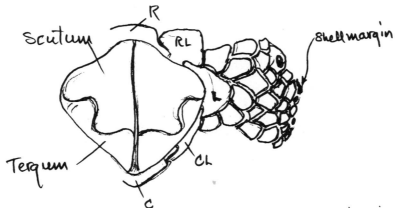

This is one of the most primitive of the operculate barnacles.

Aug 1969.

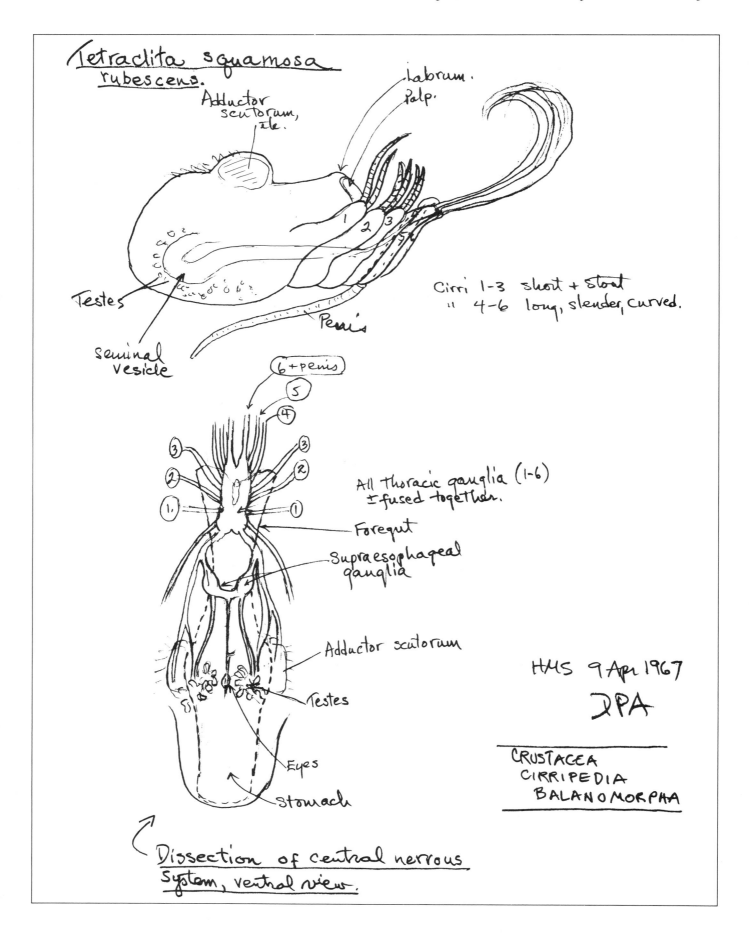

Arthropoda/Crustacea/Cirripedia/Balanomorpha

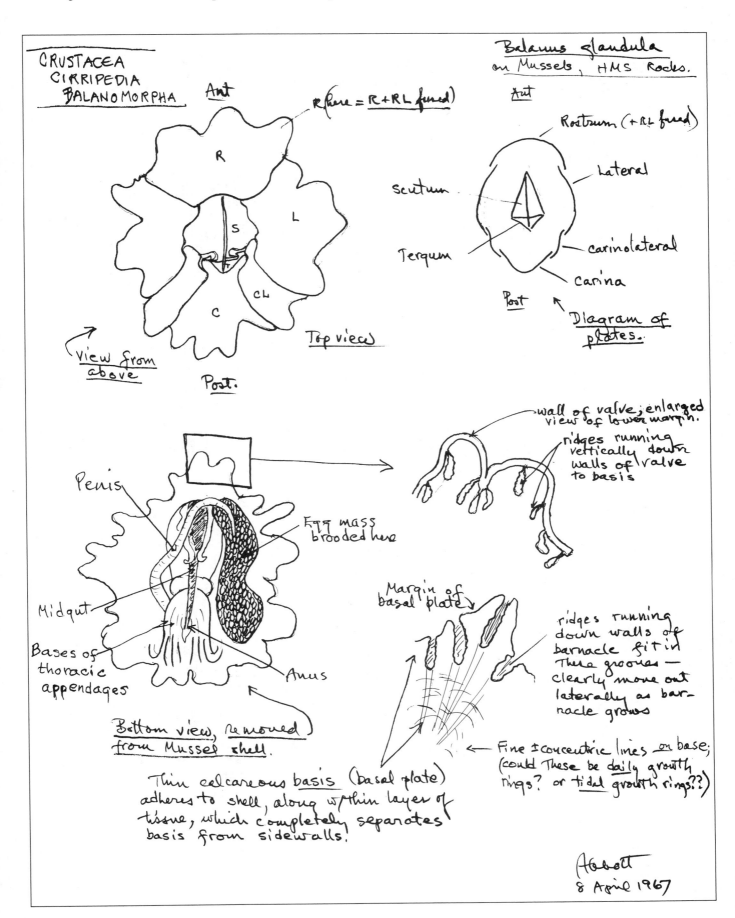

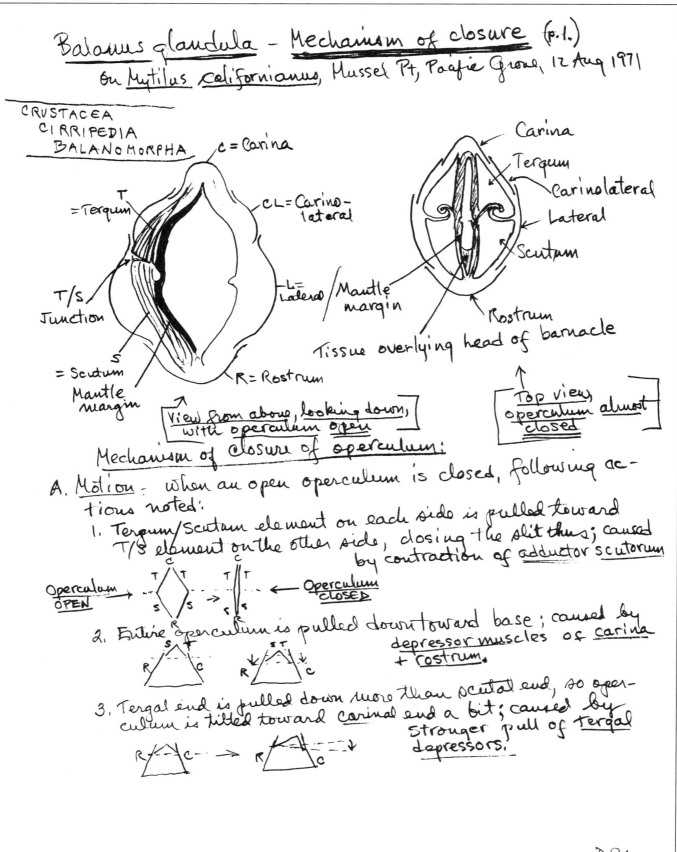

Balanus glandula — Mechanism of closure (p.1.)
On *Mytilus californianus*, Mussel Pt, Pacific Grove, 12 Aug 1971

CRUSTACEA
 CIRRIPEDIA
  BALANOMORPHA

C = Carina
CL = Carino-lateral
T = Tergum
T/S Junction
S = Scutum
L = Lateral / Mantle margin
R = Rostrum
Mantle margin

[View from above, looking down, with operculum open]

Carina
Tergum
Carinolateral
Lateral
Scutum
Rostrum
Tissue overlying head of barnacle

[Top view, operculum almost closed]

Mechanism of closure of operculum:

A. **Motion** — when an open operculum is closed, following actions noted:

1. Tergum/scutum element on each side is pulled toward T/S element on the other side, closing the slit thus; caused by contraction of <u>adductor scutorum</u>

   Operculum OPEN → ... → Operculum CLOSED

2. Entire operculum is pulled down toward base; caused by depressor muscles of <u>carina</u> + <u>rostrum</u>.

3. Tergal end is pulled down more than scutal end, so operculum is tilted toward carinal end a bit; caused by stronger pull of tergal depressors.

DPA

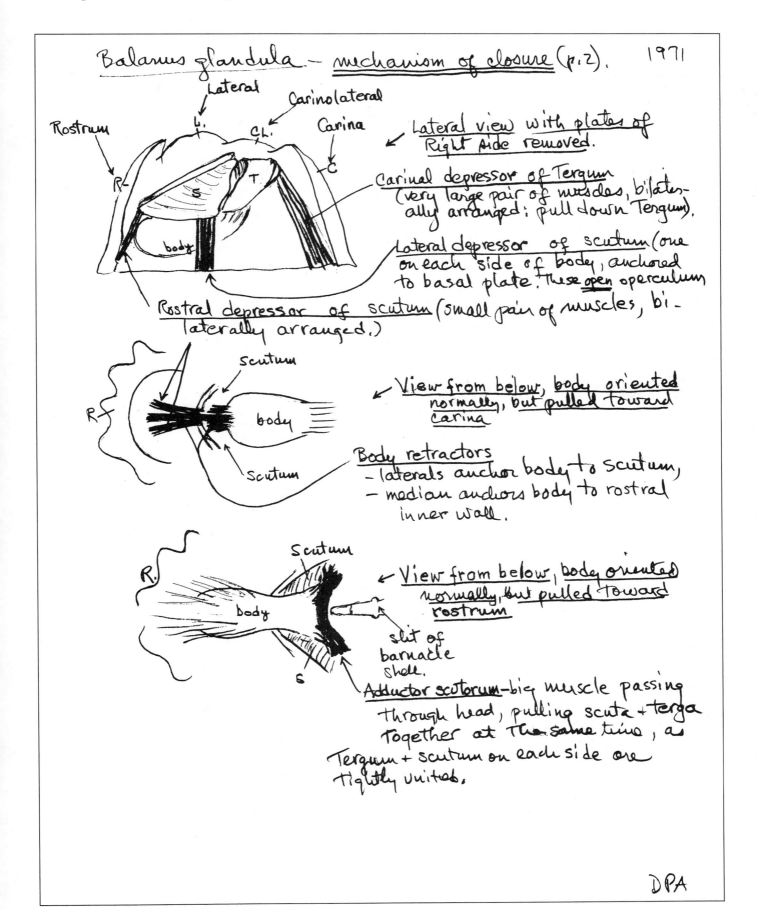

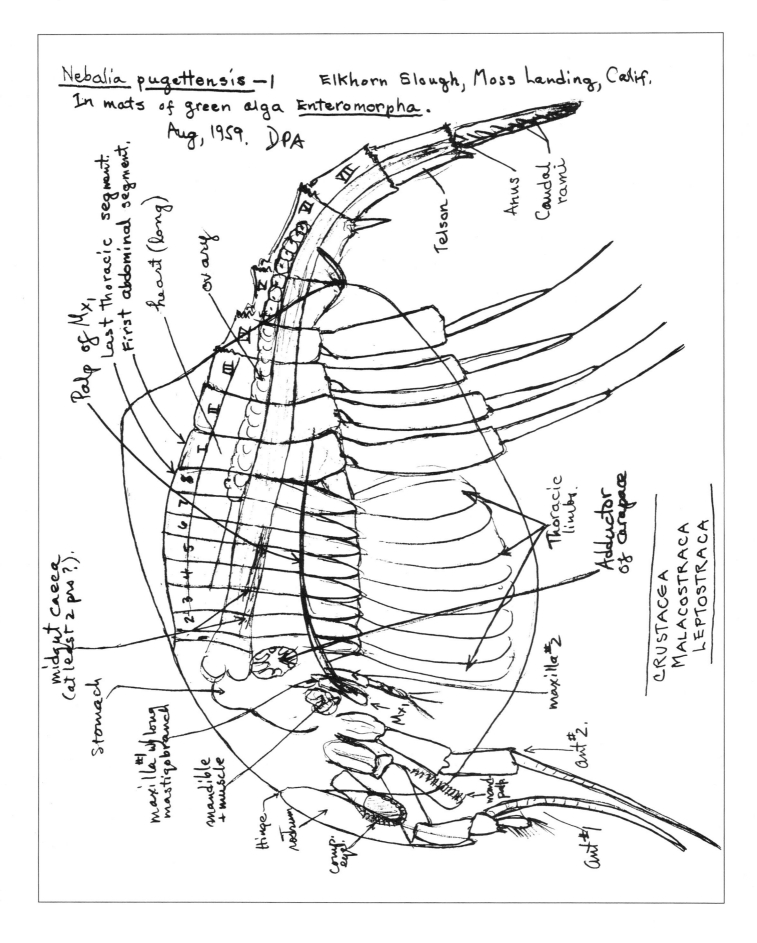

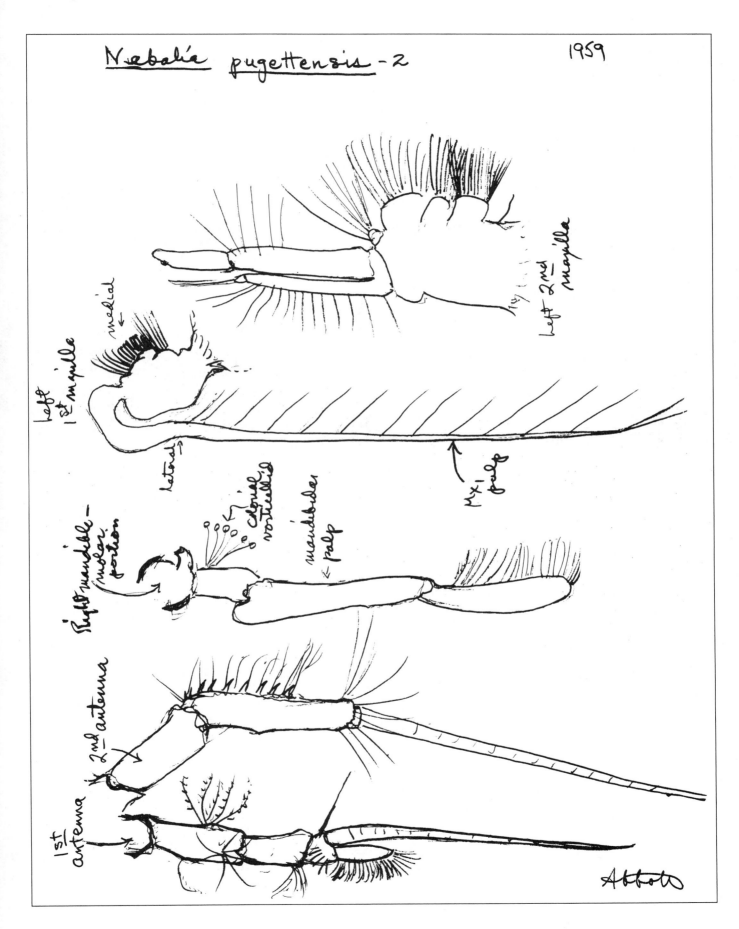

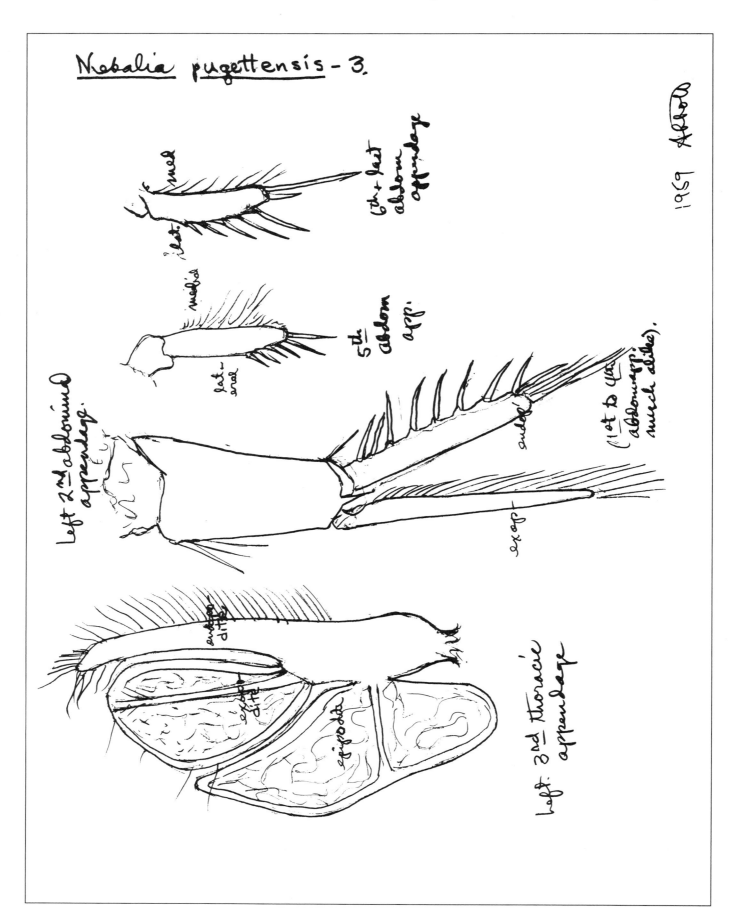

Arthropoda/Crustacea/Malacostraca/Leptostraca

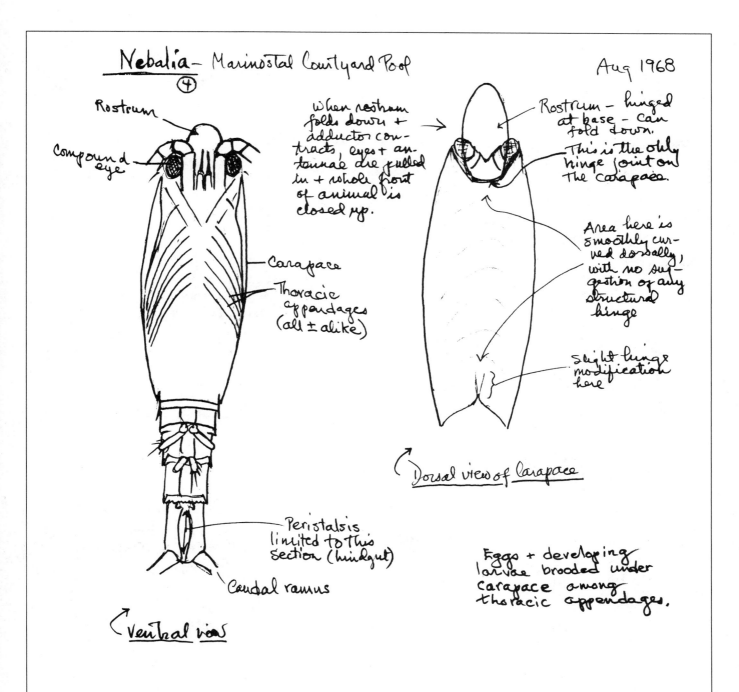

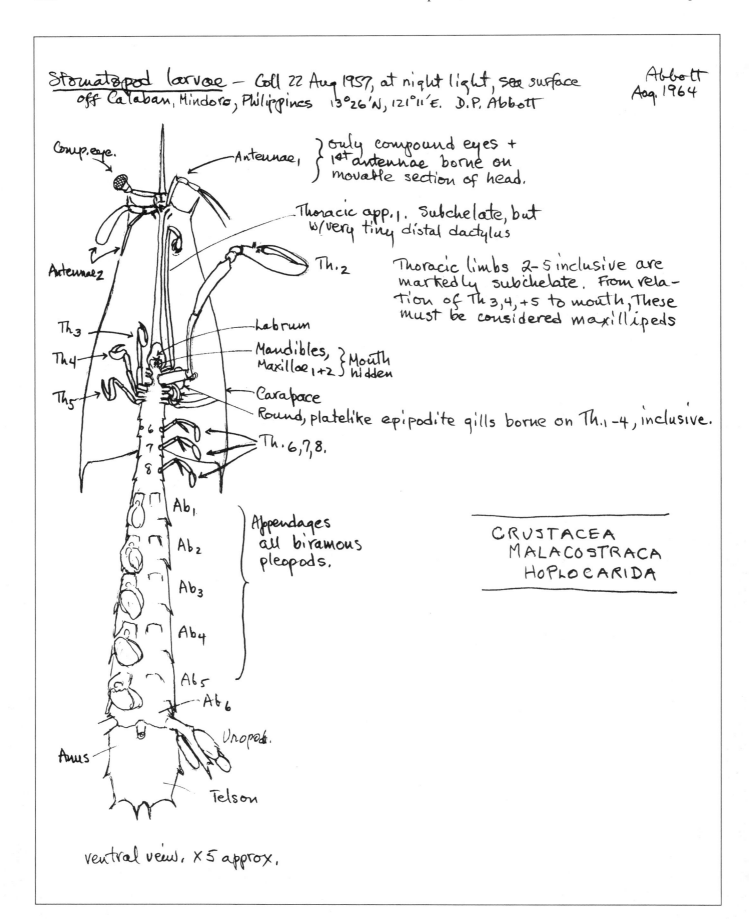

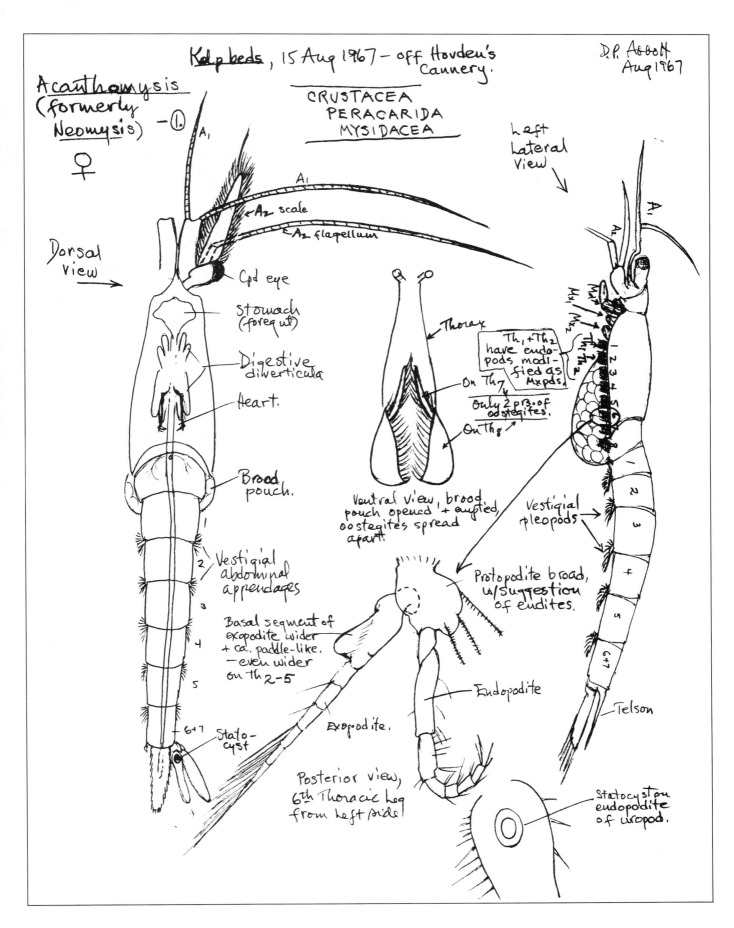

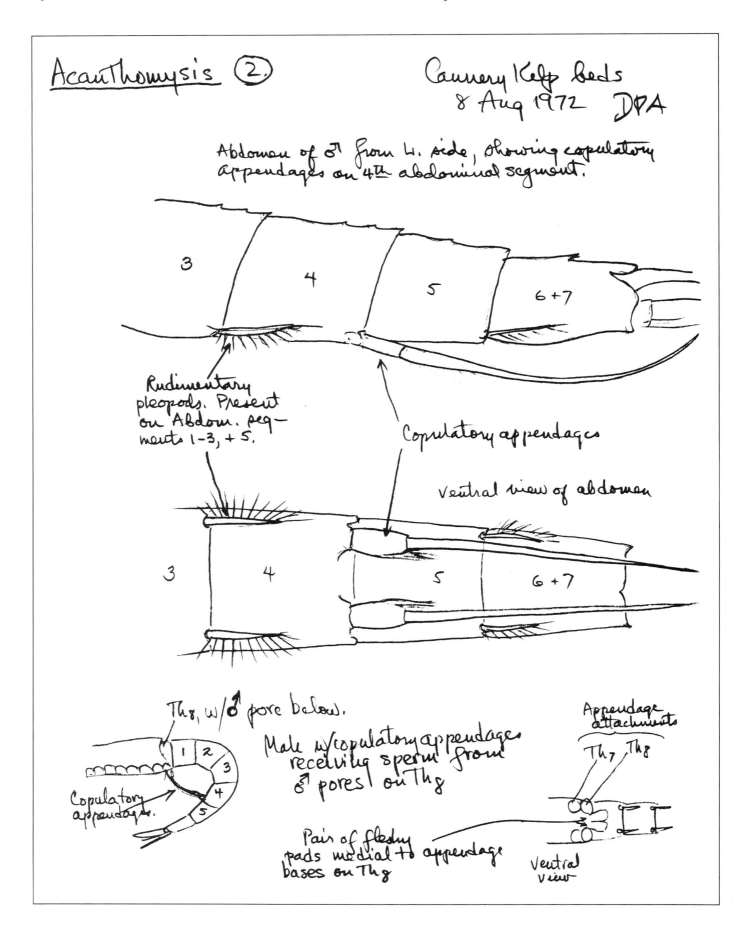

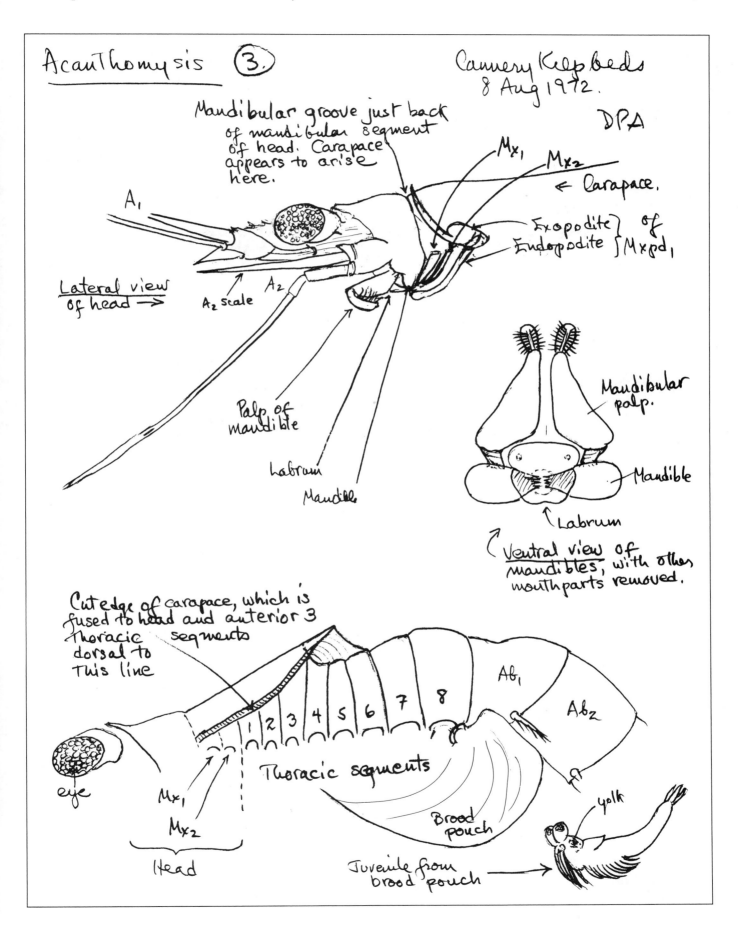

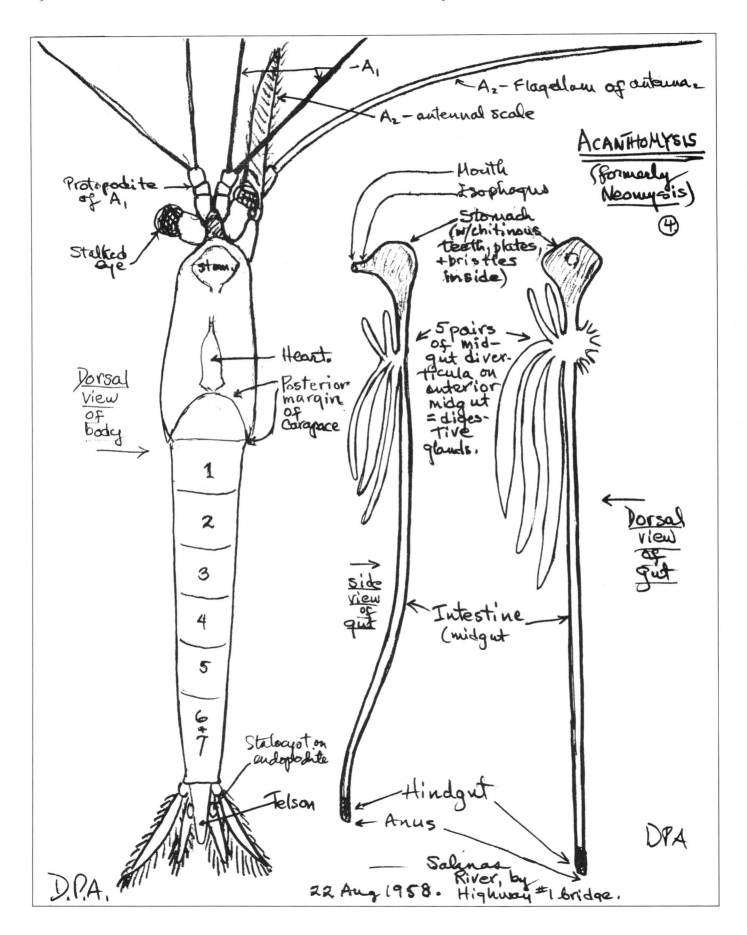

Arthropoda/Crustacea/Malacostraca/Isopoda

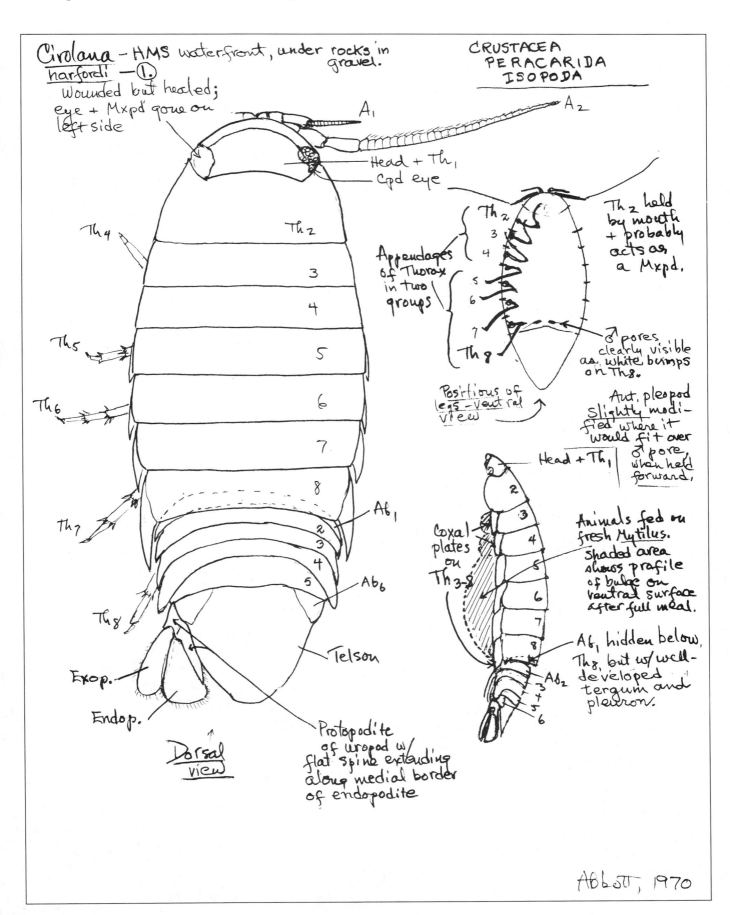

Arthropoda/Crustacea/Malacostraca/Isopoda

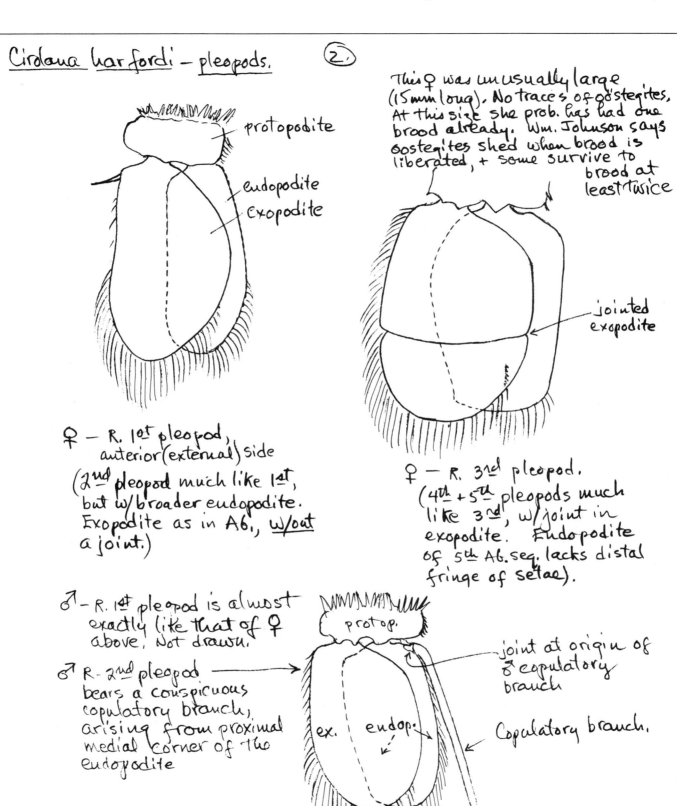

Abbott, 1970

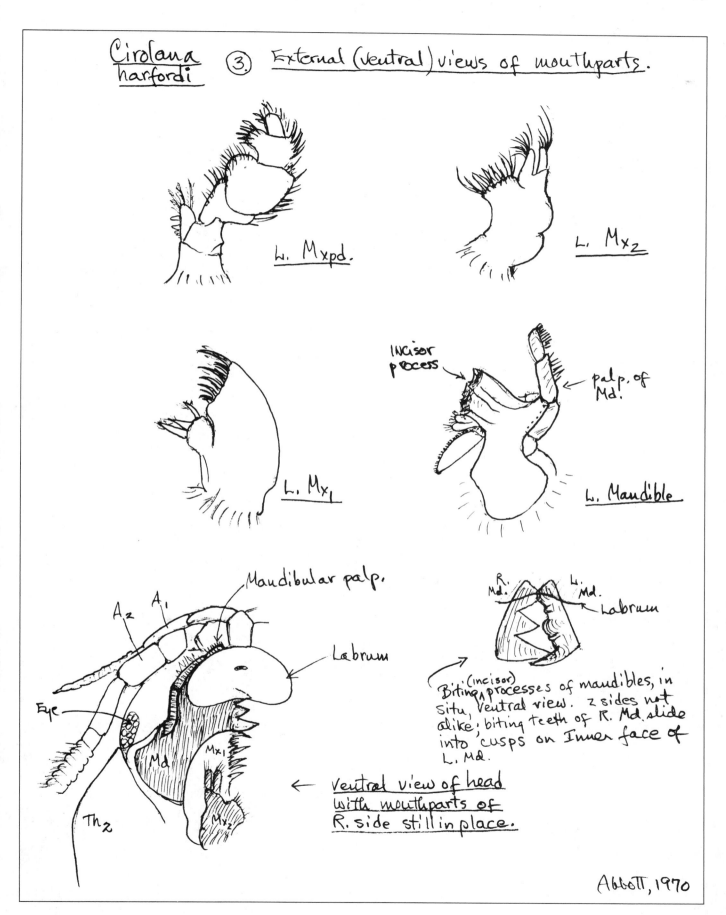

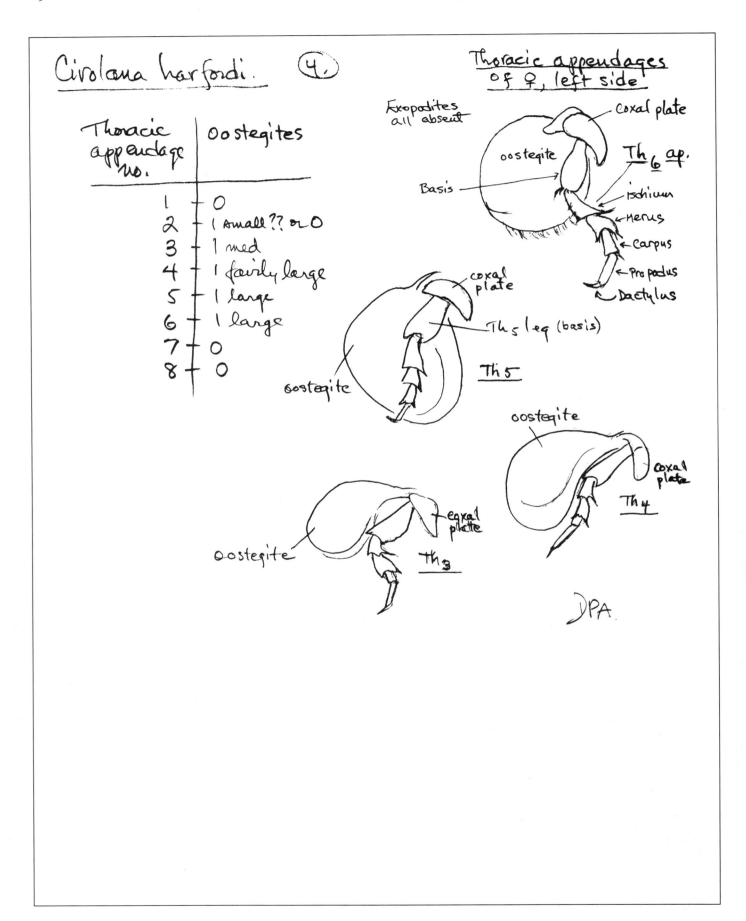

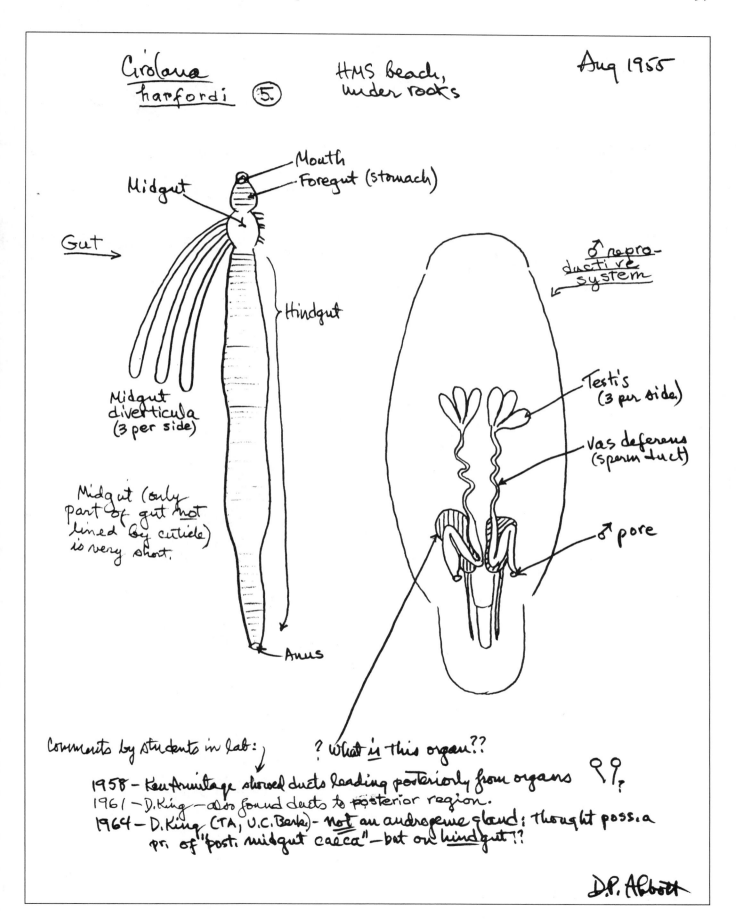

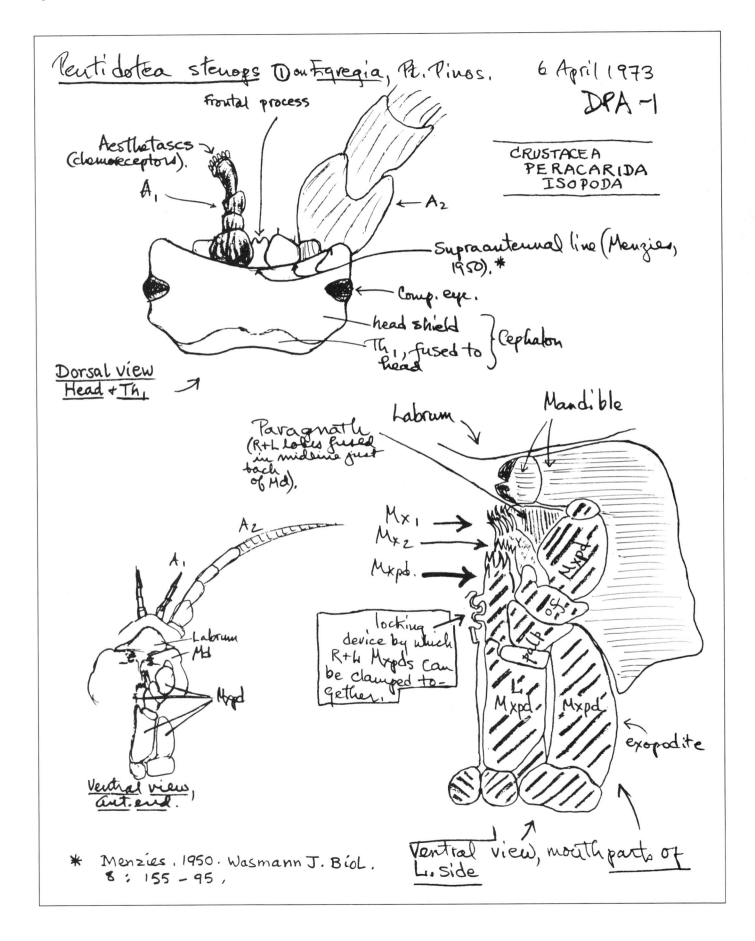

Pentidotea stenops - ♀ on *Egregia*, Pt. Pinos.   DPA 1973

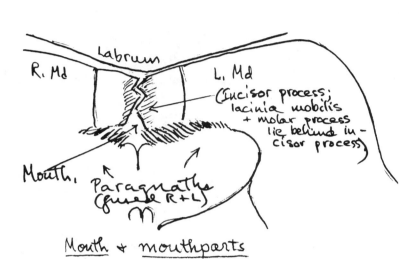

Mouth & mouthparts

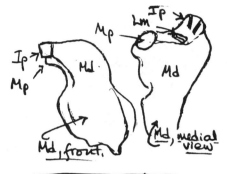

Ip = Incisor process of Md
Mp = Molar process of Md
Lm = Lacinia mobilis of Md

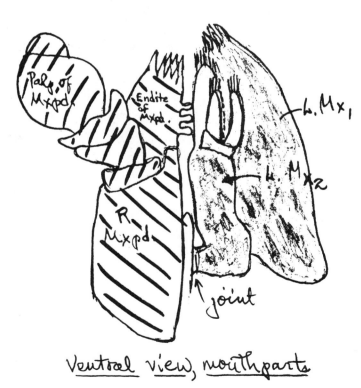

Ventral view, mouthparts

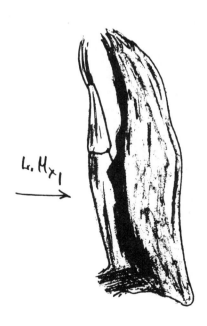

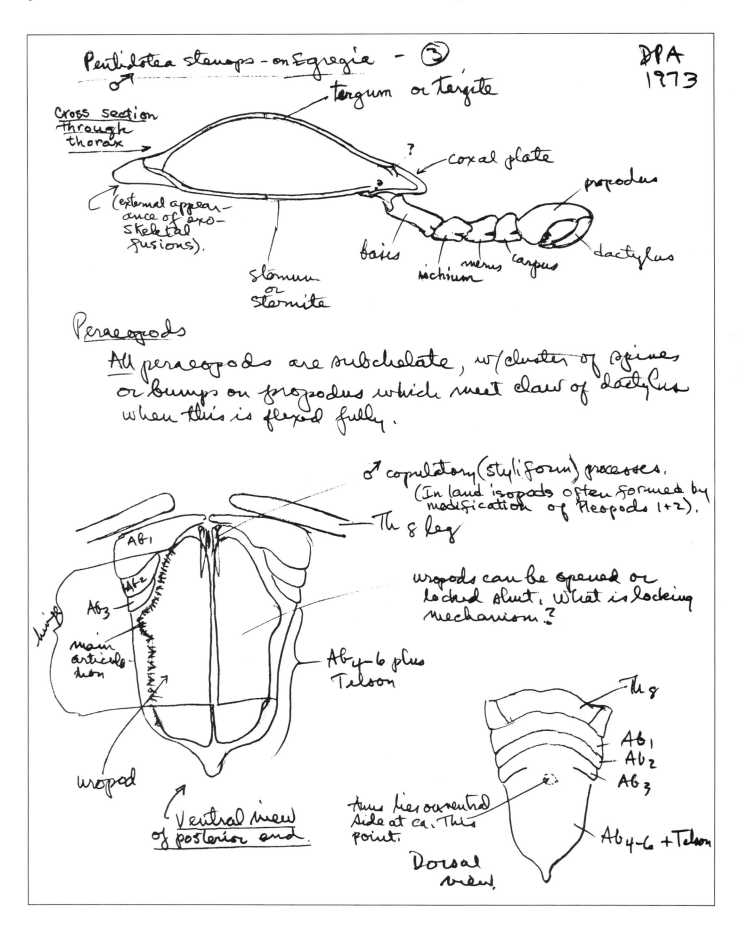

Arthropoda/Crustacea/Malacostraca/Isopoda

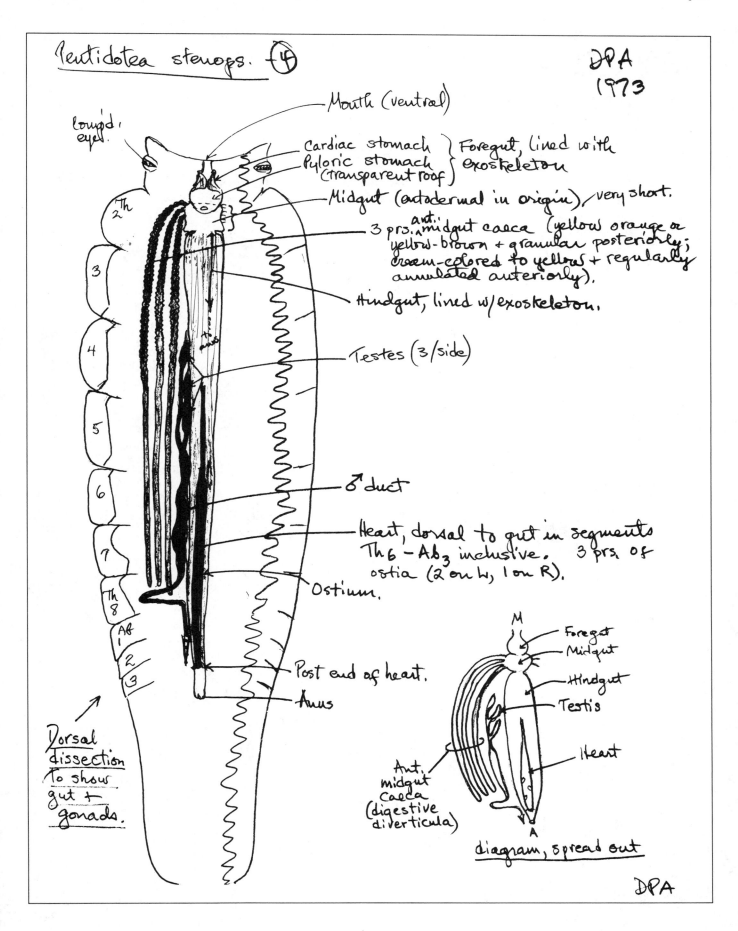

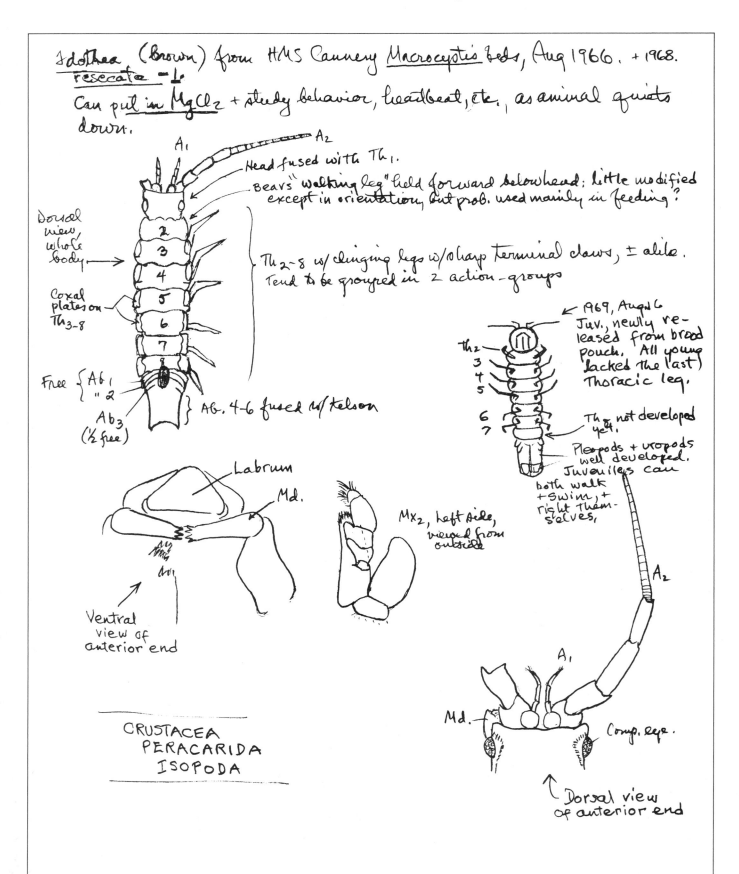

Arthropoda/Crustacea/Malacostraca/Isopoda

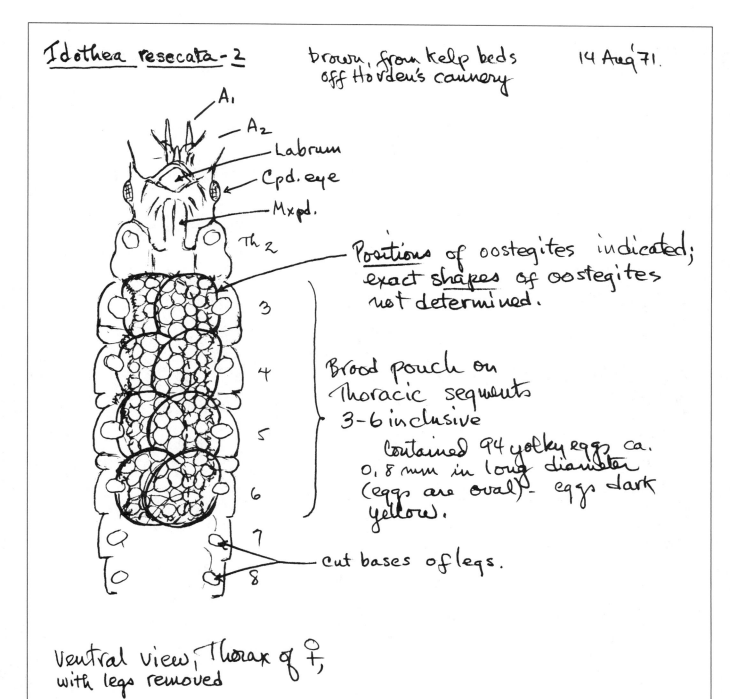

Idothea resecata - 2   brown, from kelp beds off Hovden's cannery   14 Aug '71

Labels: A_1, A_2, Labrum, Cpd. eye, Mxpd., Th_2, 3, 4, 5, 6, 7, 8

Positions of oostegites indicated; exact shapes of oostegites not determined.

Brood pouch on thoracic segments 3-6 inclusive. Contained 94 yolky eggs ca. 0.8 mm in long diameter (eggs are oval) - eggs dark yellow.

cut bases of legs.

Ventral view, Thorax of ♀, with legs removed

DPA

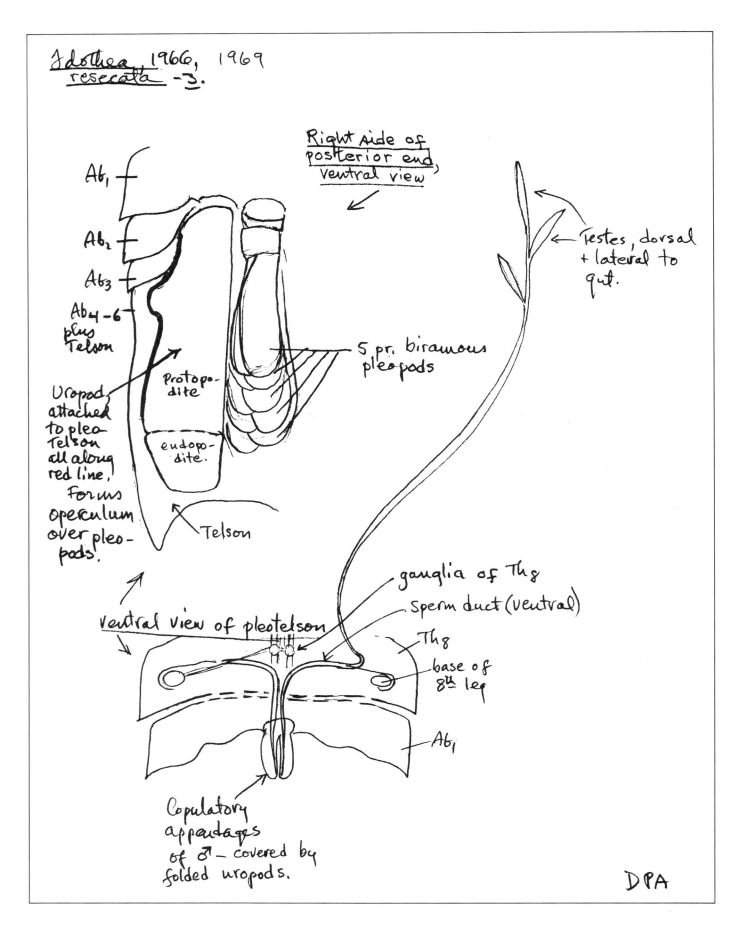

Arthropoda/Crustacea/Malacostraca/Isopoda

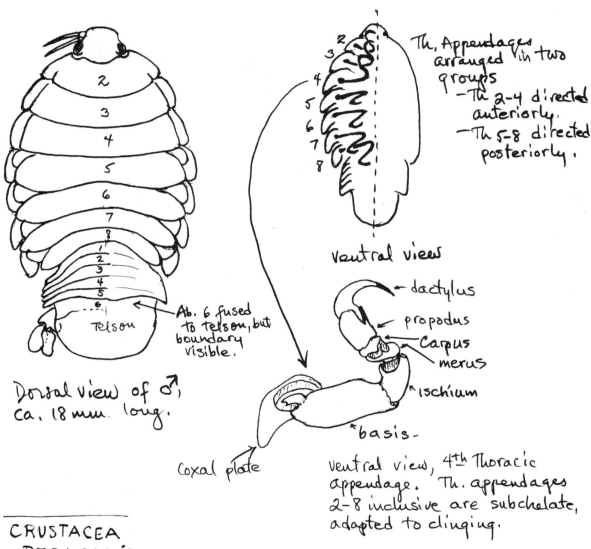

Parasitic isopods from lingcod, Monterey Bay, ca. mid-August, 1973 (coll. Scott Braude).

Dorsal view of ♂, ca. 18 mm. long.

Ab. 6 fused to telson, but boundary visible.

Th. Appendages arranged in two groups
- Th. 2-4 directed anteriorly.
- Th. 5-8 directed posteriorly.

ventral view

dactylus
propodus
carpus
merus
ischium
basis
Coxal plate

Ventral view, 4th Thoracic appendage. Th. appendages 2-8 inclusive are subchelate, adapted to clinging.

CRUSTACEA
PERACARIDA
ISOPODA

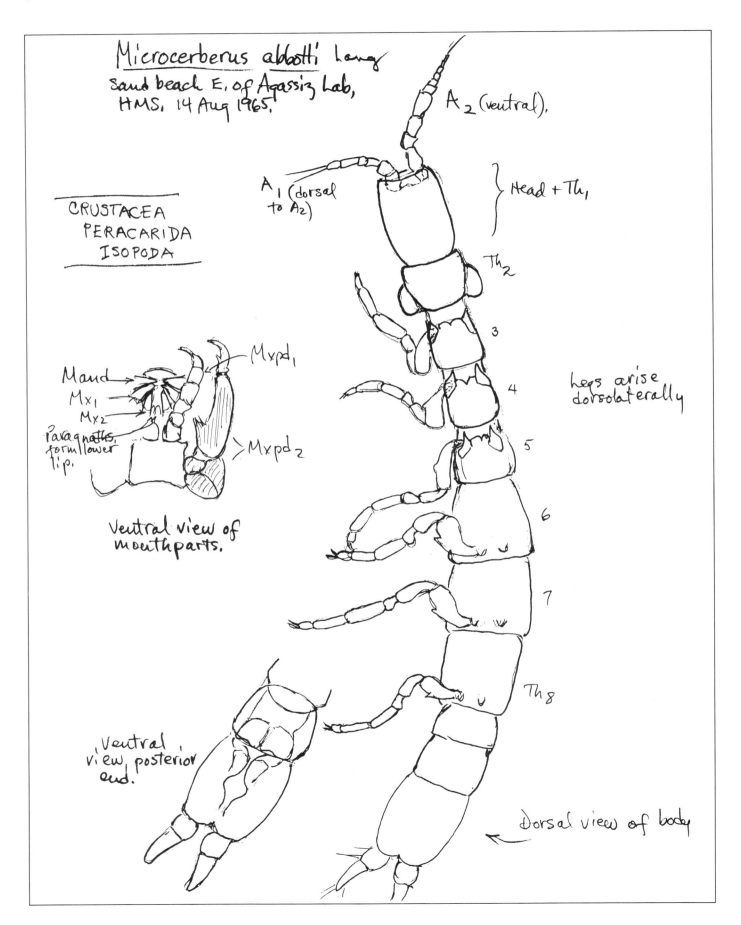

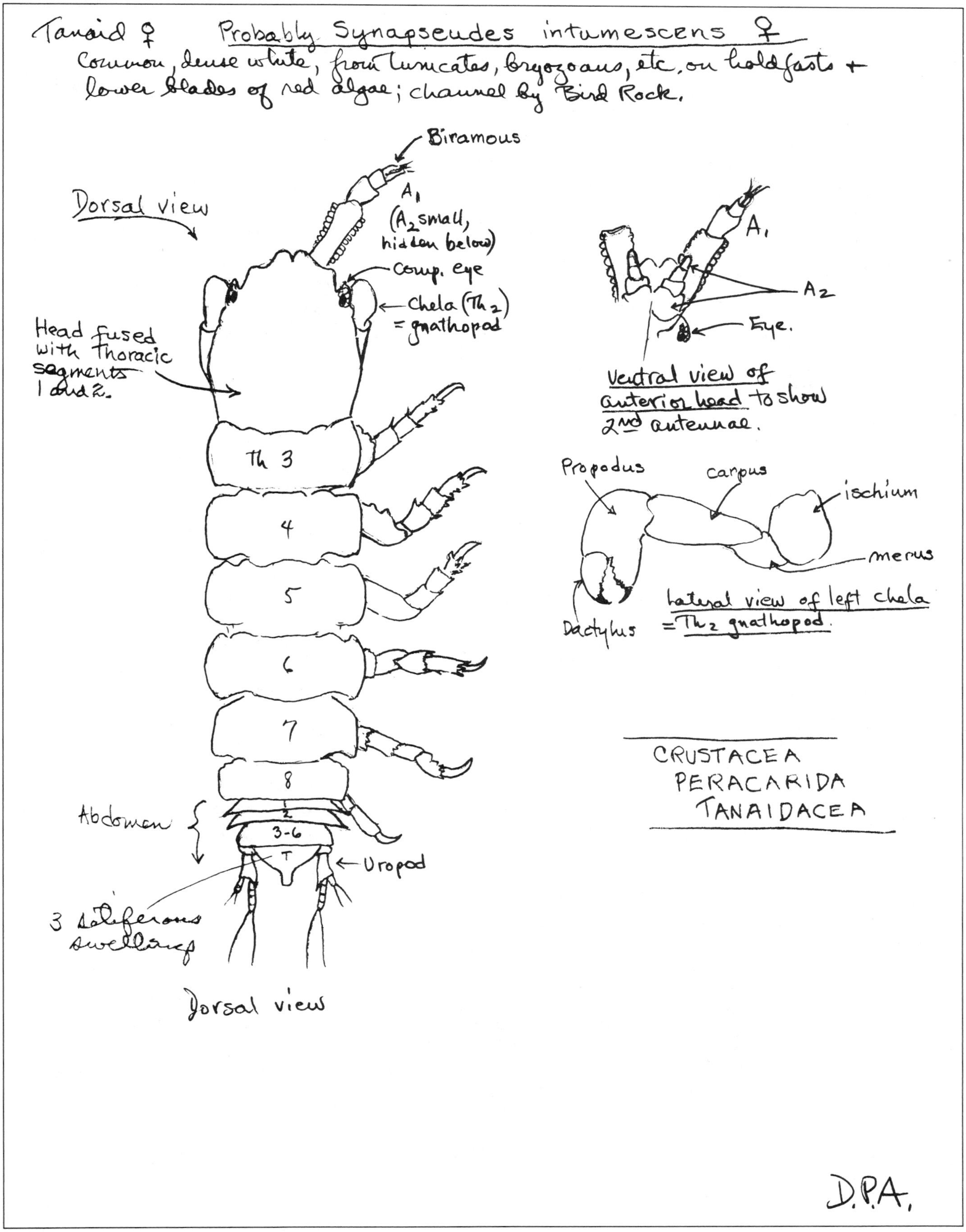

Arthropoda/Crustacea/Malacostraca/Tanaidacea

Tanaid ♀ — Probably *Synapseudes intumescens* ♀
Common, dense white, from tunicates, bryozoans, etc., on holdfasts + lower blades of red algae; channel by Bird Rock.

Dorsal view

- Biramous
- A₁
- (A₂ small, hidden below)
- Comp. eye
- Chela (Th₂) = gnathopod

Head fused with thoracic segments 1 and 2.

Th 3
4
5
6
7
8

Abdomen { 2, 3-6, T }

3 saliferous swellings

← Uropod

Dorsal view

Ventral view of anterior head to show 2nd antennae.
- A₁
- A₂
- Eye

Lateral view of left chela = Th₂ gnathopod.
- Propodus
- Carpus
- Ischium
- Merus
- Dactylus

CRUSTACEA
PERACARIDA
TANAIDACEA

D.P.A.

Tanaidacea — *Pagurapseudes* sp. — D.P. Abbott, Aug 1967

Washed w/ other mesobiota from red algae w/ hydroids + bryozoans on it, lower intertidal zones. In shells of *Barleeia* + other gastropods. 18 Aug 1967. Bird Rock area, H.M.S.

Righting

Behavior similar to that of hermit crab —
- gnathopods (Th_2) + next pr. of legs (Th_3) used <u>outside</u> of shell in walking
- all other appendages short, used for walking <u>inside</u> shell.
- Large antennae brought to mouthparts & cleaned off.

- Body dextrally coiled, like shells.
- Th_1 & $_2$ fused to head.
- Th_{3-8} w/ "walking legs"
  Th_3 — long legs.
  Th_{4-8} — short legs.
- Ab_{1-5} segments clearly delimited dorsally + laterally. No pleopods (tho' lateral bristle tufts are present).
- Ab_6 fused w/ telson
- Uropods slender.

1 ♀ w/ full "brood-pouch"

Total of 11 embryos present.

Embryos were not stuck together, but enclosed in a large membrane (single?) attached along the ventral side from Th_3 or Th_4 back to Th_6. <u>Could</u> represent fused oostegites??

CRUSTACEA
PERACARIDA
TANAIDACEA

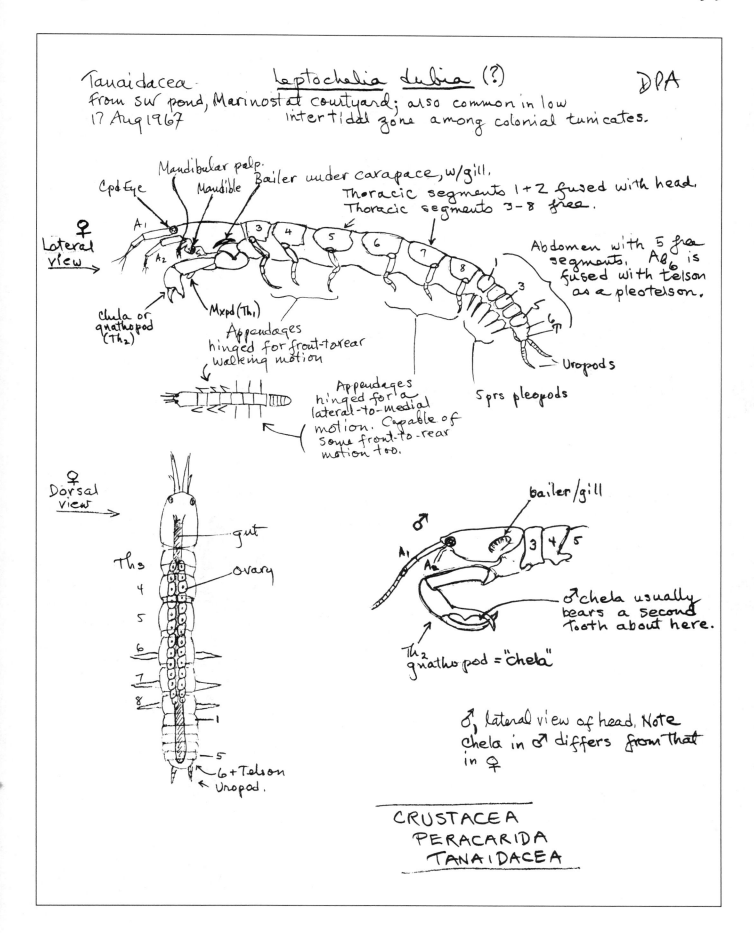

Arthropoda/Crustacea/Malacostraca/Tanaidacea

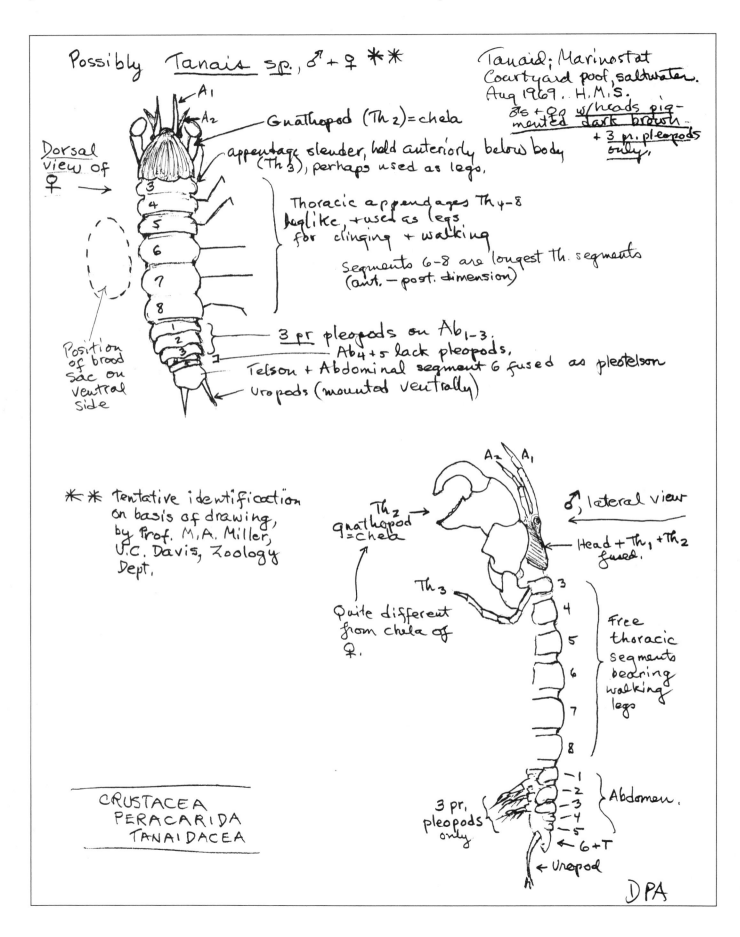

Arthropoda/Crustacea/Malacostraca/Amphipoda

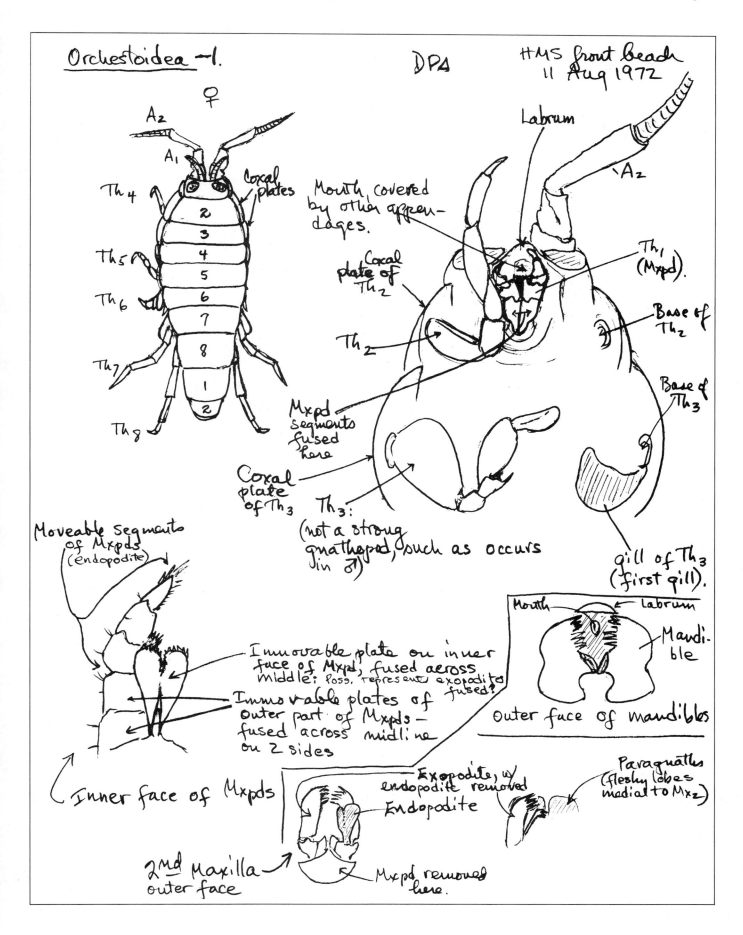

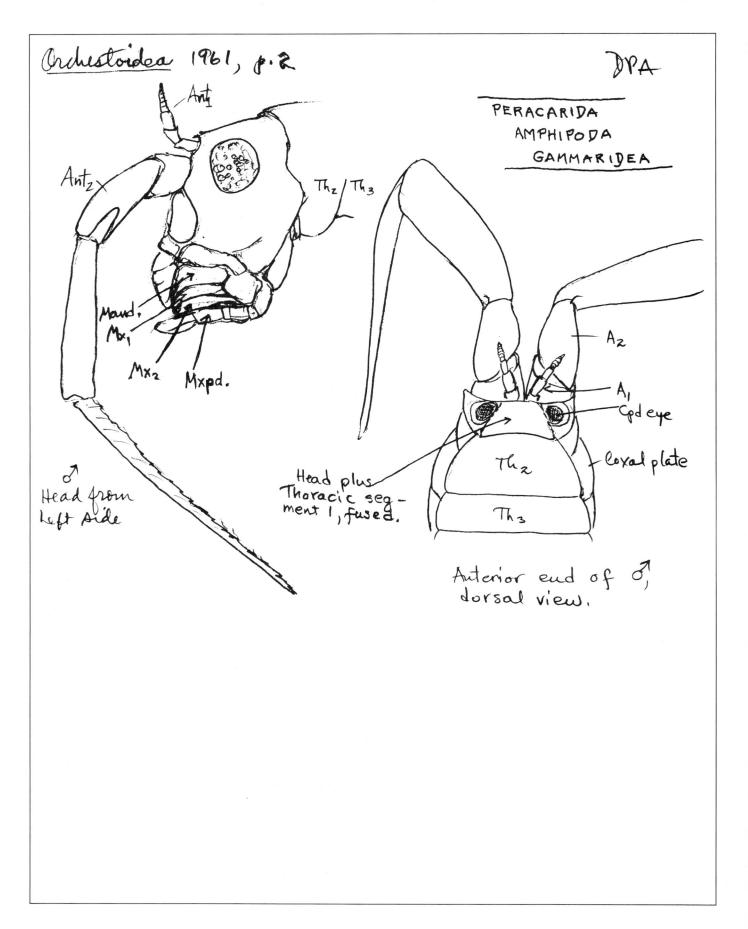

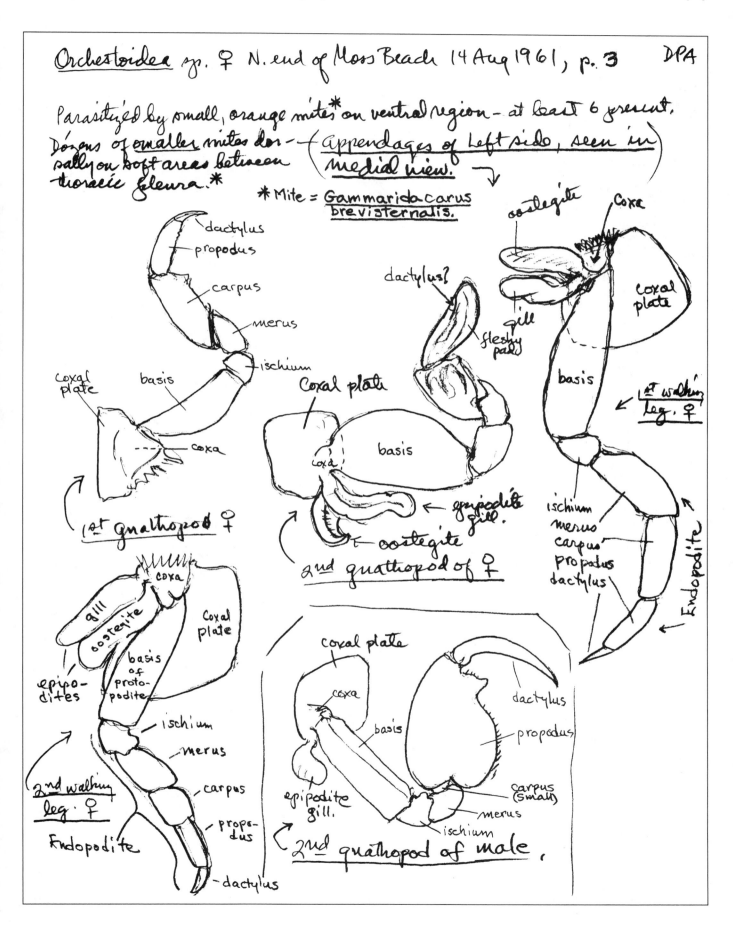

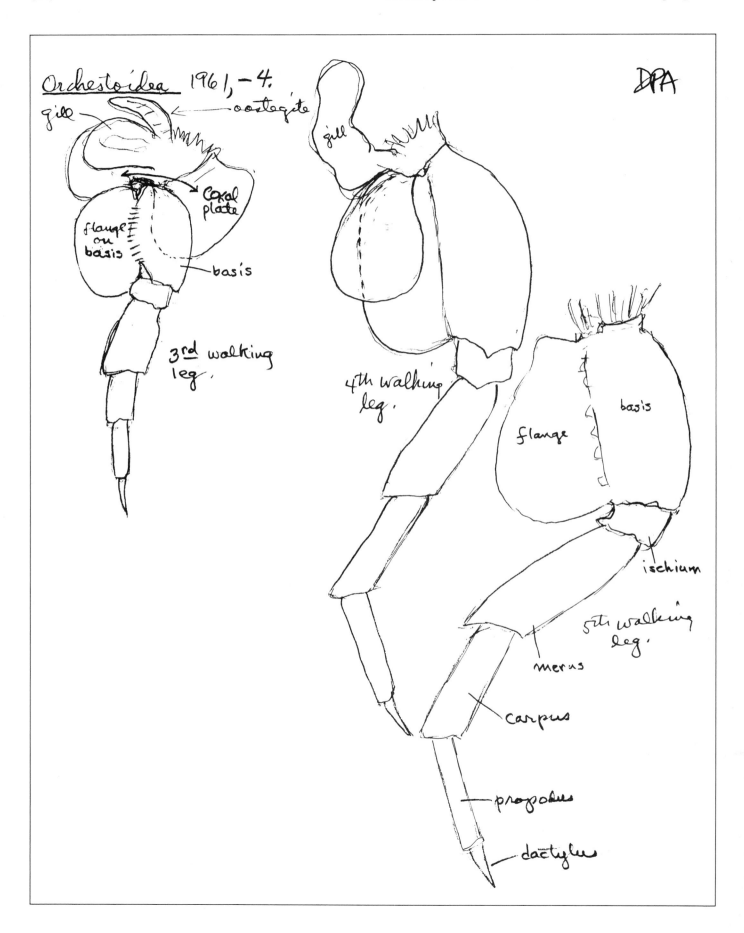

Arthropoda/Crustacea/Malacostraca/Amphipoda

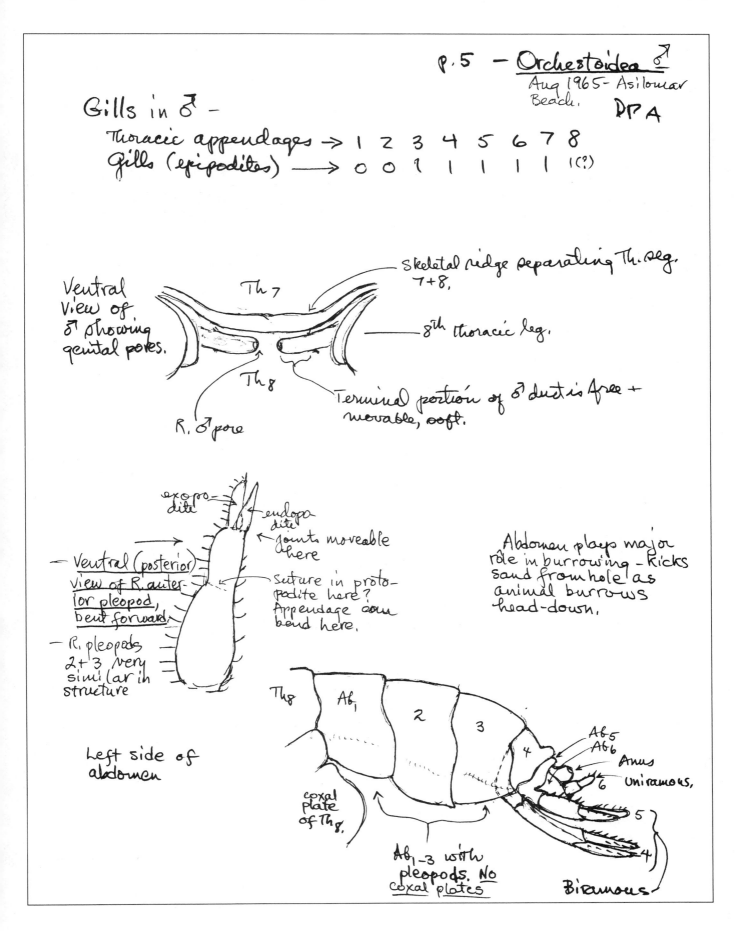

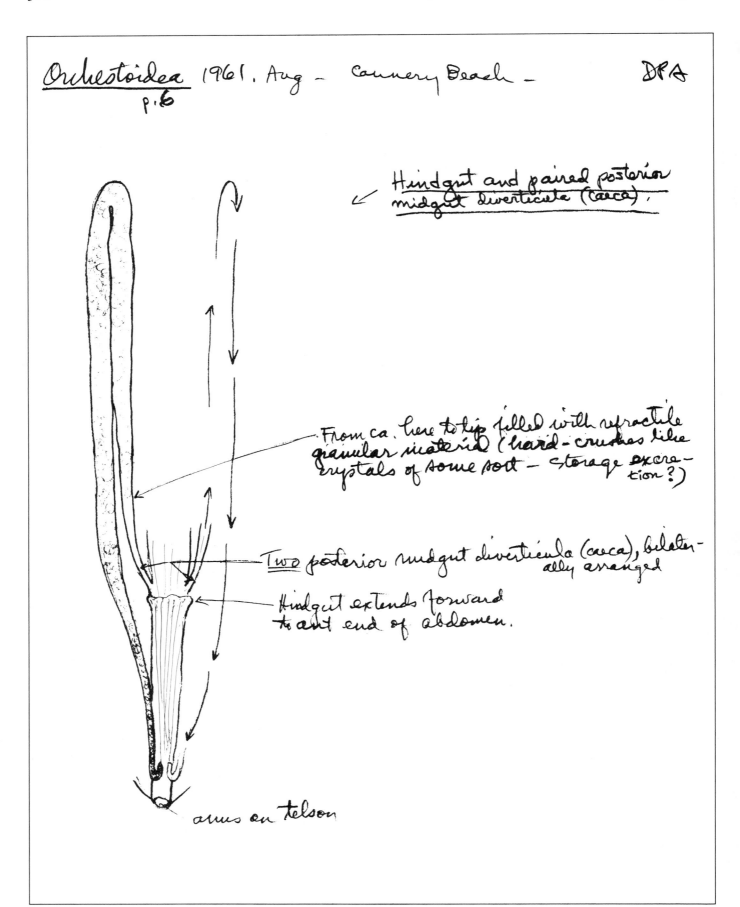

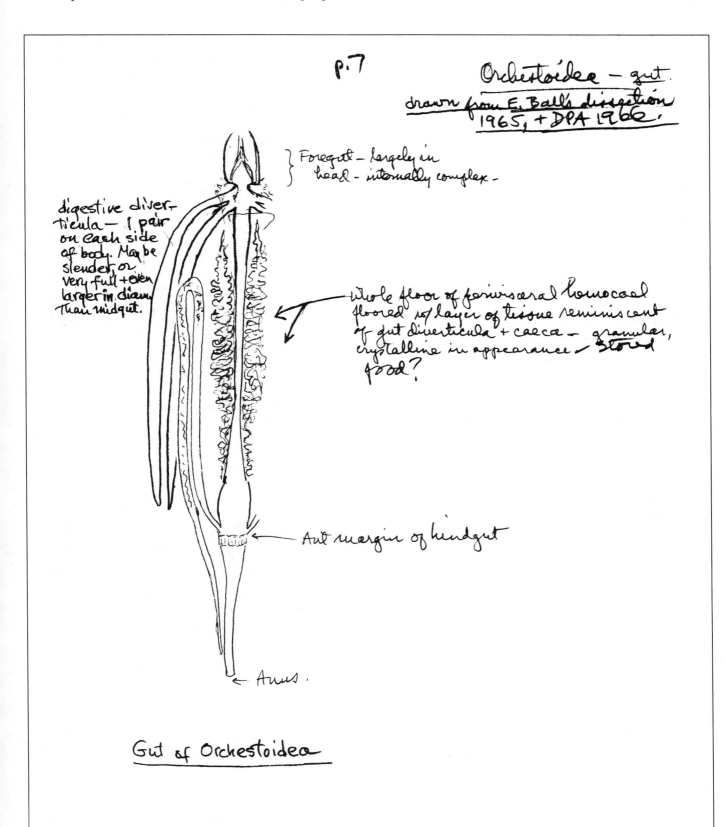

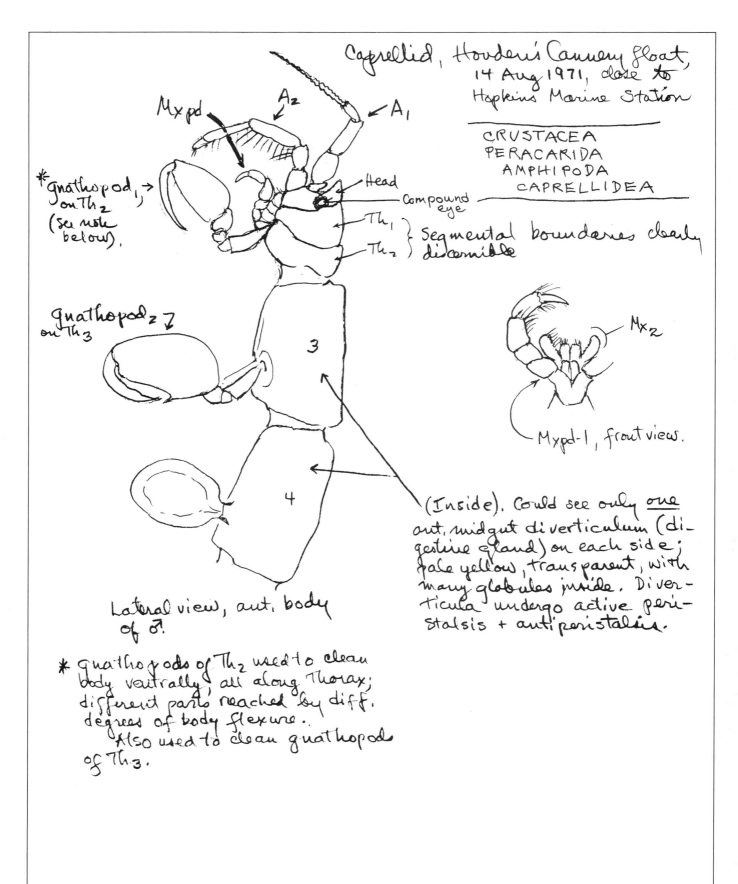

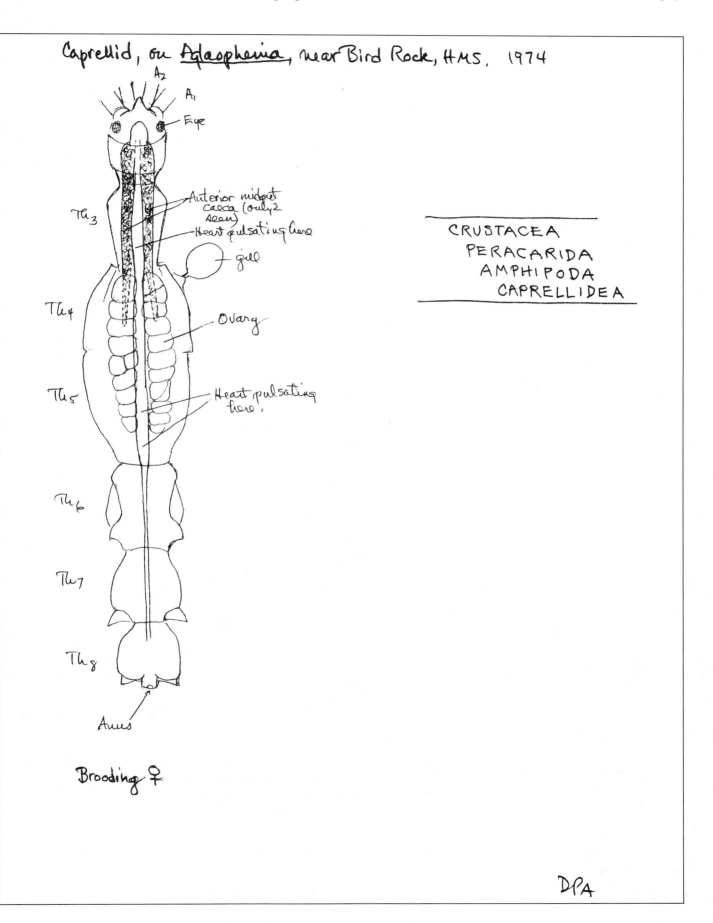

Caprellids
on *Aglaophenia* colonies

LOCOMOTION:
Body Swing:

- Attaches to *Aglaophenia* by locking legs around hydroid. Then swings its body in any & all directions.
- Body may be rigid & straight, or curved & flexible.

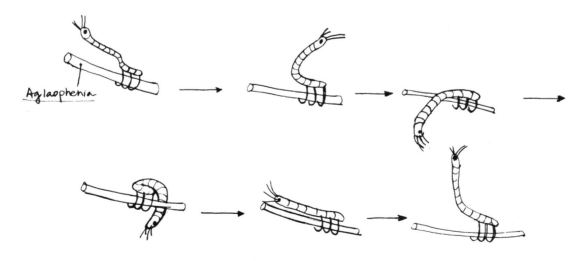

- Caprellid uses this swinging motion possibly to sense and test the immediate area. Often touches antennae to the hydroid substrate.

Eva Aladjem
14 Aug 82

Arthropoda/Crustacea/Malacostraca/Amphipoda

Walking

Walks by using legs (of Th. 6, 7, 8), balances anterior end with its gnathopods (of Th. 2 & 3).

One method:

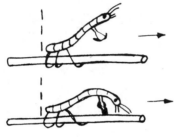

→ Upright position. Antennae extend forward. Sometimes leg pair #1 is unattached, the two hind leg pairs cling. Shown here, all three leg pairs are clinging.

→ Moves leg pair on Th_6 forward (may be ditaxic or monotaxic)

→ Balances body & clings by contacting substrate with gnathopod pair (or just one). Antennae touch substrate.

→ Brings 2nd leg pair (on Th_7) anteriorly in monotaxic or ditaxic motion.

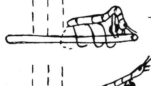

→ Brings 3rd leg pair (on Th_8) forward.

→ No need for balance with gnathopods. Has firm grip on Aglaophenia. Start over again.

↓ etc.

Many variations of these movements occur. This is just a basic scheme.
Can hang on Aglaophenia in any orientation — upward, underneath it, to the sides, etc.

E.V.A. 8/14/82

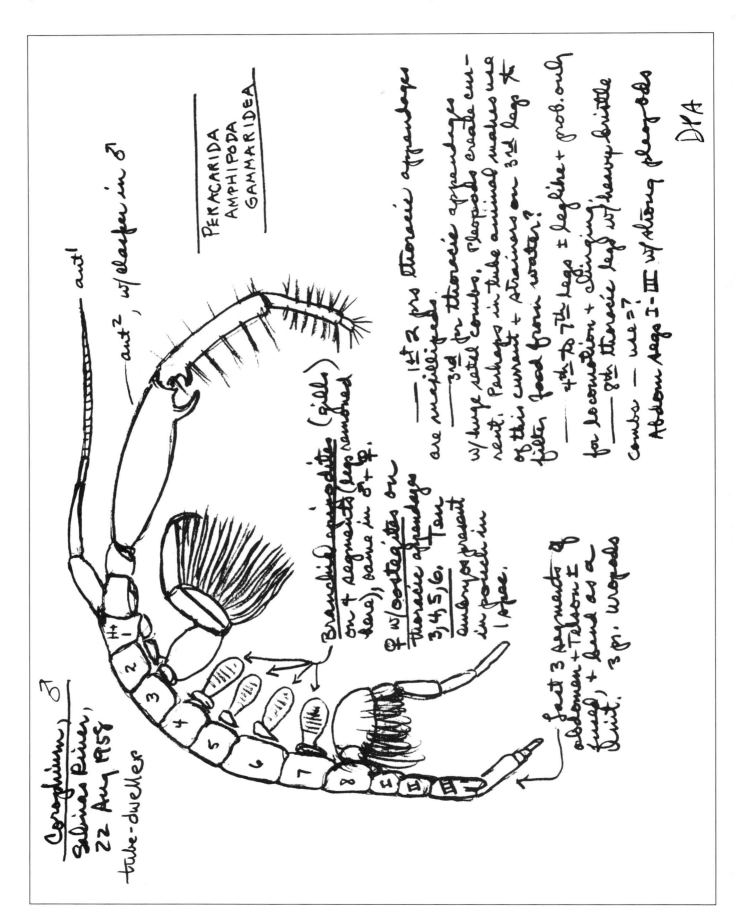

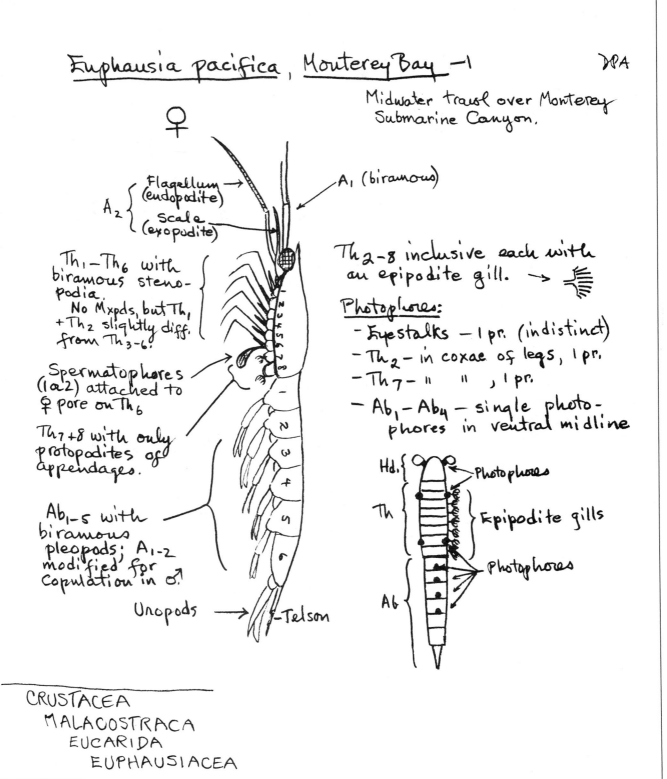

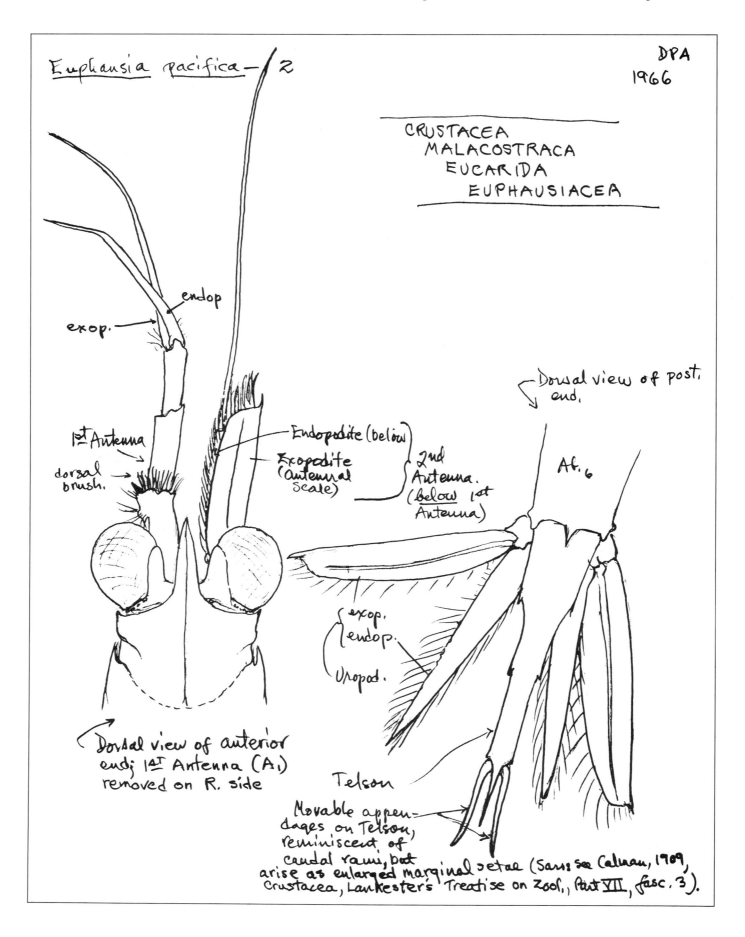

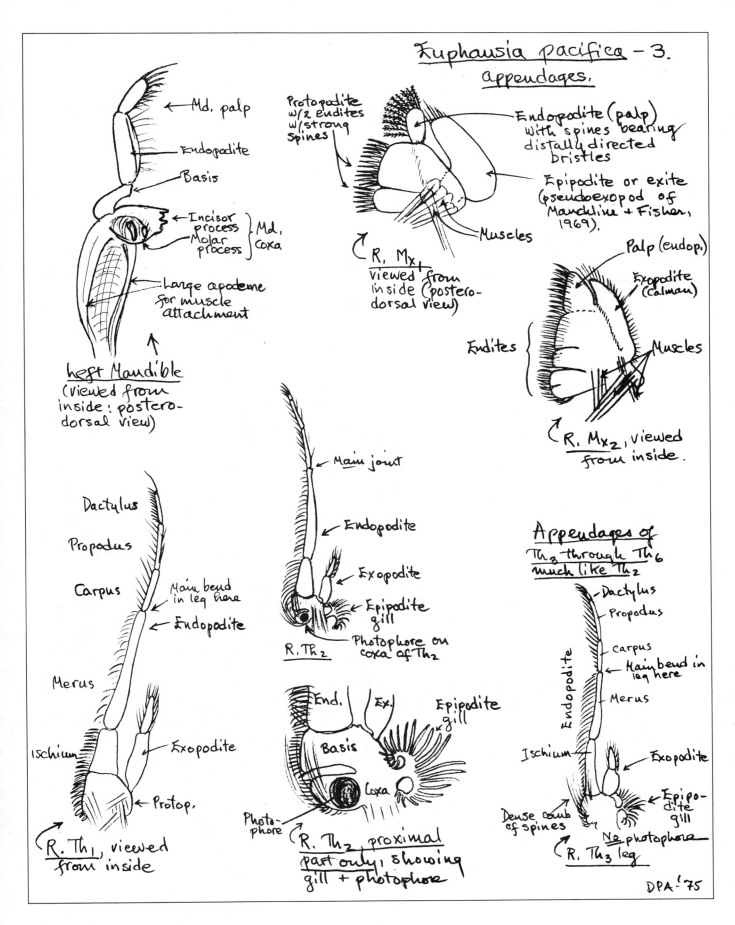

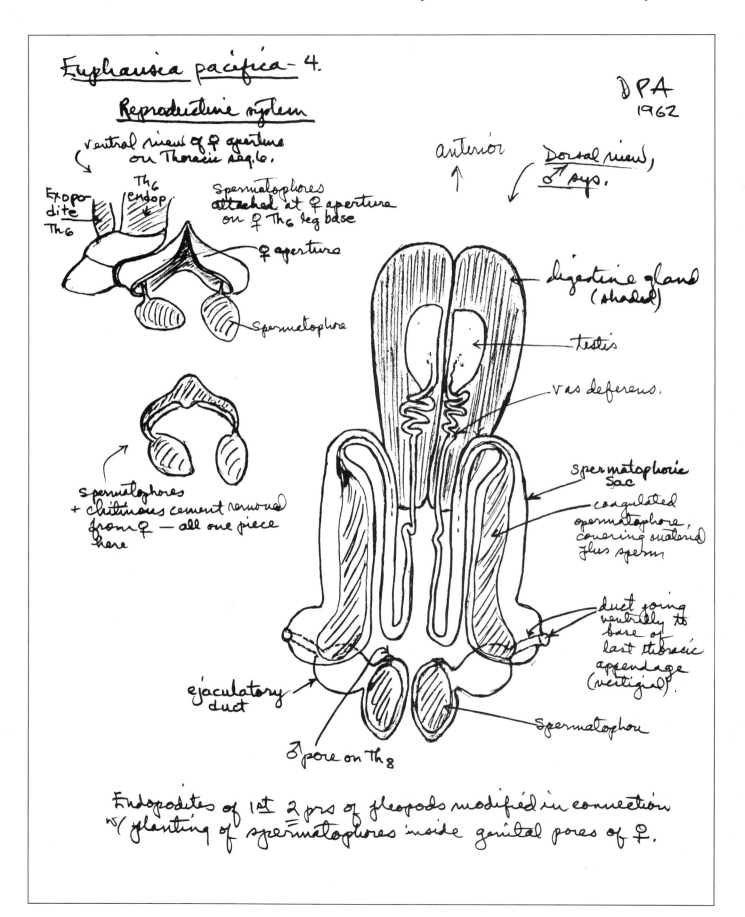

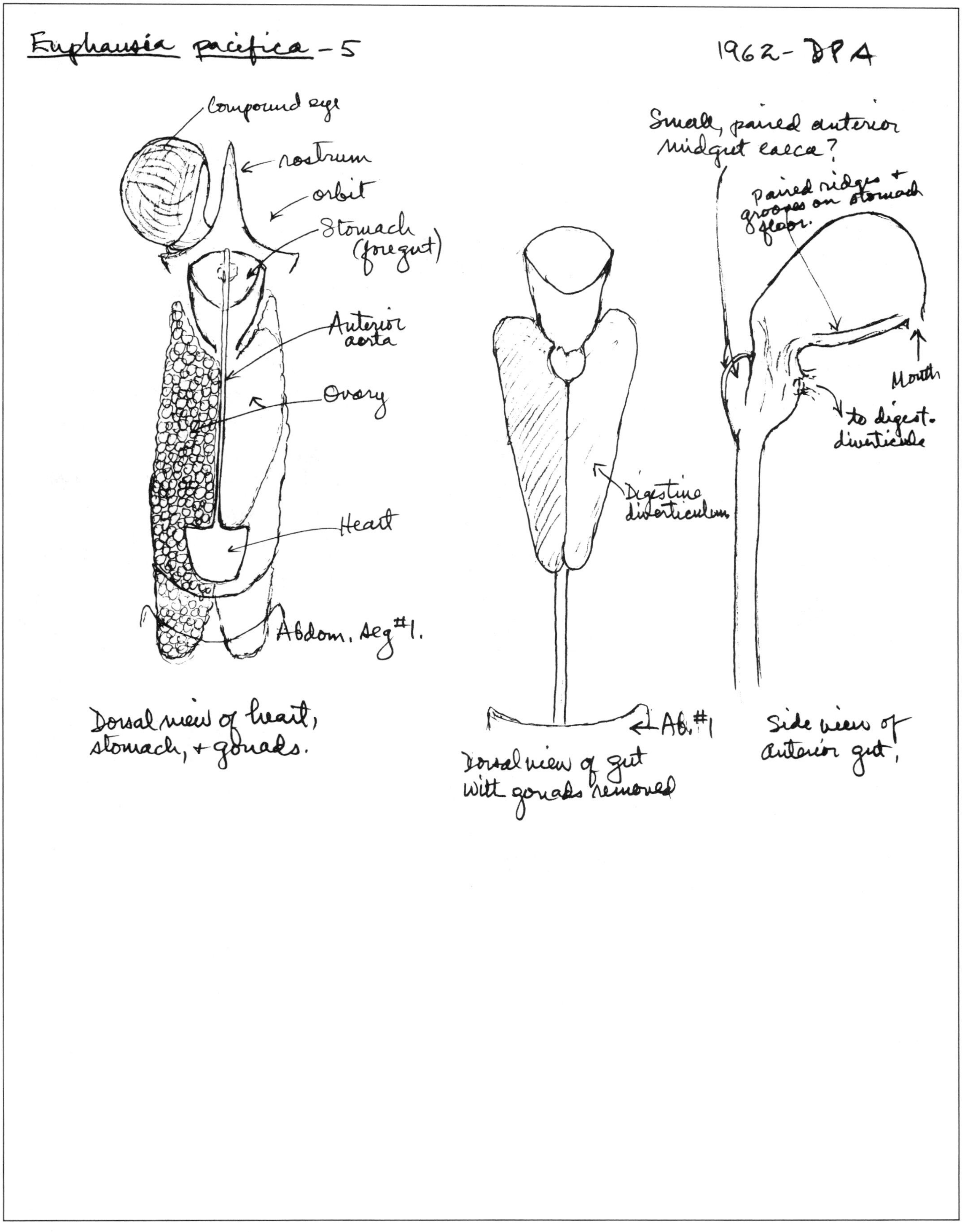

Arthropoda/Crustacea/Malacostraca/Euphausiacea

Euphausia pacifica - 5

1962 - DPA

Labels on left figure:
- compound eye
- rostrum
- orbit
- Stomach (foregut)
- Anterior aorta
- Ovary
- Heart
- Abdom. seg #1

Dorsal view of heart, stomach, + gonads.

Labels on middle figure:
- Digestive diverticulum
- Ab. #1

Dorsal view of gut with gonads removed

Labels on right figure:
- Small, paired anterior midgut caeca?
- paired ridges + grooves on stomach floor.
- Mouth
- to digest. diverticule

Side view of anterior gut.

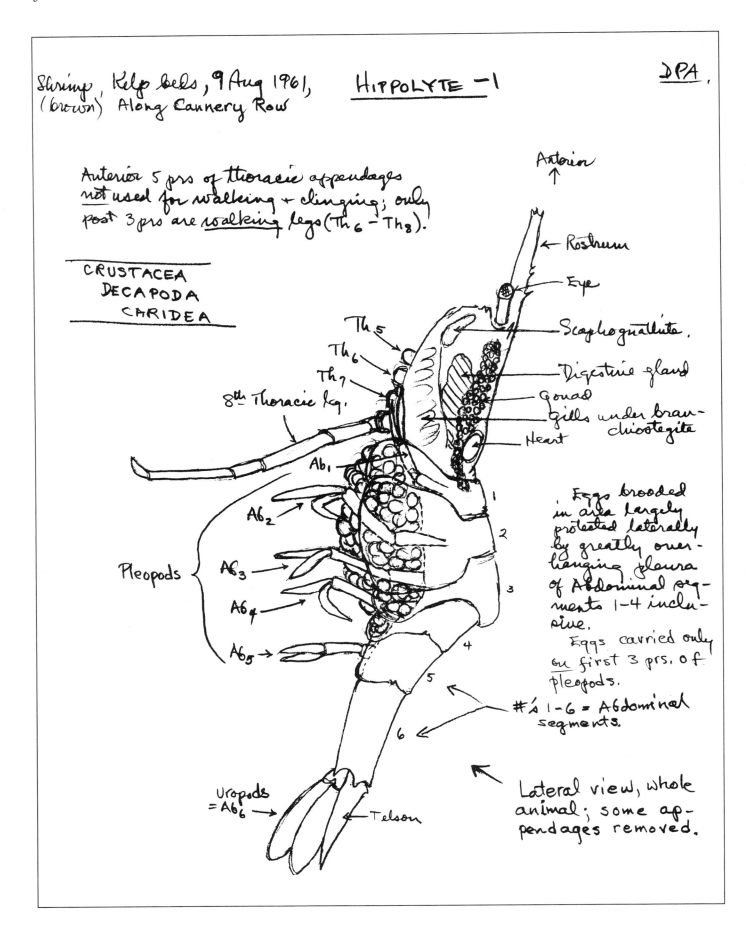

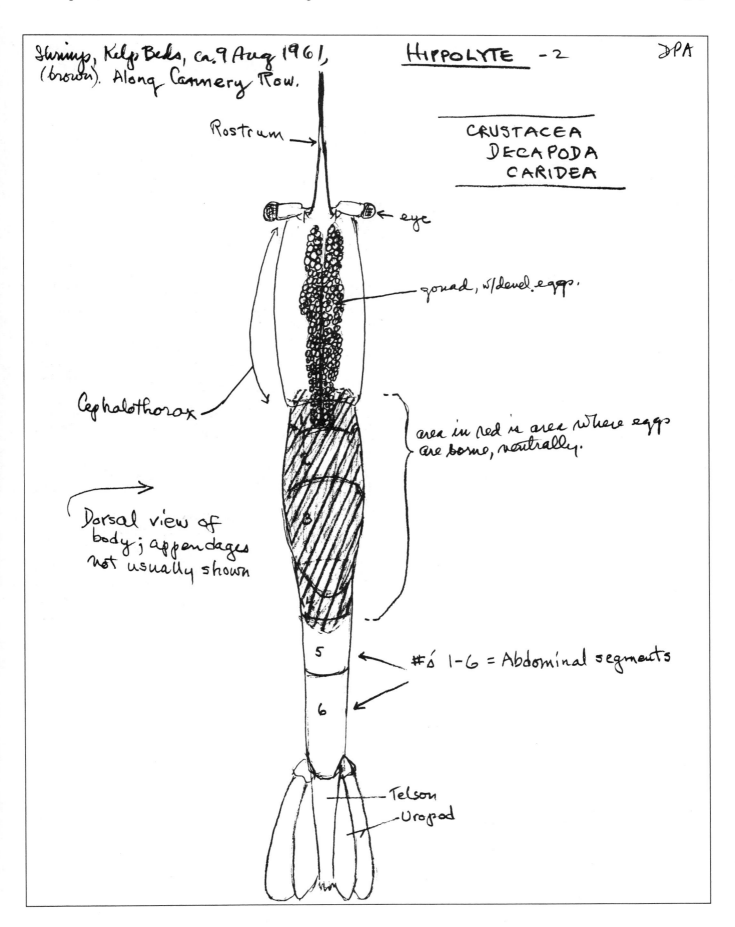

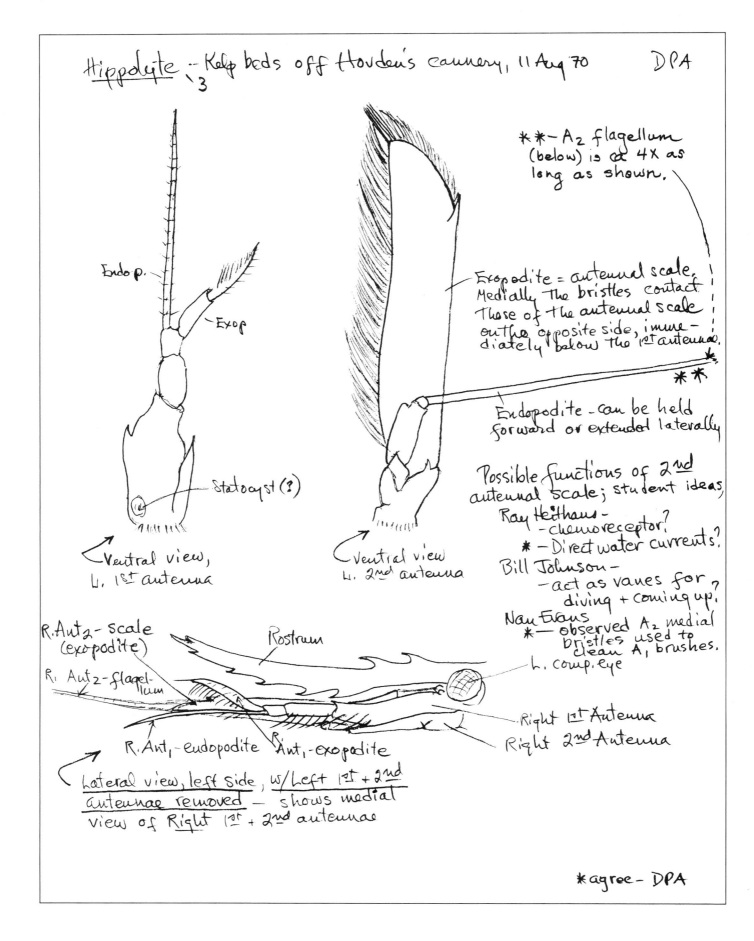

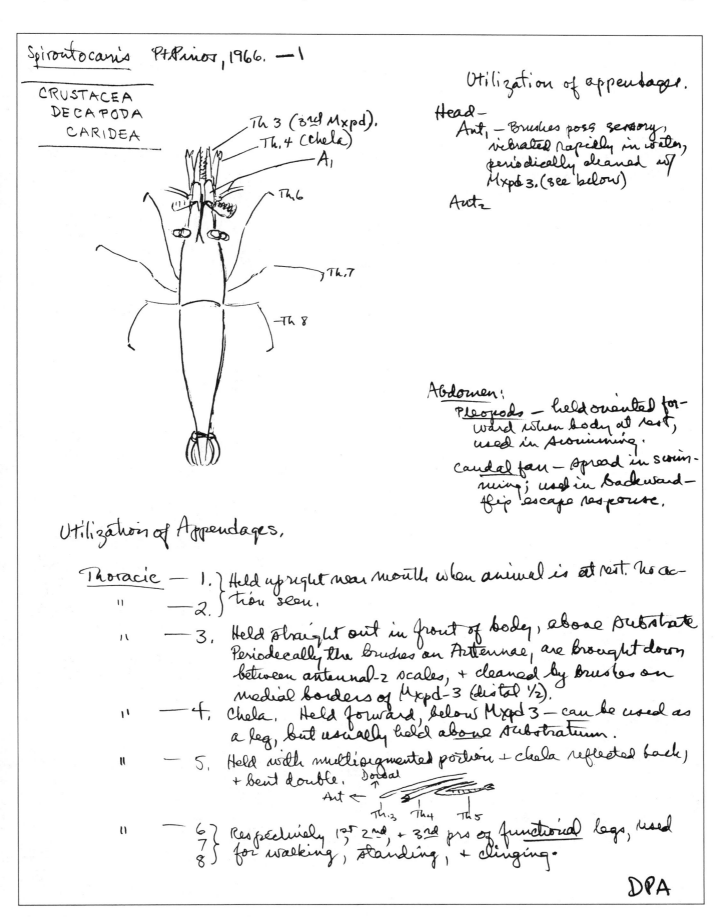

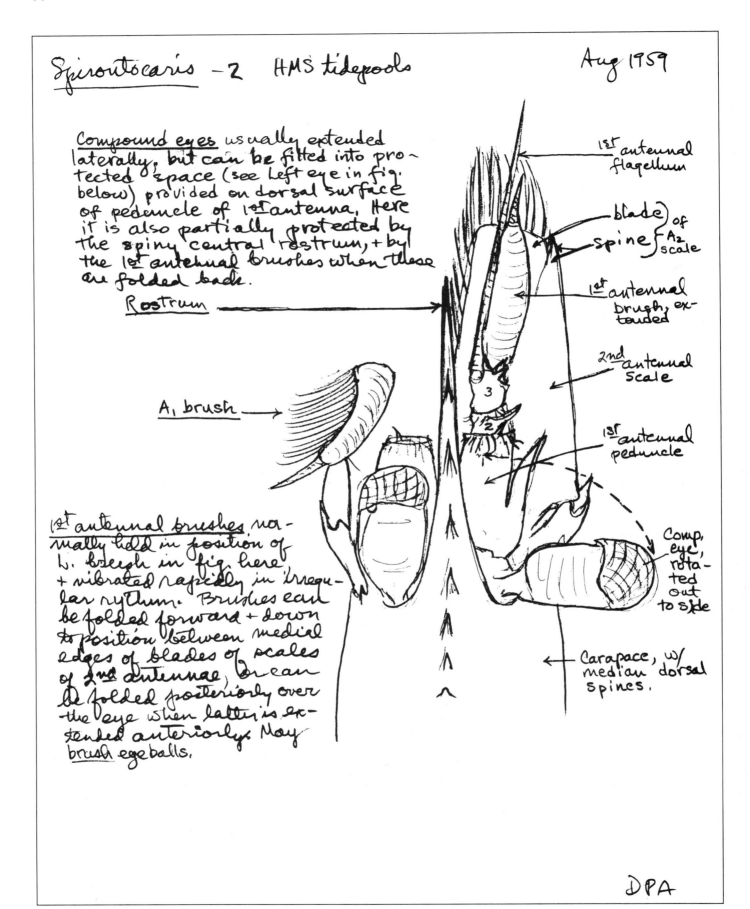

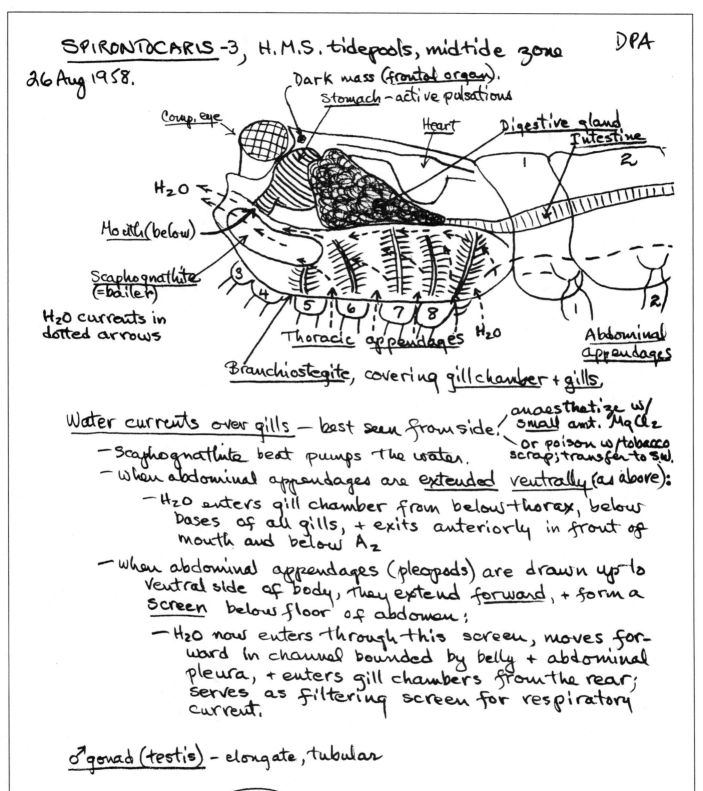

SPIRONTOCARIS - 3, H.M.S. tidepools, midtide zone
26 Aug 1958. DPA

Water currents over gills — best seen from side.

- Scaphognathite beat pumps the water.
- When abdominal appendages are <u>extended ventrally</u> (as above):
  - H_2O enters gill chamber from below thorax, below bases of all gills, + exits anteriorly in front of mouth and below A_2
- When abdominal appendages (pleopods) are drawn up to ventral side of body, they extend <u>forward</u>, + form a <u>screen</u> below floor of abdomen:
  - H_2O now enters through this screen, moves forward in channel bounded by belly + abdominal pleura, + enters gill chambers from the rear; serves as filtering screen for respiratory current.

anaesthetize w/ small amt. $MgCl_2$ or poison w/ tobacco scrap; transfer to s.w.

♂ gonad (testis) — elongate, tubular

Anterior ←   ← Base of last (8th) thoracic appendage.

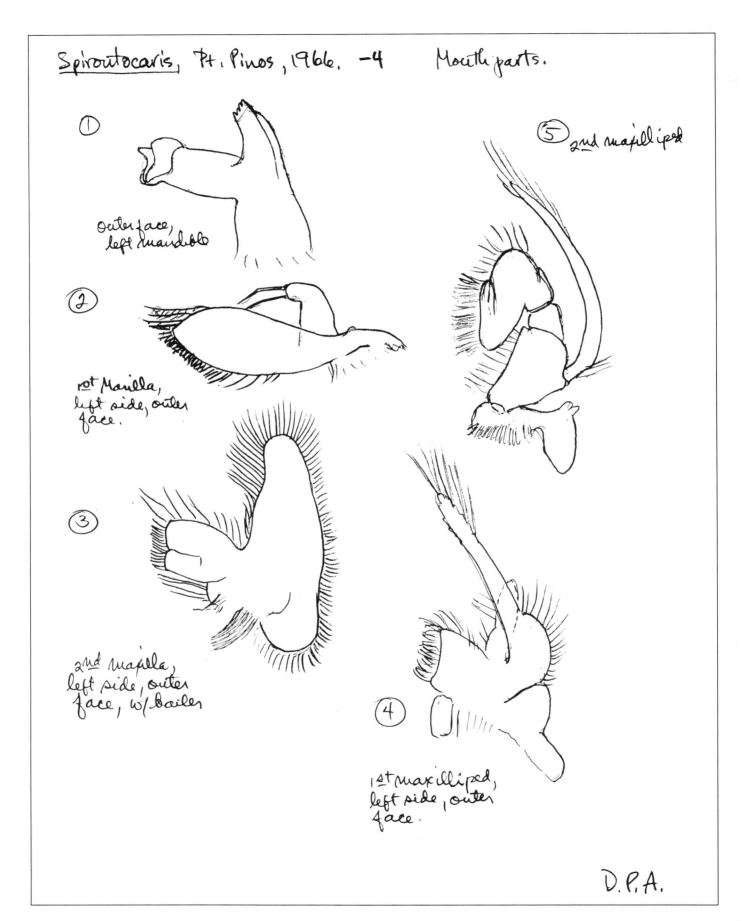

Arthropoda/Crustacea/Malacostraca/Decapoda

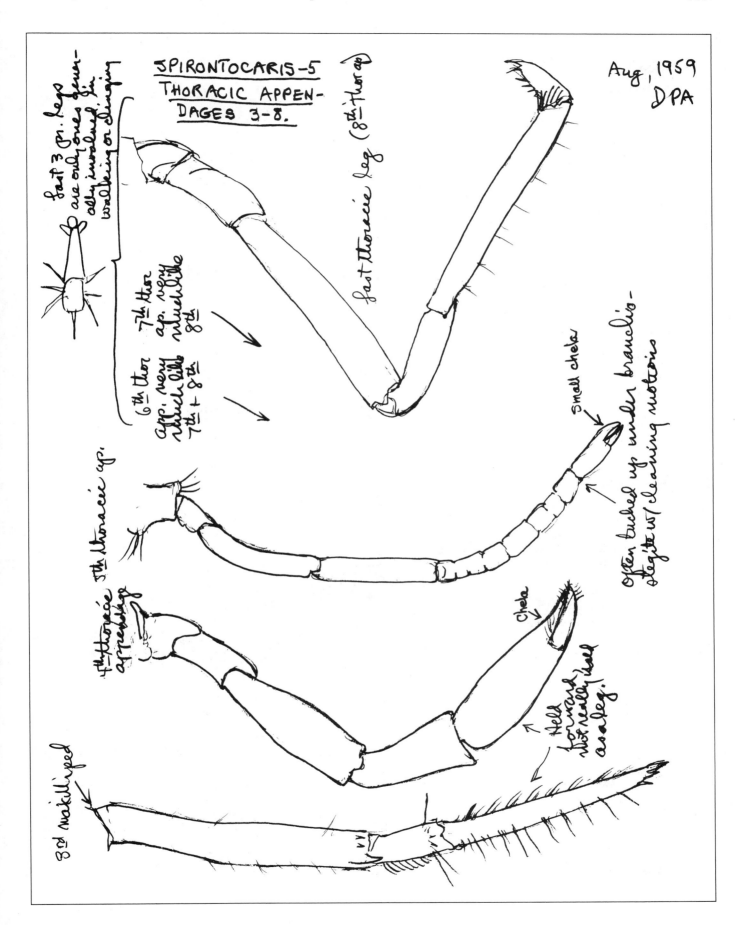

Emerita -1 19 Aug 1969 DPA
analoga

Ant. end, animal partly buried in sand.

CRUSTACEA DECAPODA
ANOMURA
HIPPIDEA

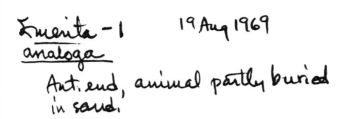

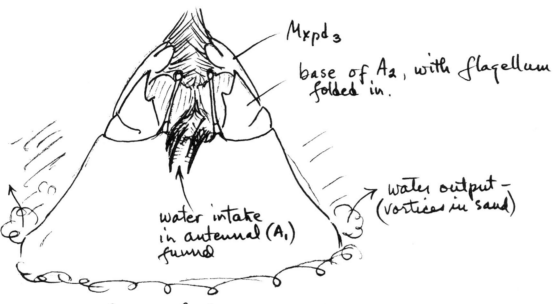

Mxpd₃

base of A₂, with flagellum folded in.

water output — (vortices in sand)

water intake in antennal (A₁) funnel

Funnel formed by A₁ protruded above sand. Eyes may or may not be elevated at same time.

gaps in bristle screen may allow some sand to enter dorsally + ventrally.

Up to a c.c. of sand can be pushed into funnel, + presumably accumulates medial to Mxpd₃. Then animal blows upward strongly, ejecting a fountain of sand; readjusts body under sand, opens funnel, + begins circulating H₂O again.

* water usually drawn down into A₁ funnel, + water ejected (which may form vortices in sand, antero-laterally to carapace). But whole current may be reversed for a period.

**

* Typical behavior on coarse sand.
** In very fine sand, A₁ + eyes may remain buried, + all water is drawn from below, creating 2 vortices:

Arthropoda/Crustacea/Malacostraca/Decapoda

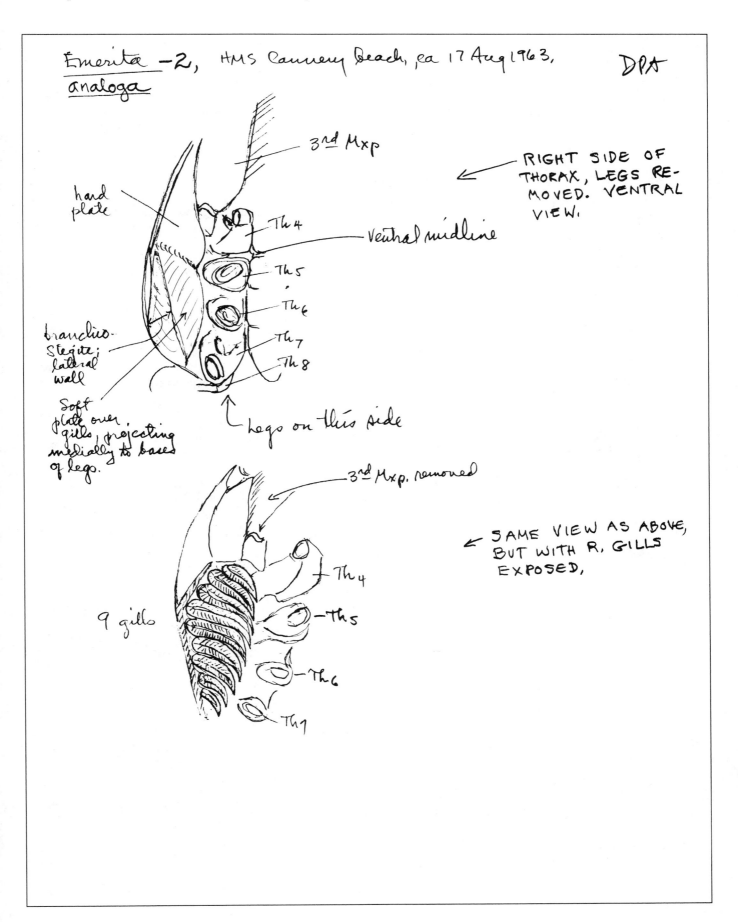

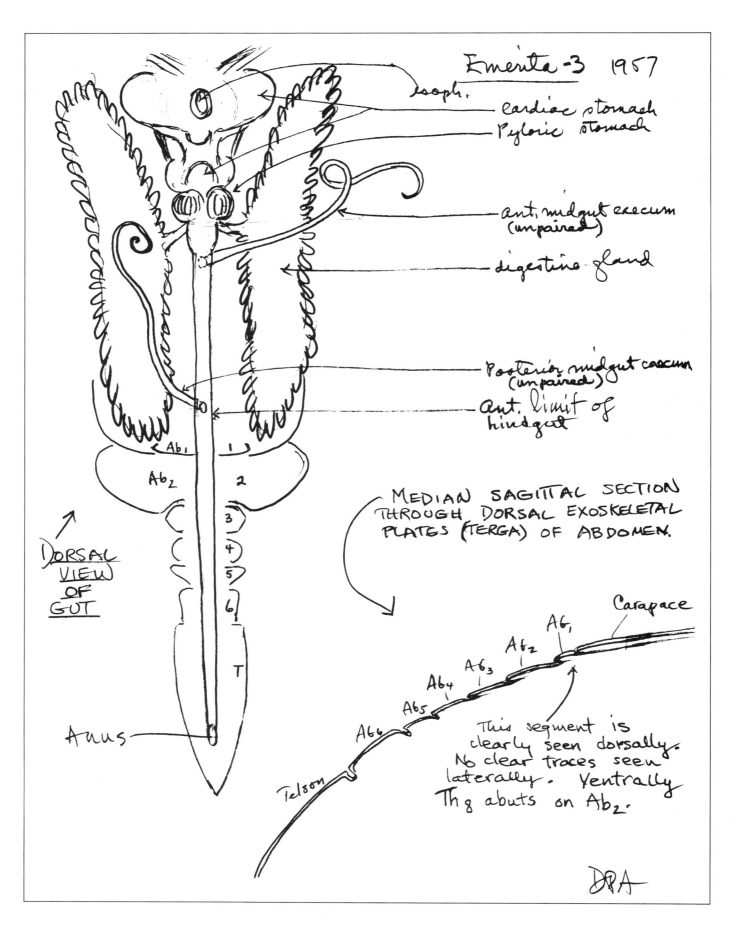

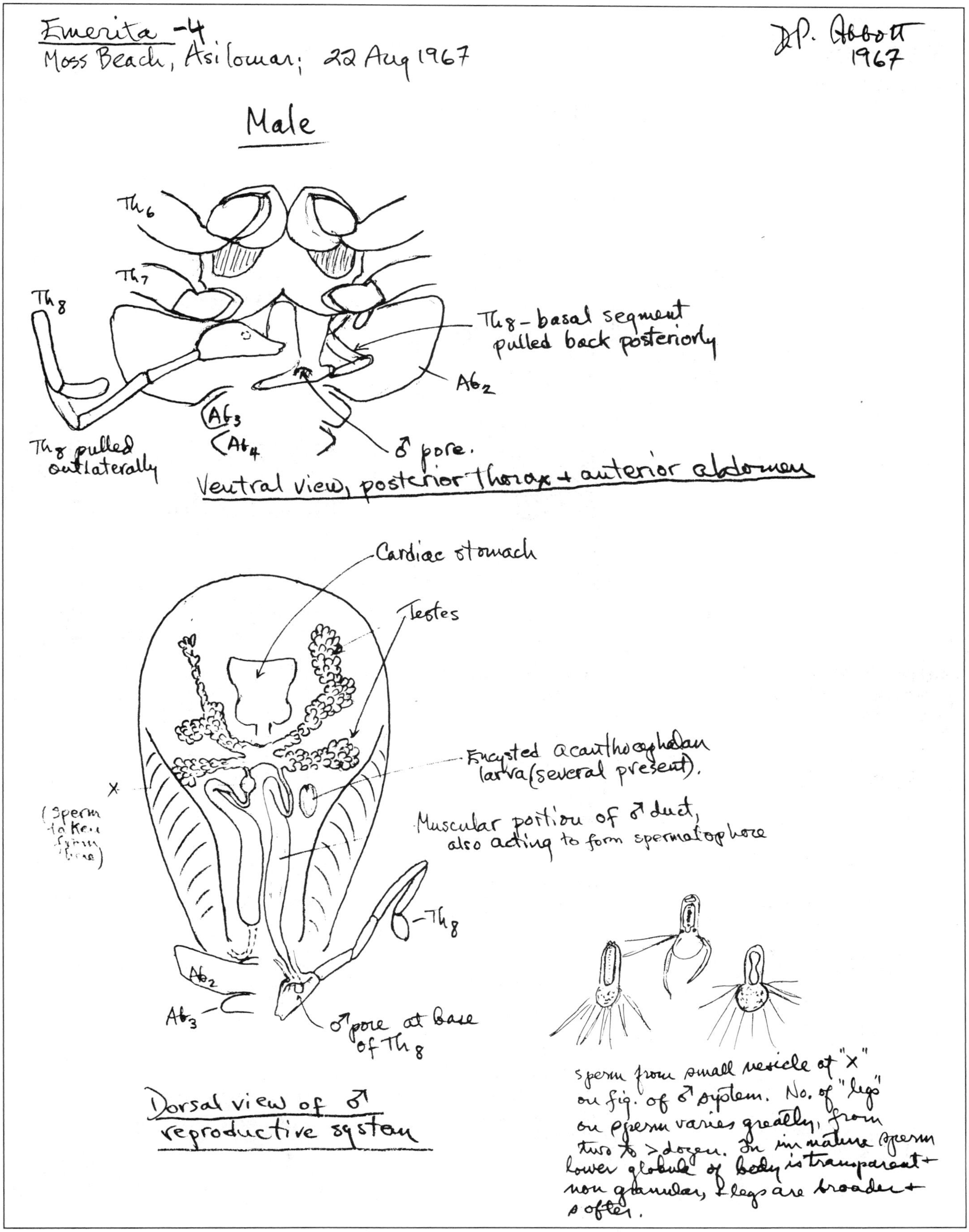

Arthropoda/Crustacea/Malacostraca/Decapoda

Emerita -4
Moss Beach, Asilomar; 22 Aug 1967

D.P. Abbott
1967

Male

Th$_6$, Th$_7$, Th$_8$

Th$_8$ pulled out laterally

Th$_8$ – basal segment pulled back posteriorly

Ab$_2$, Ab$_3$, Ab$_4$

♂ pore.

Ventral view, posterior thorax + anterior abdomen

— Cardiac stomach
— Testes
— Encysted acanthocephalan larva (several present).
— Muscular portion of ♂ duct, also acting to form spermatophore

X (sperm to be drawn (below))

Th$_8$
Ab$_2$
Ab$_3$
♂ pore at base of Th$_8$

Dorsal view of ♂ reproductive system

Sperm from small vesicle at "X" on fig. of ♂ system. No. of "legs" on sperm varies greatly, from two to > dozen. In immature sperm lower globule of body is transparent + non granular, + legs are broader + softer.

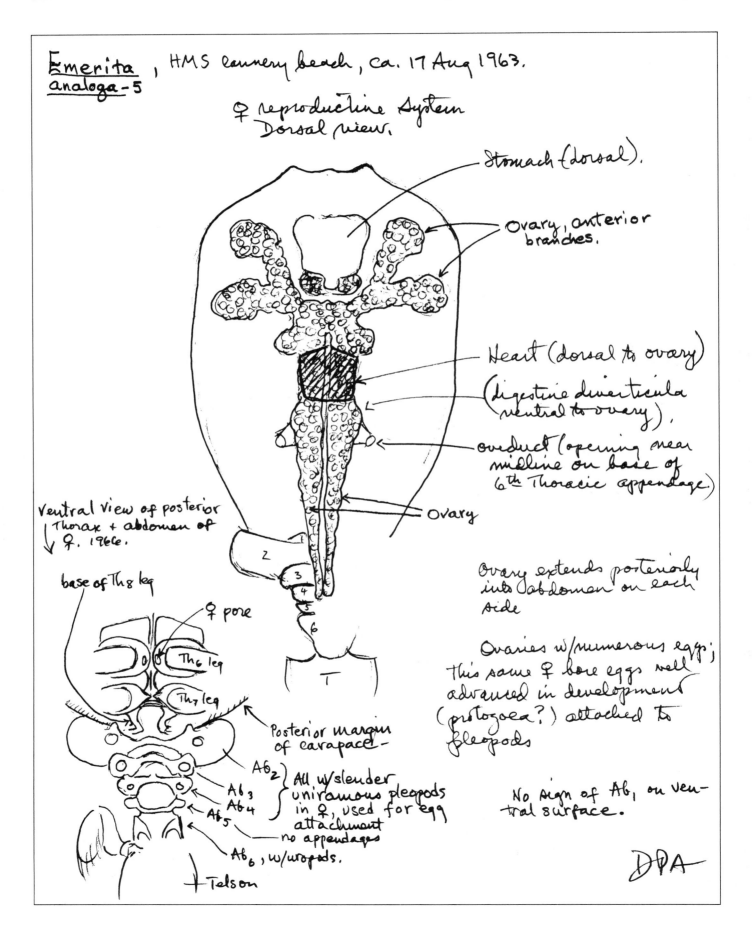

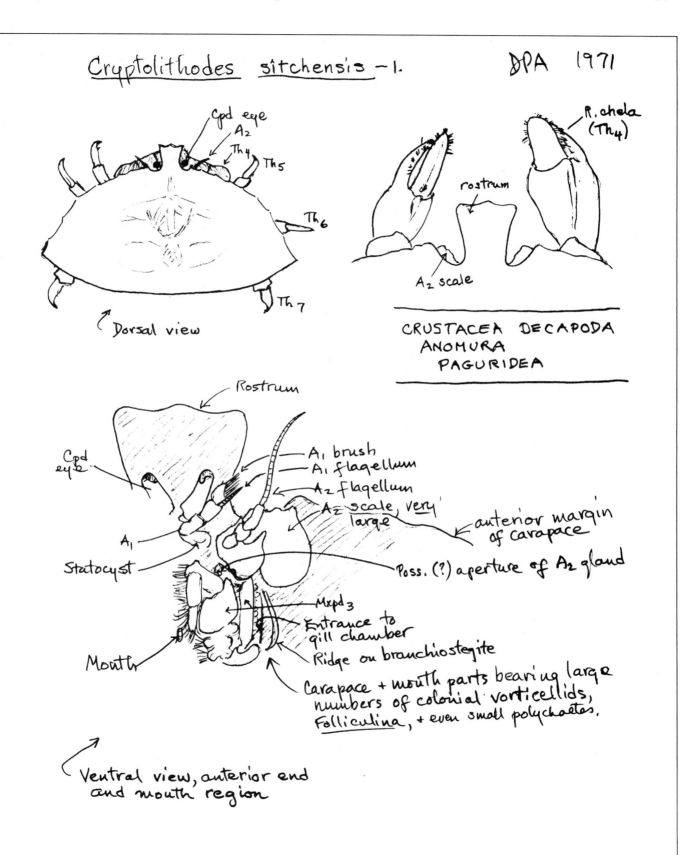

Cryptolithodes sitchensis - 2
DPA 1971

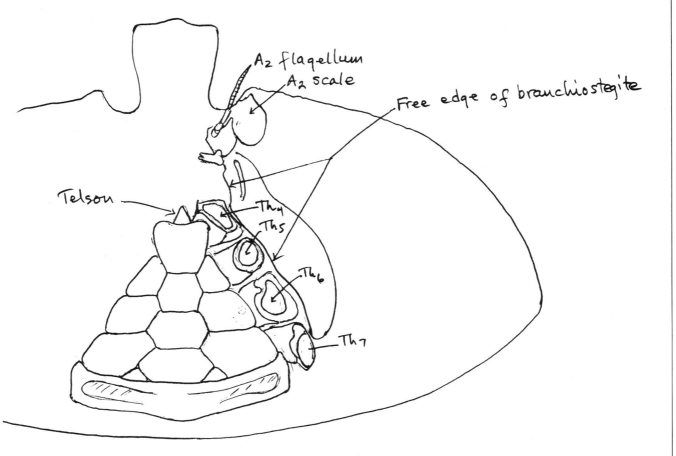

Ventral view - legs of L. side removed

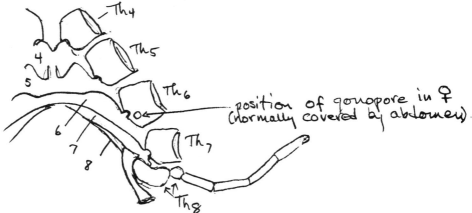

Ventral view, Th. leg bases + sternites, abdomen removed.

Arthropoda/Crustacea/Malacostraca/Decapoda

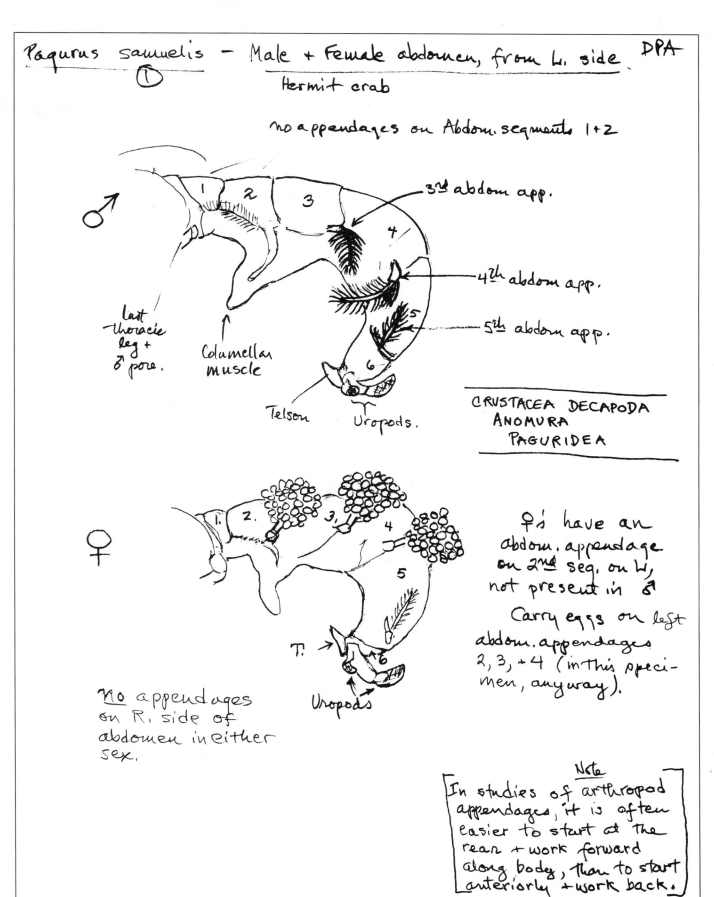

Pagurus samuelis (2)

DPA

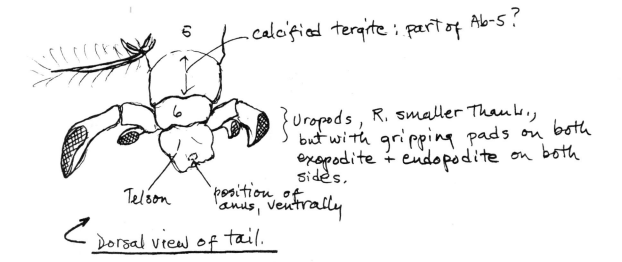

Dorsal view of tail.

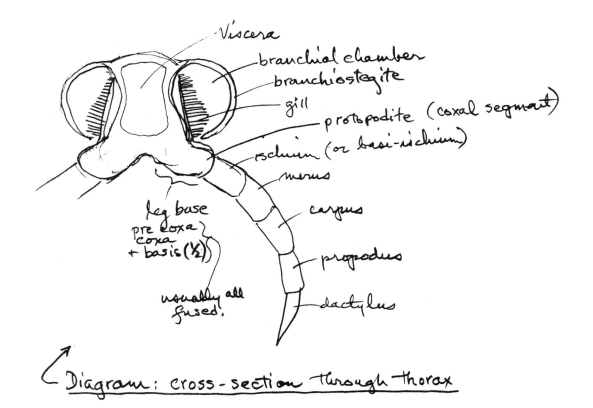

Diagram: cross-section through thorax

Arthropoda/Crustacea/Malacostraca/Decapoda

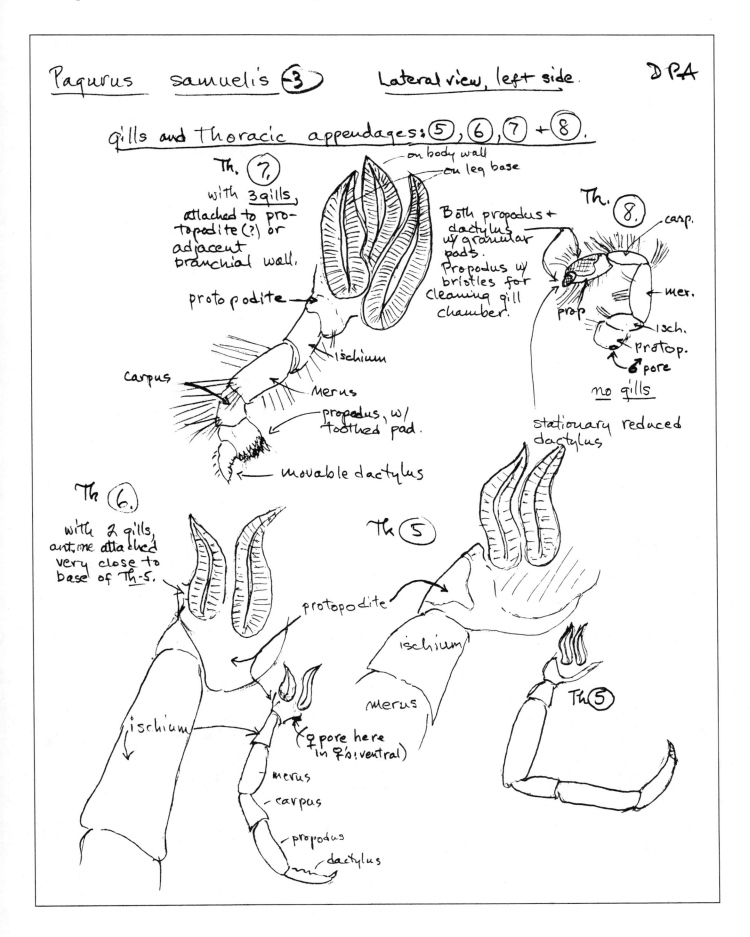

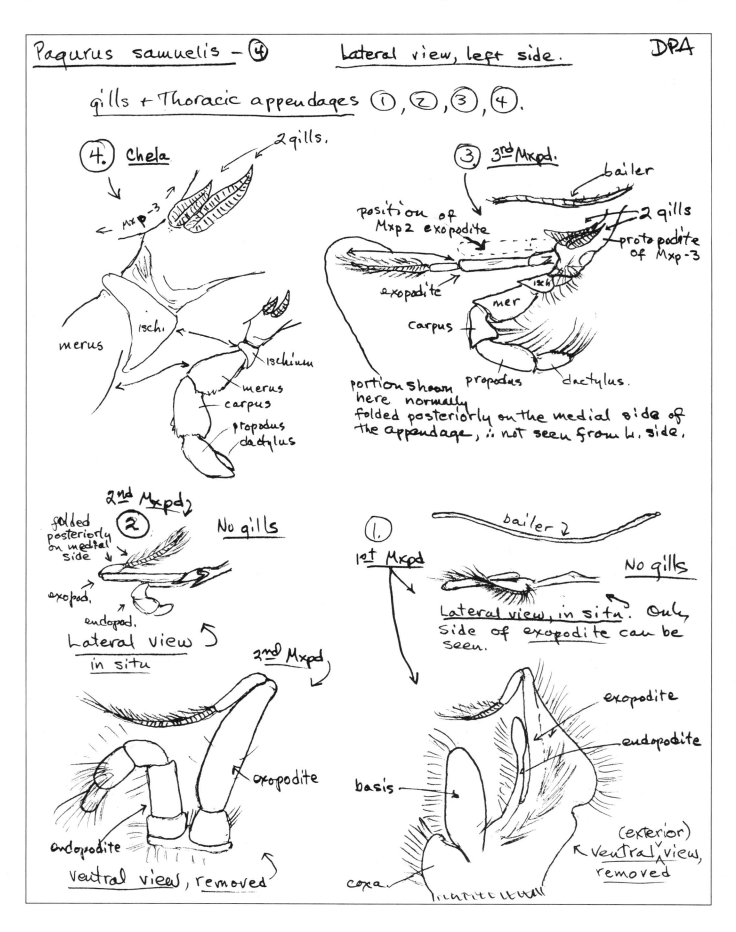

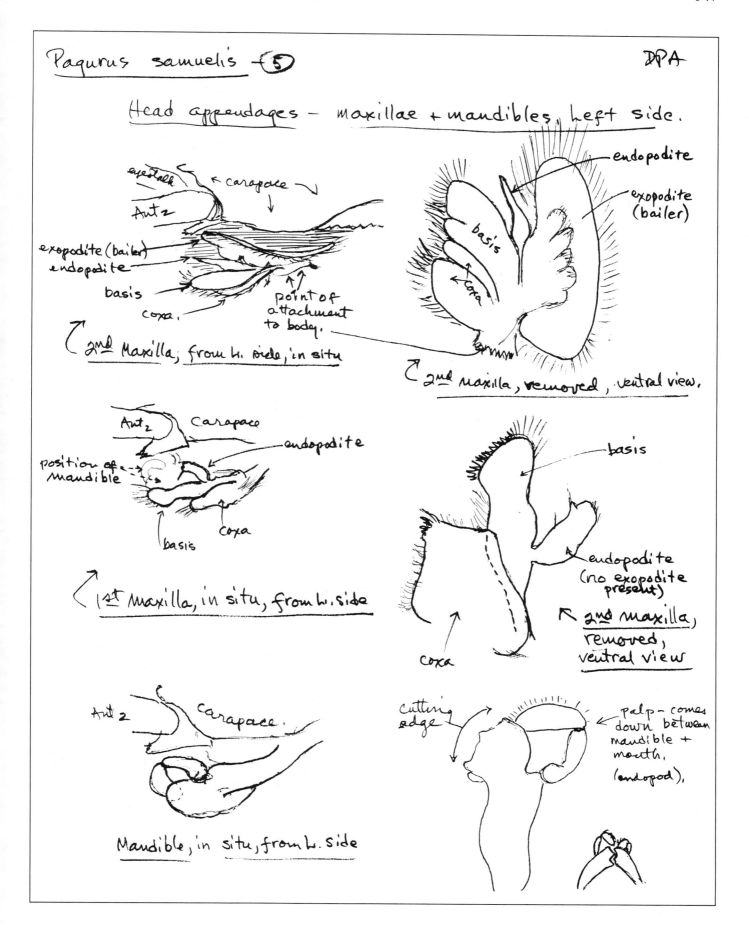

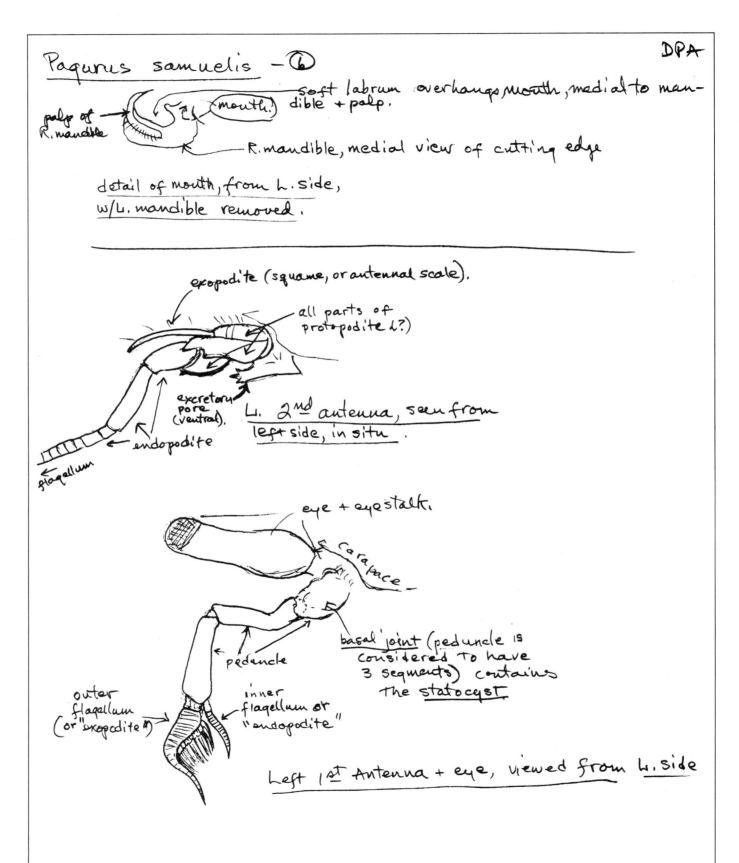

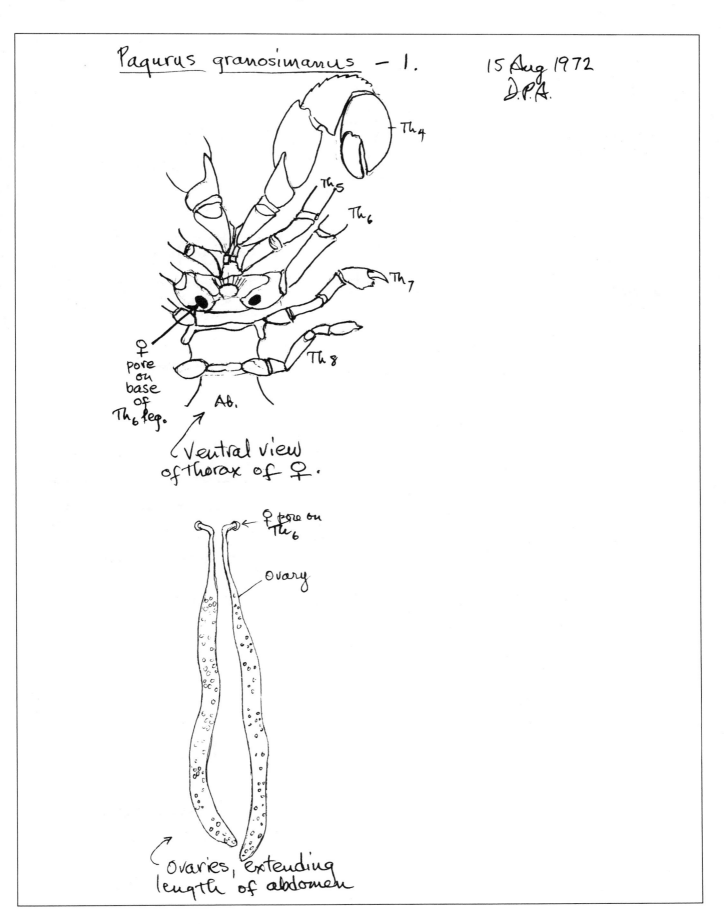

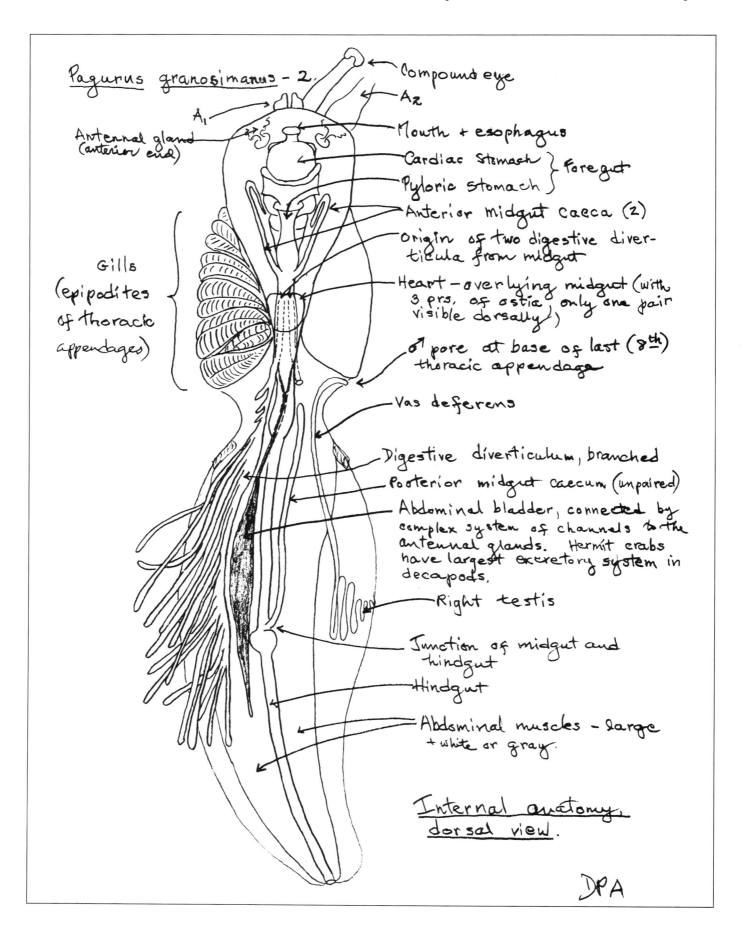

Arthropoda/Crustacea/Malacostraca/Decapoda

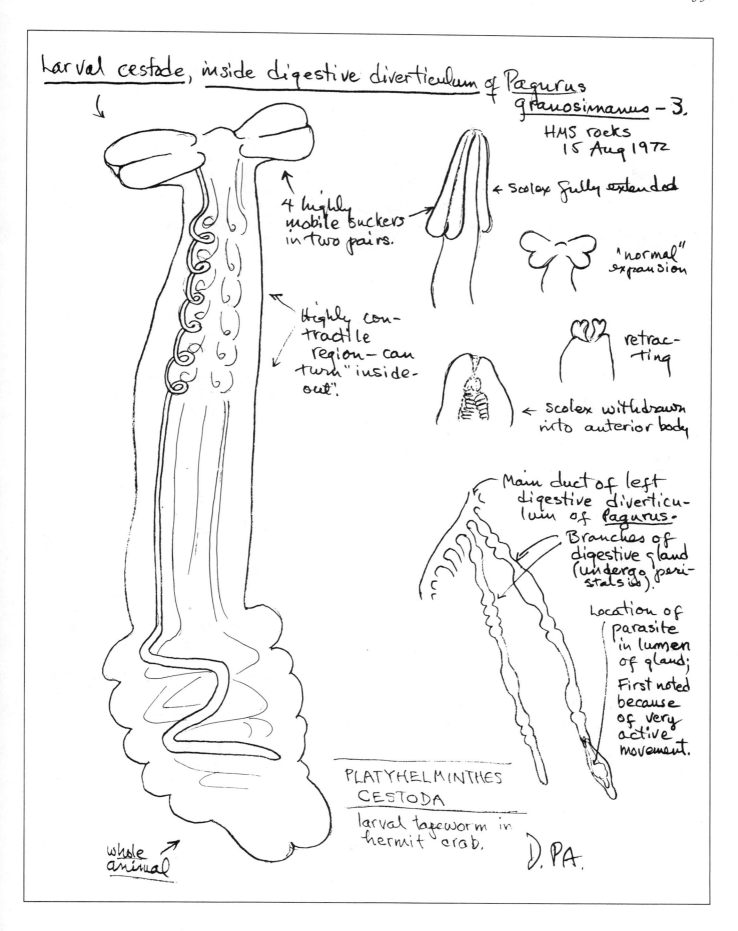

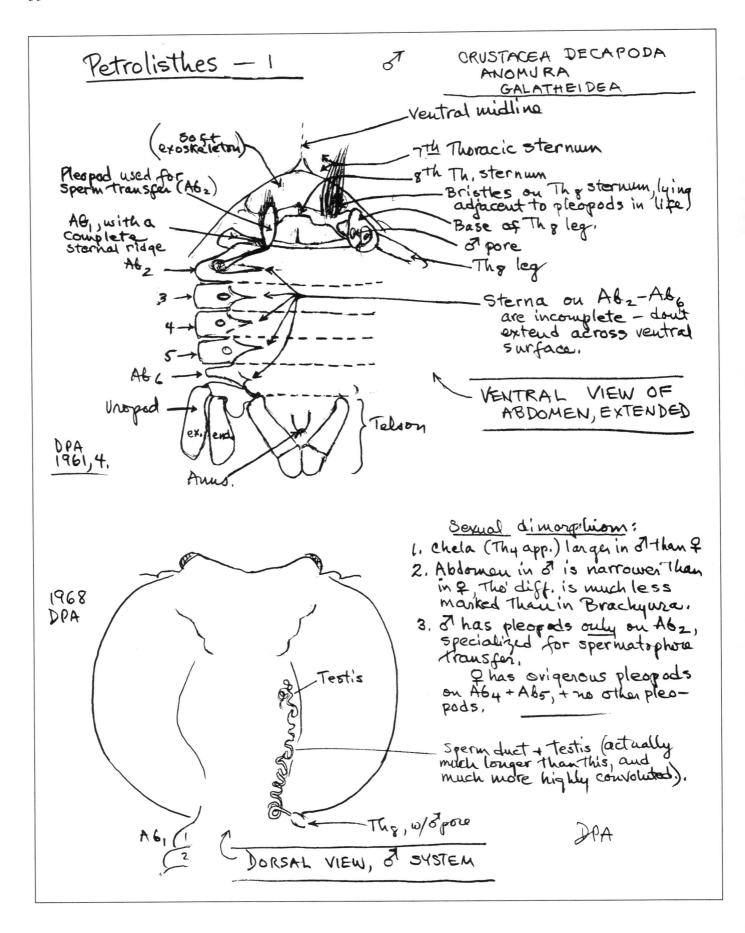

Arthropoda/Crustacea/Malacostraca/Decapoda

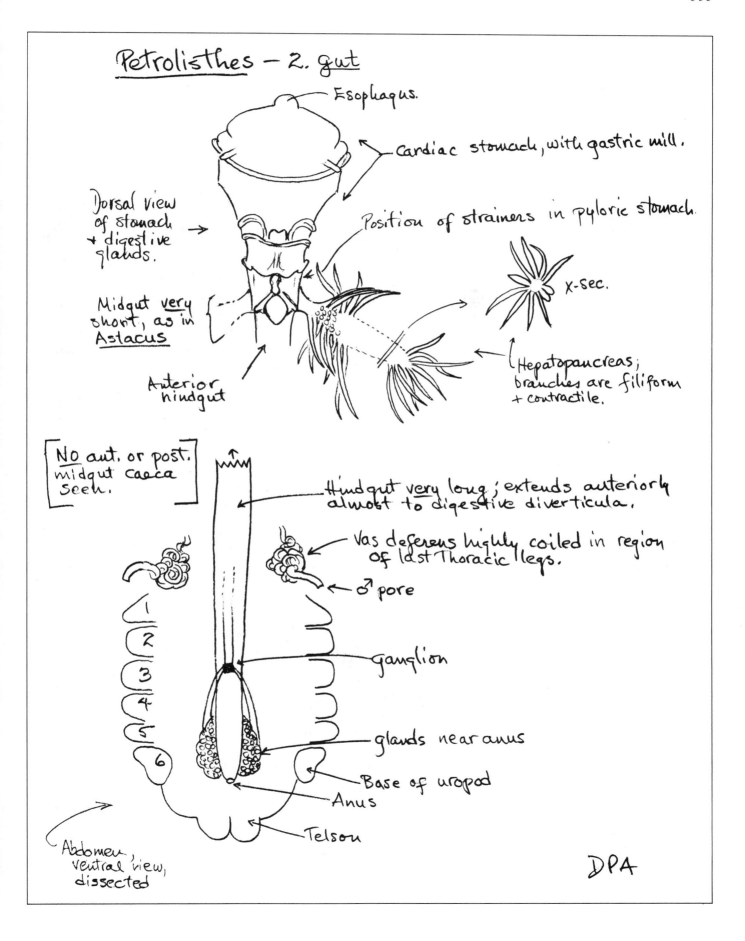

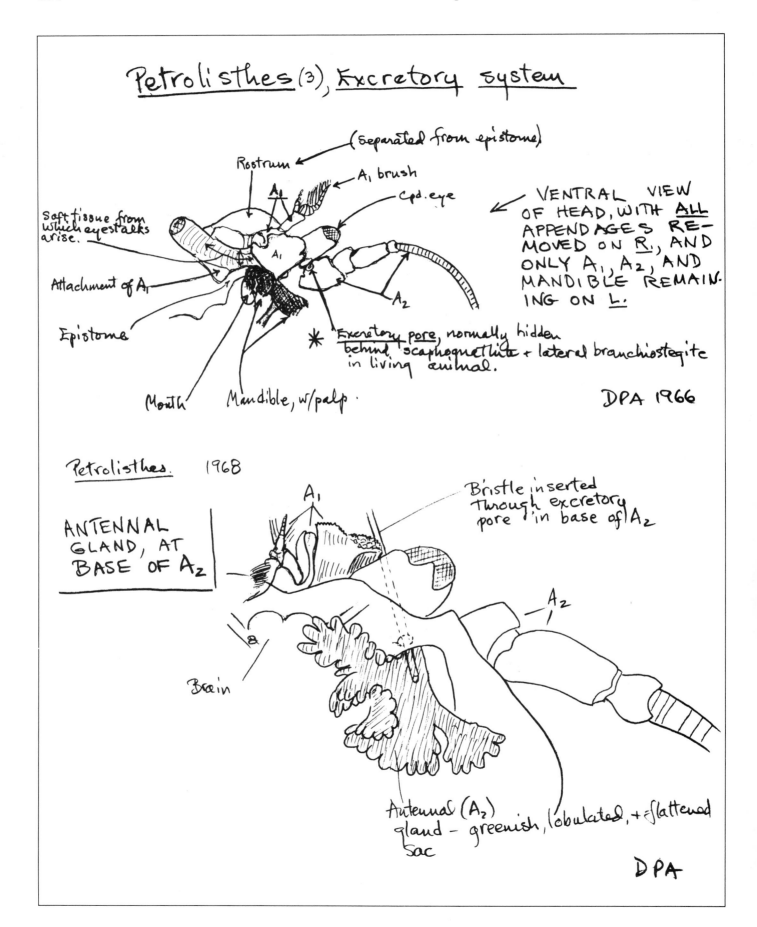

Arthropoda/Crustacea/Malacostraca/Decapoda

Petrolisthes — Asilomar Rocks 15 Aug 1961 – 4.   DPA.

1. Several individuals examined bearing a heavy growth of a colonial vorticellid on appendages (tip to base), anterior + posterior dorsal regions of cephalothorax.

2. Carmine added to water shows water enters branchial chamber posteriorly + along nearly whole ventral margin of branchiostegite, exiting anteriorly in flow which comes out past or thru 1st antennae. Latter are continually being flicked rapidly about in the excurrent water flow.
   (no reversal of water flow noted)

3. Much of water entering branchial chamber is filtered coarsely by bristles; posteriorly these lie along margins of abdominal pleura; laterally these projecting bristles lie on the ant. & post. margins of the protopodites of the walking legs (i.e. between the bases of adjacent legs), + along the ventral border of the branchiostegites (tho' there are fewer here).

4. Carmine in water (finer material) passed easily into the branchial chamber. Action of posterior (last) thoracic appendages well seen. Legs are capable of reaching + scrubbing (w/ terminal brushes).
   Appendages scrub: (1.) Bristles at bases of walking legs + margin of branchiostegite, (2) groove at posterior margin of carapace, (3) whole inner surface of branchial chamber.

5. Antennae (1st) are cleaned by bristle patch on the inner face of the distal 2 segments (propodus, dactylus), of the maxilliped (endopodite).

6. Almost no movements of, or use of, 2nd antennae noted; kept directed latero-posteriorly.

7. "Sensory tufts" on ventro-medial face of propodus of chela sometimes used to brush eyes + A_2 bases. Some tufts inhabited by amphipods.

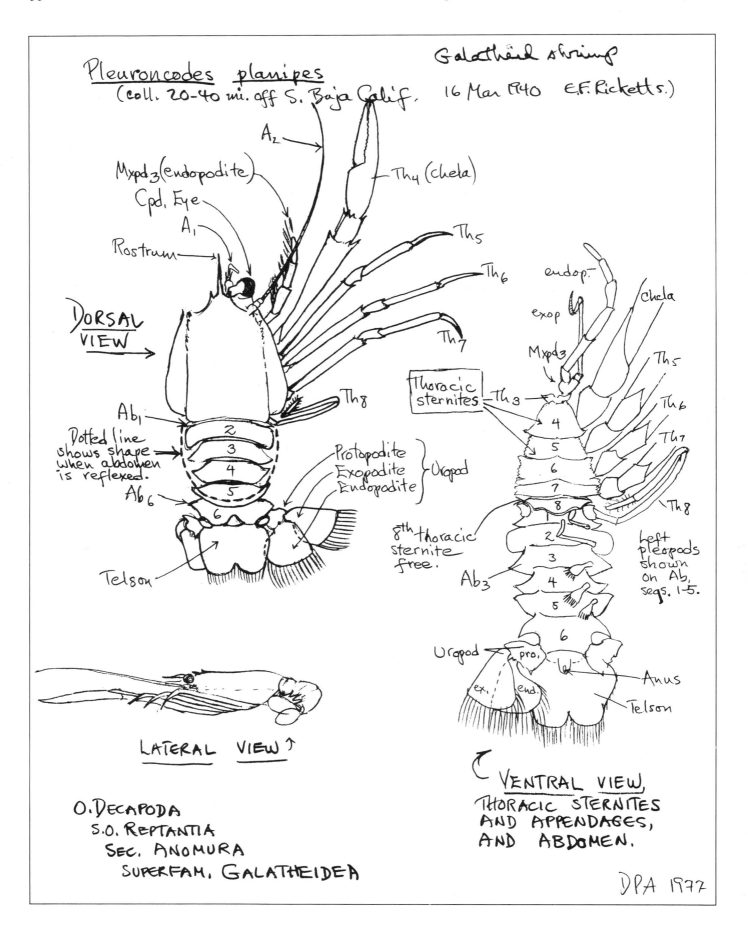

Arthropoda/Crustacea/Malacostraca/Decapoda

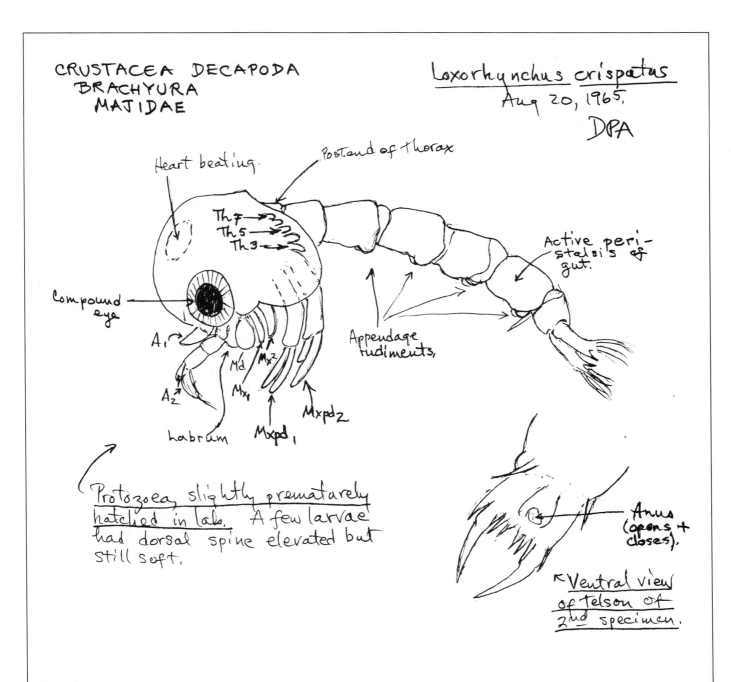

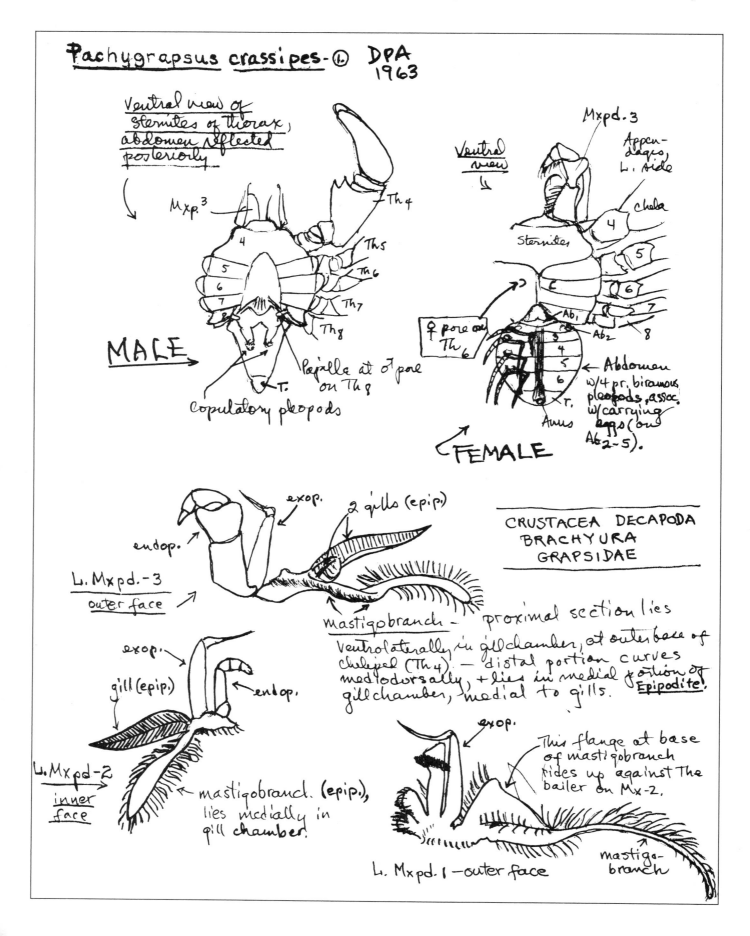

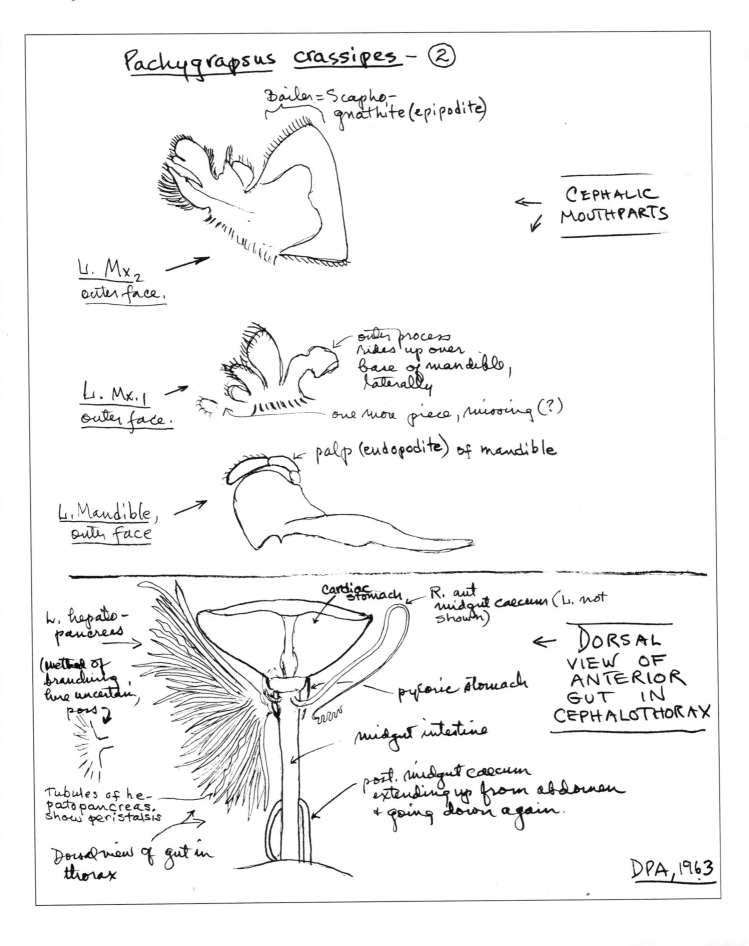

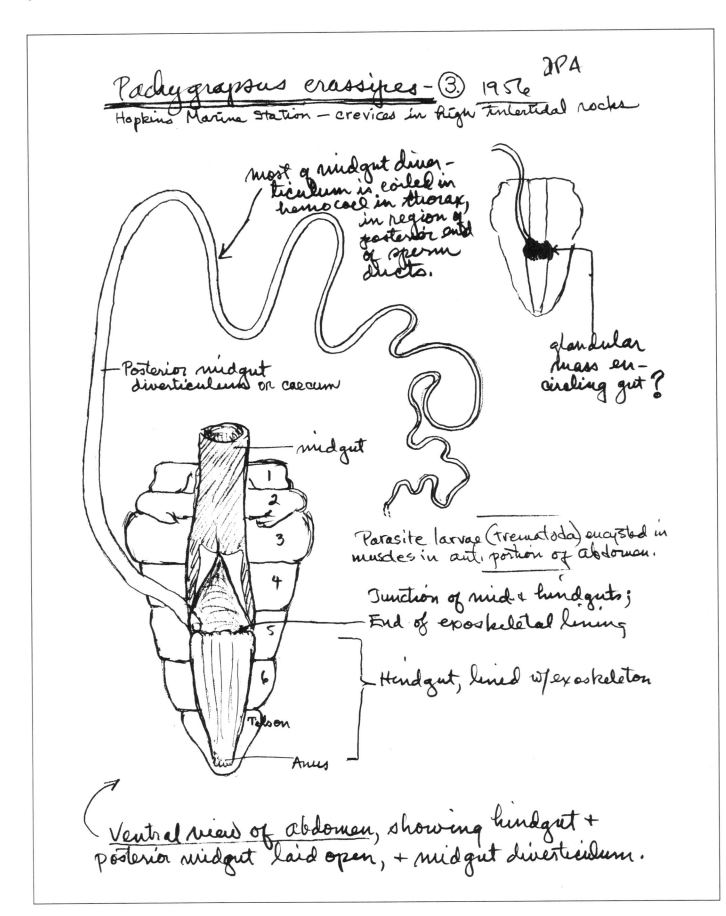

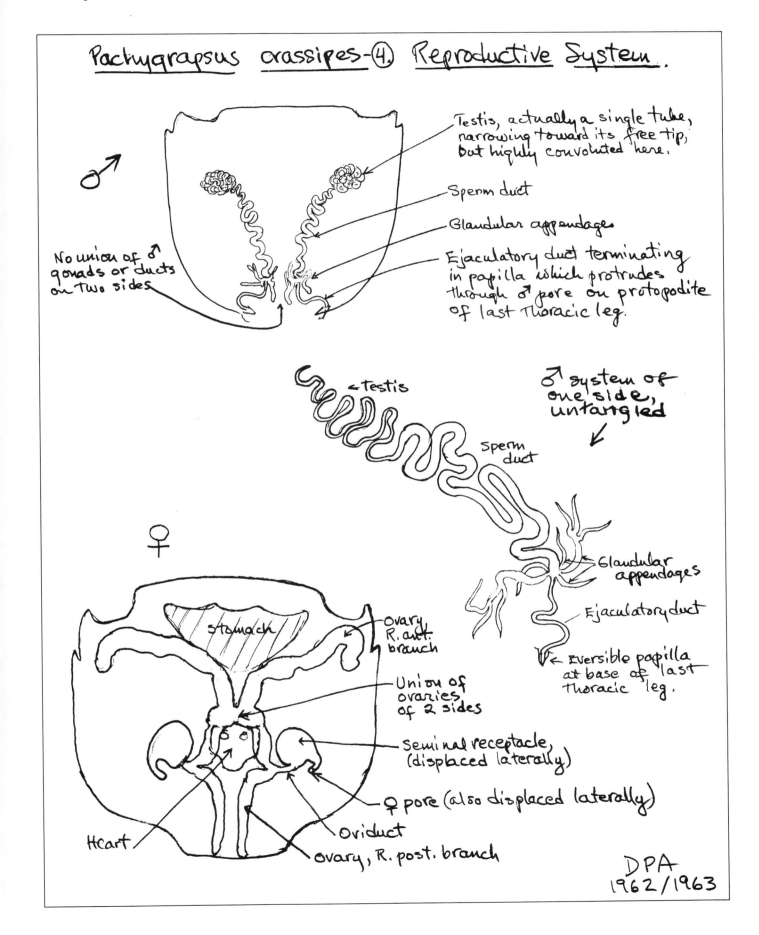

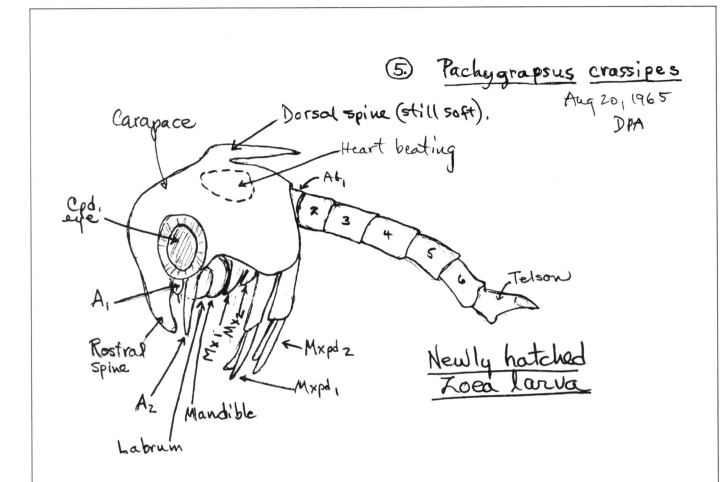

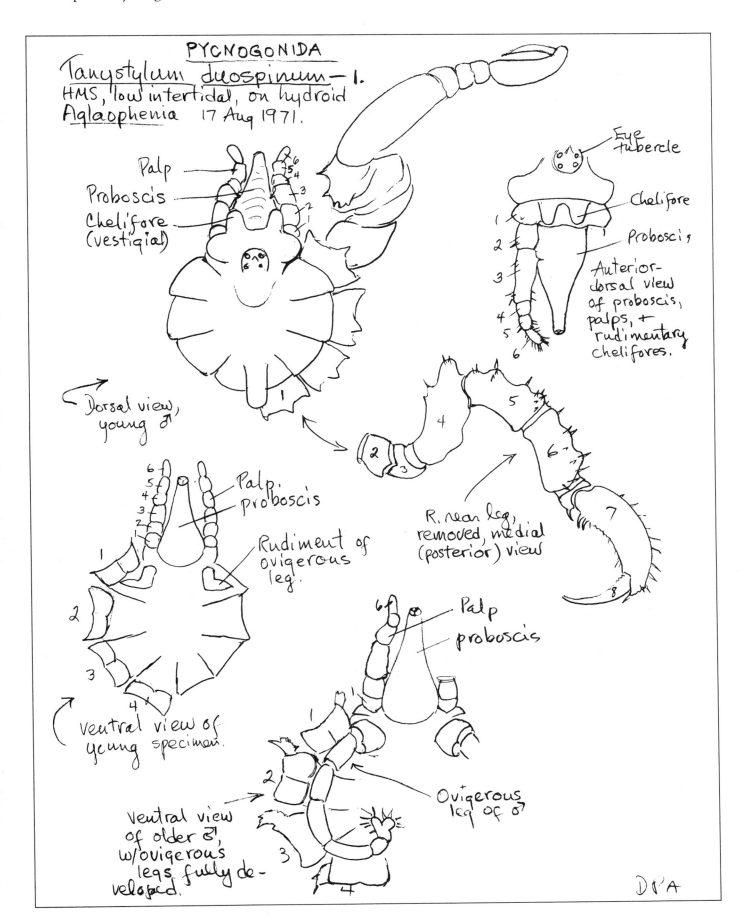

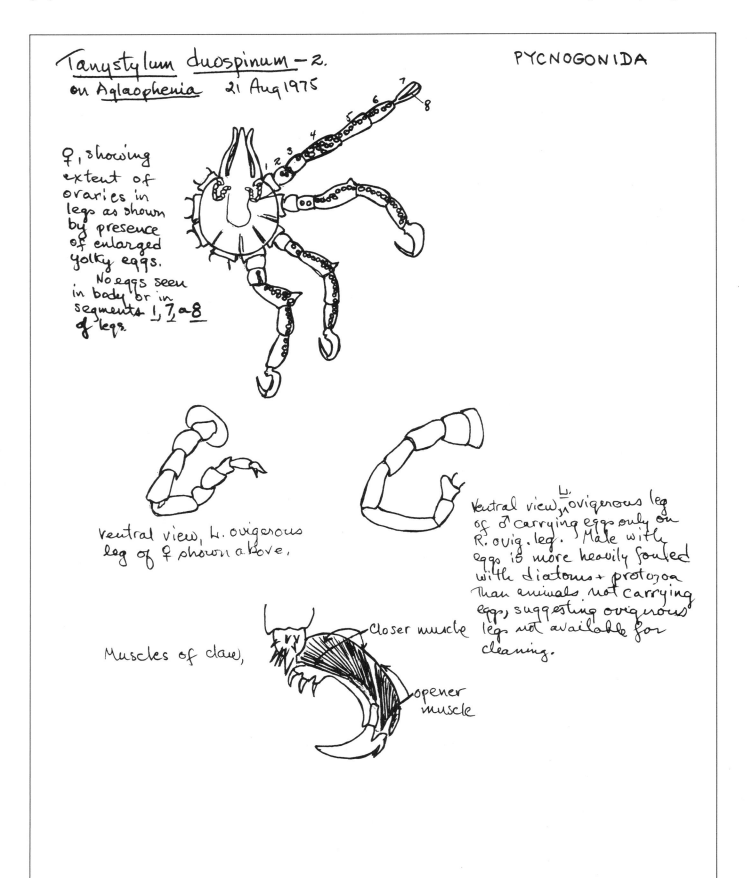

Arthropoda/Pycnogonida

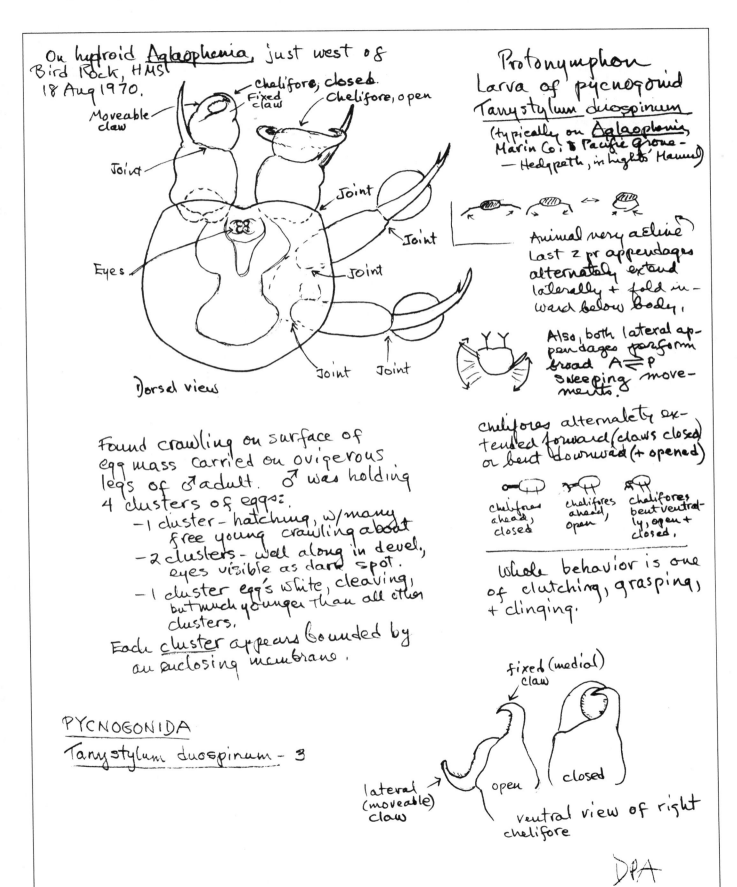

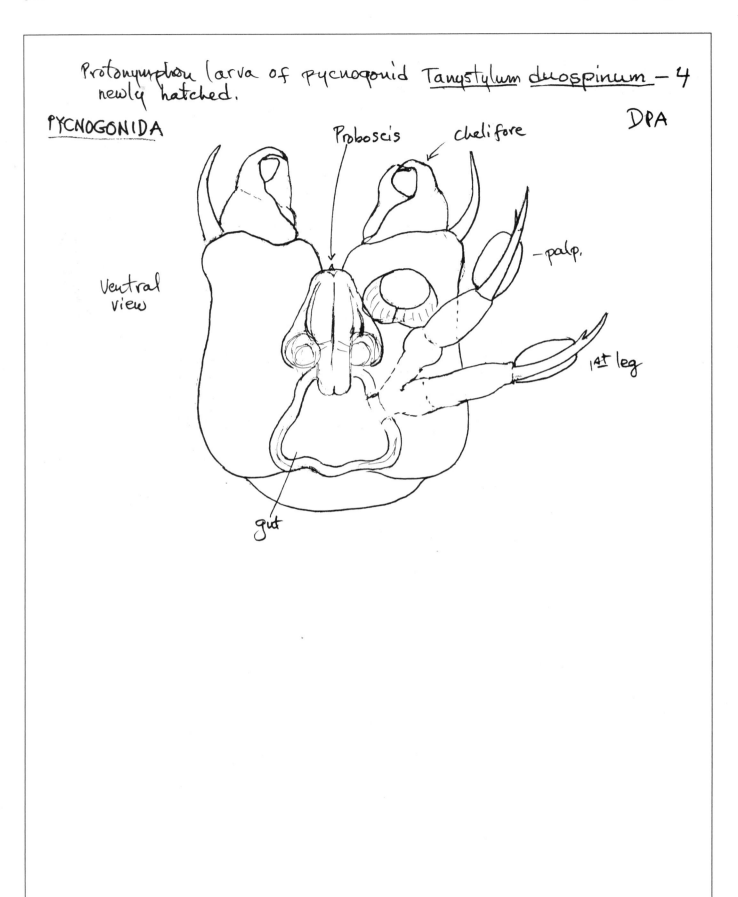

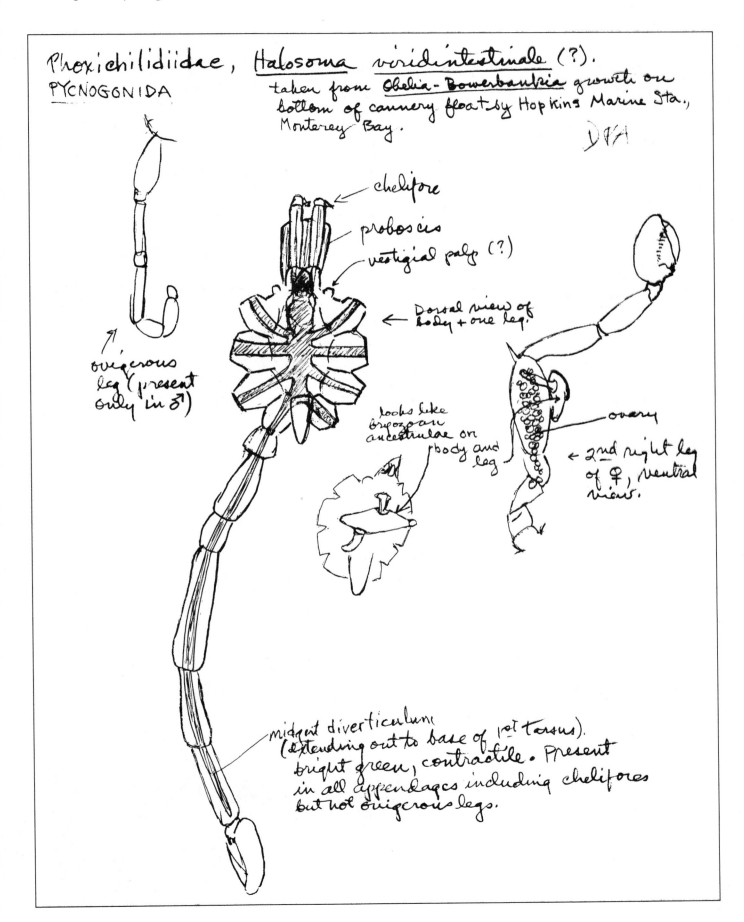

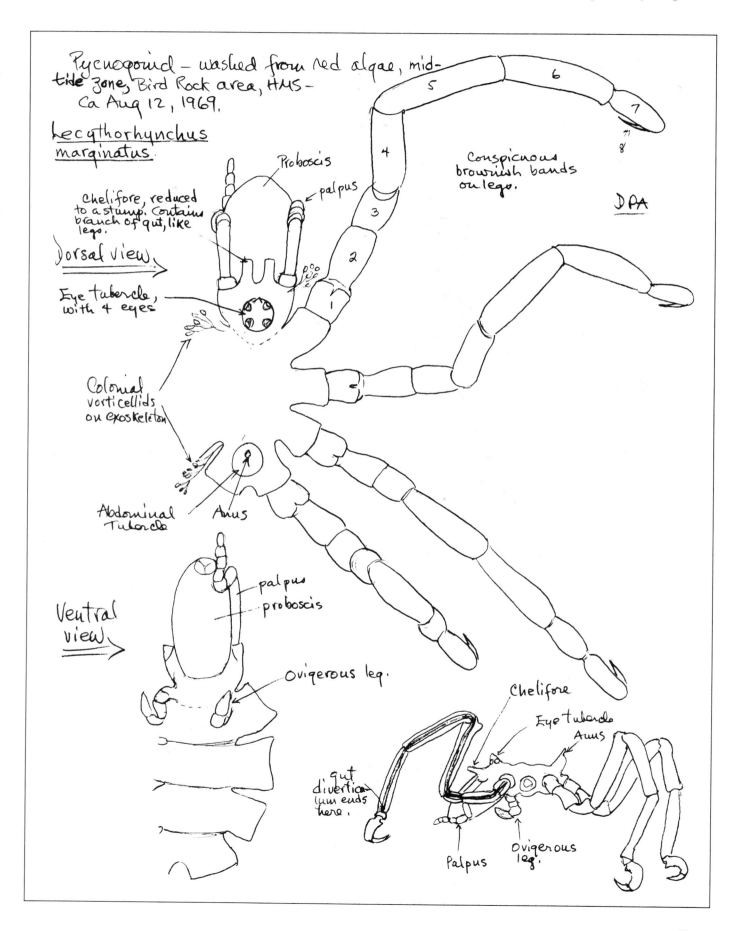

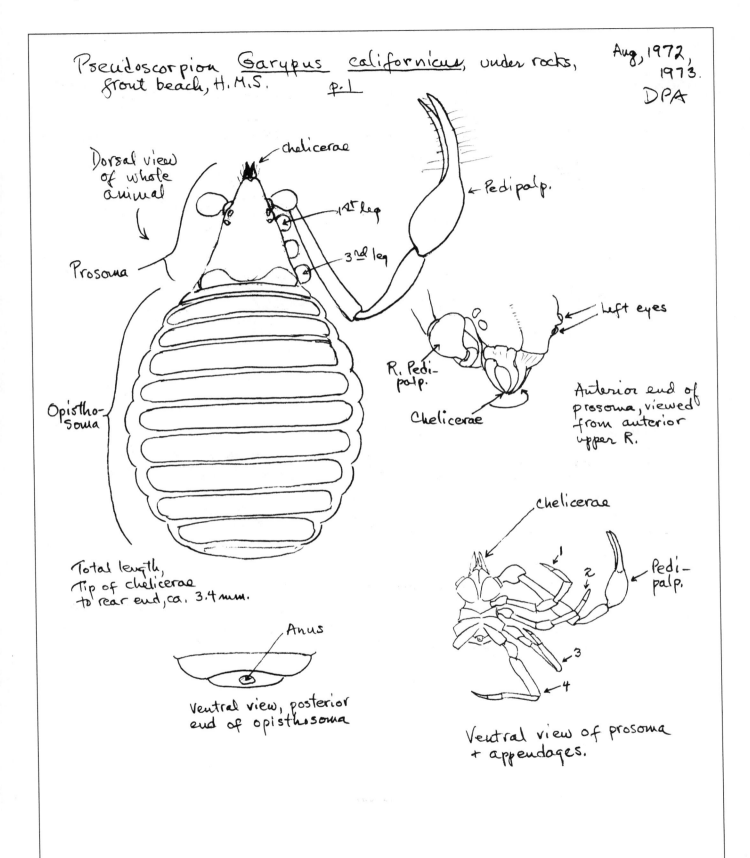

Pseudoscorpion

Garypus californicus DPA
West beach, HMS, under splash zone rocks. 17 Aug '72.
p.2

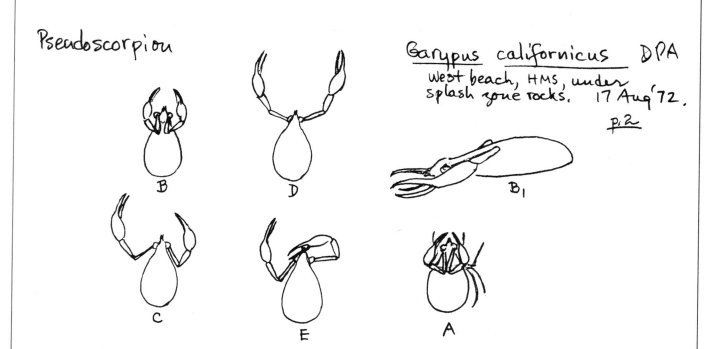

Typical stances — all dorsal views except B₁.

A. "Closed up" stance, typical following disturbance; immobile, pedipalps drawn back maximally with "elbows" meeting dorsally, + chelae protecting body anteriorly.

B-B₁ — Beginning to open up (dorsal + side views)

C — Turning

D — Typical stance when ambulating forward

E. Cleaning chelae. This is done by running fingers of chela through chelicerae, which are worked as fingers are drawn through them (from base to tip).

Arthropoda/Arachnida/Acarina

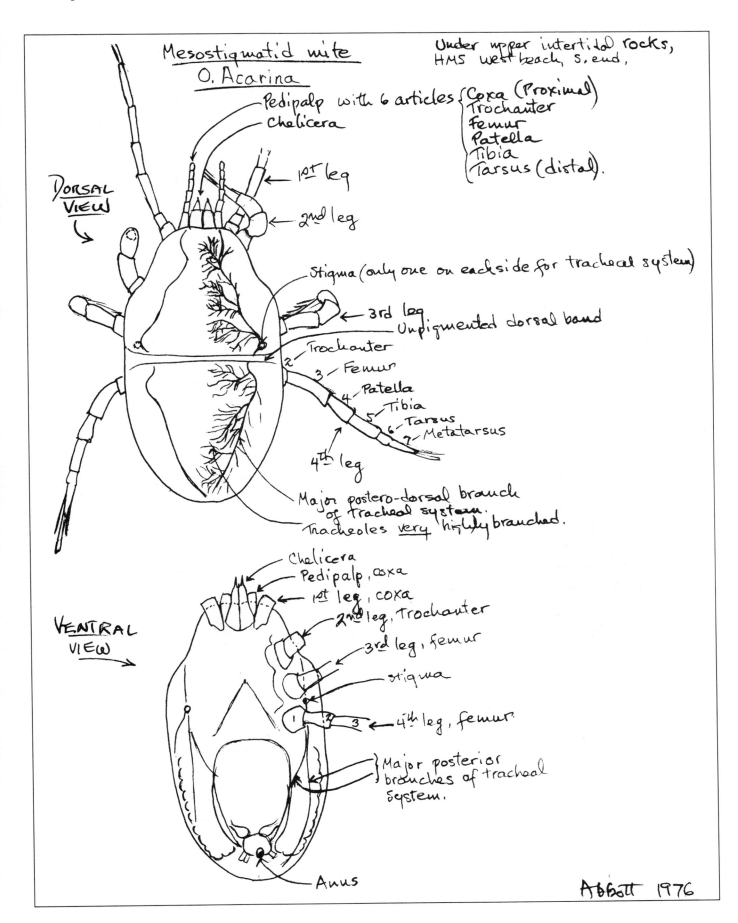

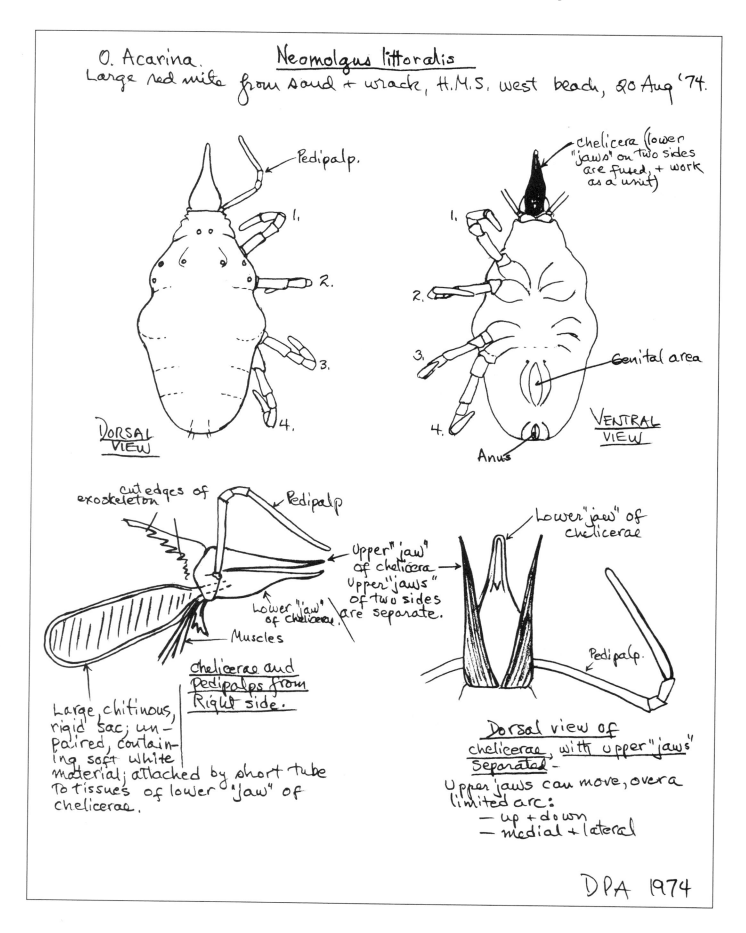

Arthropoda/Tardigrada

Tardigrade
Pt. Pinos, upper intertidal near USNPG School (Haderlie) transect, on green alga *Blidingia*, Coll. Algae class.

16 Aug 1973

DPA

Sea cucumbers are a good place to start with echinoderms—there's a high ratio of meat to bones.

In asteroids and ophiuroids, you find very little stereotyped behavior; there's a tremendous variability. In opisthobranchs, you find stereotyped behavior.

The mollusks are a splendid group of animals, spread throughout the biosphere and almost as well understood phylogenetically as vertebrates.

In the early treatment of mollusks, we get a heavy dose of terms, but we'll see these same structures time and time again all through the several groups of mollusks.

The evolutionary origin of bivalves is becoming less controversial all the time; we must go back before the scaphopods, before the gastropods, but not before the cephalopods, to get to their origins.

The arthropods have diversified musculature; they have a huge area of the body over which muscles can shift in the course of evolving. Their exoskeleton can take on an infinite variety of forms, and they have done a magnificent job of exploiting this.

In studies of arthropod appendages, it is often easier to start at the rear and work forward along the body, than to start anteriorly and work back.

I'll bet that in 500 million years, the crustaceans will be holding their own in the sea.

If all the copepods in the world disappeared, there would be a vast reorganization of all life in the sea.

When you are about to dissect a snail, like Littorina, *take several—the first ones usually turn to chowder.*

DRAWINGS BY STUDENTS

EVA ALADJEM
Amphiodia occidentalis, pp. 112, 113, 114, 115
Artemia sp., pp. 263, 264
caprellid sp., pp. 320, 321

ILZE K. BERZINS
Synoicum parfustis, p. 120
Perophora annectens, pp. 121, 122

VICKI BUCHSBAUM
syllid sp., p. 230

GALEN HOWARD HILGARD
Eucopella sp., p. 17
Membranipora tuberculata, p. 61
Bowerbankia sp., p. 78
Leptasterias sp., p. 96
Littorina planaxis, pp. 138, 139
Archidoris montereyensis, p. 170
terebellid sp., pp. 240, 241
Phragmatopoma californica, p. 243

MICHAEL IMPERATO
Olivella biplicata, p. 153

JAMES WATANABE
Aglaophenia sp., p. 12

LANI WEST
Nucella emarginata, pp. 147, 148

The coral Balanophyllia *is an animal draped over a very spiny chair—its own skeleton.*

In some acoel flatworms, the gut is not a cavity, it is a mush of cells; food forces a cavity when the animal eats.

The planula of Aglaophenia *is a long snaky thing. It crawls, settles, and sends up a stalk. It is beautifully engineered for getting along in a wave-swept area.*

If you look at Aglaophenia, *peek inside the leaves of one of its corbulae.*

In opisthobranchs, the penis retracts into a little sheath; without the sheath, it would drag in the sand and get scratched.

The stomach is up in the head in an amphipod—that's the only place left for it.

I've always felt that this conchostracan has a lovely face—not the kind of face I'd fall in love with, but truly lovely.

Cadulus *is like the Marx Brothers—blood rushes from its head to its foot, takes a look, and then rushes back to its head.*

One of the flatworms was named by D. P. Costello, of Woods Hole. I worked some with him and we always thought of publishing a paper together—by Abbott and Costello.

This has been a beautiful, cool, luminous universe in which to have spent some time. I feel fortunate.

INDEX

Acanthina spirata, 144–45
acanthocephalan larva, with *Emerita*, 339
Acanthomysis (=*Neomysis*), 289–92
Acarina, 371–72; see also 313
Acmaea, 135, 136
Acoela, 55
Actaeon; see *Rictaxis*
Actiniaria, 43–45
actinotroch larva, 86
actinula larvae, 4
Adula (=*Botula*) *californiensis*, 177–78
Aeolidacea, 172–74
Aequorea, 18
Aglaja (=*Navanax*), 160–62
Aglaophenia, 12–13; with caprellids, 319–21; with ostracod, 269; with pycnogonids, 363–65
Alcyonacea, 38–39
Alcyonaria, 37–42
alga: with *Phoronis*, 85; with *Perophora annectens*, 121; with Hesionidae, 224
alga, brown; see *Egregia*, *Macrocystis*
alga, coralline (red alga), 75
alga, green, 124, 284, 373
alga, red: with pycnogonid, 368; with *Synapseudes*, 307; (*Plocamium*), with stauromedusa, 22; (*Rhodymenia*), with *Thalamoporella*, 62; substrate for *Eucopella*, 17; pycnogonid washed from, 368
Amphiodia occidentalis, 110–15
Amphipholis squamata, with orthonectids *Rhopalura ophiocomae*, 108
Amphipoda, 311–22; eating *Tubularia*, 3; with *Polyclinum*, 118; in gut of *Pollicipes*, 277; see also 355
Anasca, 58–67
anemone; see Actiniara, Corallimorpharia
Anisodoris nobilis, 166
Annelida, 220–48; see also Polychaeta
Anomalodesmata, 206–7
Anomura, 336–56
Anostraca, 260–64
Anthomastus ritteri, 38, 39
Anthomedusae, 1–11
Anthopleura elegantissima, comparison with *Corynactis*, 46
Anthozoa, 37–49
Aplacophora, 215

Arachnida, 369–72
Archaeogastropoda, 127–36
Archidoris montereyensis, 170
Artemia, 260–64; eaten by *Pollicipes*, 277
Arthropoda, 260–373
Articulata, 81–84
Ascidia ceratodes, 124–25; with *Holoporella*, 71; with *Eupentacta*, 100; with *Lamellaria*, 141
Ascidiacea, 118–23
Ascomycetes (lichen), on *Acanthina*, 144
Ascophora, 68–73
Astacus, 353
Asteroidea, 87–98
Astropecten, 87–89
Atolla, 32–33
Aurelia aurita, 28

Balanomorpha, 279–83
Balanophyllia elegans, 48, 49
Balanus glandula, 281–83
Balanus tintinnabulum, with *Hiatella*, 205
Barentsia, 80
Barleeia, with *Pagurapseudes*, 308
barnacles; see Cirripedia
Beroë cucumis, 50–51
Bivalvia, 176–207
Blidingia (green alga), with tardigrade, 373
Bossea (red alga); see alga, coralline
Botryllus, 123
Botula, comparison with *Lithophoga*, 179; see also *Adula californiensis*
Bowerbankia, 76–78; with *Halosoma*, 367
Brachiopoda, Articulata, 81–84
Brachyura, 357–62
Branchiopoda, 260–68
Branchiura, on sand dab, 274
Bryozoa, 58–78; ancestrula on *Halosoma*, 367; boring, 155; associated with *Synapseudes*, 307
Bugula californica, 64–65

Cadulus fusiformis, 216–18
Calliarthron (red alga); see alga, coralline
Calycella (?), with *Eudendrium*, 5
Campanulariidae; see *Eucopella*
caprellids, 318–21; on *Eucopella*, 17

Cardium; see *Clinocardium*
Caridea, 328–35
Catophragmus, 279
Celleporaria; see *Holoporella brunnea*
Cephalaspidea, 156–63
cercariae (larvae of Trematoda, Platyhelminthes), 151, 176
Ceriodaphnia?, 267–68
Cestoda, 351; larva, with *Pagurus*, 351
Chaetoderma erudita, 215
Cheilostomata, 58–73
chiton; see Polyplacophora
Chlorella (green alga), tunicate feeding on, 124
Chondrophora, 9
Chordata, 118–26
Chrysaora, 23, 24
Chthamaloidea, 279
ciliates, peritrich (Protozoa): with *Dodecaceria*, 232
ciliates, vorticellid (Protozoa): with cladoceran, 267; with *Cryptolithodes*, 341; with *Lumbrinereis*, 229; with pycnogonid, 368; with *Tricellaria*, 66; with *Nebalia*, 285; with *Petrolisthes*, 355
Cirolana harfordi, 293–97
Cirratulidae, 231–33
Cirriformia, 231
Cirripedia, 275–83
Cladocera, 267–68
Cladonema, 11
clam; see Bivalvia
Clavelina, near *Phoronis*, 85
Clinocardium nuttallii (=*Cardium nuttallii*), 186–89
Clione? (a pteropod), 175
Clytia (?), with *Eudendrium*, 5
Cnidaria (Coelenterata), 1–49
Coelenterata, 1–49
Conchostraca, 265–66
Conus californicus, 155
Copepoda, 271–73; parasitic in *Amphiodia*, 110; see also *Tigriopus*
coral; see Madreporaria
Corallimorpharia, 46–47
corixids, cladocerans dipnetted with, 267
Coronatae, 32–36
Corophium, 322
Corynactis, 46, 47; reminiscent of *Balanophyllia*, 46
Crisia, 74

Crustacea, 260–362
Cryptolithodes sitchensis, 341–42
Cryptomya californica, 202
Ctenophora, 50–54
Ctenostomata, 75–78
Cucumaria curata, 99
Cyanoplax hartwegi, 210
Cyclostomata, 74

Decapoda, 328–62
Dendraster excentricus, larvae, 116–17
Dendronotacea, 171
Dendrostomum (= *Themiste*) *zostericola*, 249–51
Dentalium (= *D. neohexagonum*), 219
diatoms, 3, 16, 66, 364
Dodecaceria tubes, 79
Dodecaceria fewkesi, 232–33
doliolid tunicate, 126
Doliolidae, 126
Doridacea, 166–69
Dorvillea moniloceras, 226–27
Dorvilleidae, 226–27

Echinodermata, 87–117
Echinoidea, 116–17
Ectoprocta (Bryozoa), 58–78
Egregia (brown alga), with *Pentidotea*, 298–300
Emerita analoga, 336–40
Enteromorpha (green alga), with *Nebalia*, 284
Entoprocta, 79–80; loxosomid, with *Lumbrinereis*, 229
Epiactis prolifera, 44–45
Eucarida, 323–62
Eucopella (Family Campanulariidae), 15–17
Eudendrium californicum, 5–7
Eupentacta quinquesemita, 100–3
Euphausia pacifica, 323–27
Euphausiacea, 323–27
Eutonina, 19, 20

Fissurella volcano, 128–29
Flabellinopsis iodinea, 174
flatworm; see Platyhelminthes
Flustrella corniculata, 75
Folliculina (Protozoa): with *Cryptolithodes*, 341; with *Ischnochiton*, 214
Forcipulatida, 94–98

galatheid shrimp, 356
Galatheidea, 352–56

Gammaridacarus brevisternalis (a mite), with *Orchestoidea*, 313
Gammaridea, 311–17, 322
Garypus californicus, 369–70
Gastropoda, 127–75; associated with *Pagurapseudes*, 308
Gorgonacea, 37
Granulina, 154
Grapsidae, 358
Gymnosomata (Pteropoda), 175

Haliotis, 127
Halosoma viridintestinale, 367
Halosydna brevisetosa, 220–22
Haminoea, 163
Harpacticoida (=Harpacticoidea); see *Tigriopus californicus*
Henricia, 91
Hermissenda crassicornis, 172–73
Hesionidae, 224–25
Hesperonoë, 223
Heterodonta, 192–205
Hiatella arctica, 205
Hinnites giganteus, 183
Hippidea, 336–40
Hippodiplosia insculpta, 70
Hippolyte, 328–30
Holoporella brunnea, 71–73
Holothuroidea, 99–107
Hoplocarida; see stomatopod larva
hydroids, 3, 5; with *Pagurapseudes*, 308; see also Hydrozoa
Hydrozoa, 1–21

Idotea (*Pentidotea*) *resecata*, 302–4
Idotea (= *Pentidotea*) *stenops*, 298–301
Irus lamellifer, 190–91
Ischnochiton mertensii (= *Lepidozona mertensii*), with triclad flatworm, 214
isopod, parasitic on lingcod, 305
Isopoda, 293–306

jellyfish; see Hydrozoa, Scyphozoa

Katharina tunicata, 212–13
Kellia laperousii, 192
kelp beds (*Macrocystis* beds), 11, 14, 15, 20, 21, 50, 58, 291, 302, 303, 328, 329, 330

Lamellaria, 141–43
Lamellariacea, 141–43
Laqueus californicus, 84

larva: acanthocephalan, 339; actinotroch of phoronid, 86; actinula of *Tubularia*, 4; cestode, 351; cestode, with *Pagurus*, 351; nauplius of *Pollicipes polymerus*, 278; parasitic trematode, 360; phoronid, 86; planula of *Aglaophenia*, 12; planula, in *Eudendrium*, 6; planula, in *Plumularia*, 14; protonymphon, of *Tanystylum duospinum*, 365–66; protozoea of *Loxorhynchus crispatus*, 357; stomatopod, 288; zoea, of *Pachygrapsus*, 362; of *Dendraster excentricus*, 116–17
Lecythorhynchus marginatus, 368
Lepadomorpha, 275–78
Lepas, comparison with *Pollicipes*, 277
Lepidozona mertenzii; see *Ishnochiton mertensii*
Leptasterias, 94–96
Leptochelia dubia (?), 309
Leptomedusae, 12–20
Leptostraca, 284–87
Leptosynapta, 104–7
Limnomedusae, 21
limpet; see Archaeogastropoda
lingcod, parasitic isopods on, 305
Lithophaga, 179–80; comparison with *Botula*, 179
Littorina, 136
Littorina planaxis, 137–39
Lottia, 135, 136
Loxorhynchus crispatus, 357
loxosomid entoprocts, with *Lumbrinereis*, 229
Luidia, 90
Lumbrinereidae, 228–29
Lumbrinereis, 228–29
Lyonsia californica, 206

Macoma secta, 195–97
Macrocystis (brown alga), with *Eucopella*, 15
Macrocystis beds; see kelp beds
Madreporaria (=Scleractinia), 48–49
Majidae, protozoea larva, 357
Malacostraca, 284–362
Manania (?), 22
Megatebennus bimaculatus, 130–31
Melibe leonina, 171
Membranipora, 58–61; compared with *Phidolopora*, 68
Mesogastropoda, 136–43
mesostigmatid mite, 372
Mesozoa (Orthonectida), 108

Index

Metridium, 43
Microcerberus abbotti, 306
mite, 313, 371–72
Mollusca, 127–219
Mopalia muscosa, 211
mussels: with *Balanus*, 281; with *Phascolosoma*, 256; see also *Mytilus*
Myoida, 203–5
Mysidacea, 289–92
Mytiloida, 177–82
Mytilus: comparison with *Adula*, 177; comparison with *Lithophaga*, 180; defecated by *Eudendrium*, 5; fed to *Cirolana*, 293
Mytilus californianus, 181; *Balanus* on, 282
Mytilus edulis, 182; with *Botryllus*, 123; with *Eupentacta*, 100
Myxicola, 248

Nassariidae, 149
Nassarius, 149
nauplius larva, of *Pollicipes polymerus*, 278
Navanax; see *Aglaja*
Nebalia, 287
Nebalia pugettensis, 284–86
Neogastropoda, 144–55
Neomolgus littoralis, 371
Neomysis, 289–92; see also *Acanthomysis*
Notaspidea, 164–65
Notoacmea persona, 132–33
Notoplana, 56–57
Nucella emarginata, 146–48
Nuculana (=*Nucula*), 176
Nuculoida, 176
Nuda, 50–51
Nudibranchia, 166–73
Nuttallina californica, 208–9

Obelia, associated with *Halosoma*, 367
Olivella biplicata, 150–53
Ophelia (?), 234–35
Opheliidae, 234–35
Ophioplocus esmarki, 109
Ophiuroidea, 108–15
Opisthobranchia, 156–75
Orchestoidea, 311–17
Orthonectida, 108
ostracod, 270
ostracod (Podocopa), 269
Ostracoda, 269–70

Pachygrapsus crassipes, 358–61; zoea larva, 362

Pagurapseudes, 308
Paguridea, 341–51
Pagurus granosimanus, 349–51
Pagurus samuelis, 343–48
Palaeotaxodonta, 176
Pandora, 207
parasitic isopods, from lingcod, 305
Patellacea, 132–33, 135
Patiria miniata, 92–93
Paxillosida, 87–89
Pectinaria, 244–45
Pectinariidae, 244–45
Pedicellina, 79
Pelagia, 25–27
Penitella penita, 203–4
Pennatulacea, 40–42
Pentidotea resecata, 302–4
Pentidotea stenops, 298–301
Peracarida, 289–322
Periphylla, 34–36
peritrich ciliates (Protozoa), with *Dodecaceria*, 232
Perophora annectens, 121–22
Petricola (=*Rupellaria*), 194
Petrolisthes, 352–55
Phacellophora camtshatica, 29–31
Phascolosoma agassizii, 252–56; mixonephridia, 257–58
Phidolopora, 68–69; compared with *Membranipora*, 68
Pholadamyoida, 206–7; see also 190
phoronid larva, 86
Phoronida, 85–86
Phoronis, 85
Phragmatopoma tubes, syllid crawling among, 230
Phragmatopoma californica, 242–43
Phyllocarida, 284–87
Phyllochaetopterus, terebellid from, 240
Phyllospadix (sea grass, a vascular plant), with *Thalamoporella*, 63
phytoplankton, use with *Leptastarias*, 96; used in feeding experiments, 61, 78, 86
Pisaster giganteus, 97
planula larva, 6, 12, 13, 14
Platyasterida, 90
Platyhelminthes, 55–57, 214, 351; see also Trematoda
Pleurobrachia, 52–54
Pleurobranchaea, 164–65
Pleuroncodes planipes, 356
Plocamium (red alga), 22
Plumularia, 14
Podocopa (Ostracoda), 269

Pododesmus cepio, 184–85
Polinices, 140
Pollicipes polymerus, 275, 276, 277, 278
Polychaeta, 220–48; with *Cryptolithodes*, 341
Polycheria (?), with *Polyclinum*, 118
Polychoerus carmelensis, 55
Polycladida, 56–57
Polyclinum planum, 118–19
Polynoidae, 220–23
Polyorchis montereyensis, 8
Polyplacophora, 210–14
Priapulida (Priapula), 259
Priapulus, 259
Prosobranchia, 127–55
protonymphon larva, 356–66
Protothaca, 193
protozoa, 364; see also ciliates
protozoea larva, 357
Psammogorgia arbuscula, 37
Pseudoscorpionida, 369–70
pteropod, 175
Pterioida, 183–85
Pteriomorpha, 177–85
Pycnogonida, 363–68; associated with *Eucopella*, 17
Pycnopodia helianthoides, 98

Rathkea, 10
Rathkeidae, 10
rediae (larvae of Trematoda, Platyhelminthes), 151, 176
Renilla, 42
Rhodymenia (red alga), with *Thalamoporella*, 62
Rhopalura ophiocomae, with *Amphipholis*, 108
Rictaxis (=*Actaeon*) *punctocaelatus*, 156–59
Rupellaria; see *Petricola*

Sabellariidae, 242–43
Sabellidae, 246–48
sand dab, branchiuran on, 274
Scaphopoda, 216–19
Scleractinia; see Madreporaria
Scrupocellaria, 67
Scyphozoa, 22–36
sea cucumber; see Holothuroidea
sea squirt; see Tunicata
Semaeostomeae, 23–31
Siliqua lucida, 200–1
Sipuncula, 249–58
snail; see Gastropoda
Spinulosida, 91–93
Spirontocaris, 331–35

Stauromedusae, 22
Sternaspidae, 236–38
Sternaspis, 236–38
stomatopod larva, 288
Stomatopoda, 288
Styela montereyensis, with *Botryllus*, 123
Stylatula elongata, 40
Stylatula gracilis, 41
Syllidae, 230
Synapseudes intumescens (probably), 307
Syncoryne eximia, 1–2
Syndesmis (flatworm similar to), with *Leptosynapta*, 106
Synoicum parfustis, 120

Tanaidacea, 307–10
Tanais, 310
Tanystylum duospinum, 363–66; protonymphon larva, 365, 366
Tardigrada, 373

Tegula, 127, 134
Tentaculata, 52–54
terebellid: behavior, 240–41; juvenile, 239
Terebellidae, 239–41
Terebratalia or *Terebratulina*, 81–83
Terebripora; see *Conus*
Tetraclita squamosa, 280
Thalamoporella californica, 62–63
Thaliacea, 126
Themiste zostericola; see *Dendrostomum*
Tigriopus (Copepoda): feeding experiments with, 1, 3, 46, 47, 49, 52, 86; use with *Leptasterias*, 96
Tigriopus californicus, 271–73
Trematoda, parasitic larvae of, 151; in *Nuculana*, 176; in *Pachygrapsus*, 360
Tresus nuttallii, 198–99
Tricellaria occidentalis, 66
Tricladida, with *Ischnochiton*, 214
Triopha carpenteri, 167
Triopha catalinae (= *T. carpenteri*), 167

Triopha maculata, 168–69
Trochacea, 127, 134
Tubularia, comparison with *Velella*, 9
Tubularia marina, 3–4
Tunicata, 118–26
tunicates: with *Leptochelia*, 309; with *Synapseudes*, 307
Turbellaria, 55–57, 214

Urechis burrow, *Hesperonoë* from, 223

Vallentinia adhaerens, 21
Velella, 9; possible homolog in *Tubularia*, 4
Veneroida, 192–202
vorticellid ciliates; see ciliates

Zoantharia, 43–49
zoea larva, 362
Zygobranchia, 127

Library of Congress Cataloging-in-Publication Data

Abbott, Donald Putnam.
  Observing marine invertebrates.

  1. Marine invertebrates—Pacific Coast (U.S.)—Atlases.
2. Marine invertebrates—Pacific Coast (U.S.)—Anatomy—
Atlases.   I. Hilgard, Galen Howard.   II. Title.
QL155.A24   1987   592.092′632   87-9931
ISBN 0-8047-1426-6 (alk. paper)